THE FIRST KUWAIT OIL CONCESSION AGREEMENT

H.H. Shaikh Ahmad al Jaber as Subah, Ruler of Kuwait 1921–1950

THE FIRST KUWAIT OIL CONCESSION AGREEMENT

A Record of the Negotiations 1911–1934

by

Archibald H. T. Chisholm
C.B.E.(Mil.), M.A.(Oxon)

FRANK CASS : LONDON

First published 1975 in Great Britain by
FRANK CASS AND COMPANY LIMITED
67 Great Russell Street, London WC1B 3BT, England

and in United States of America by
FRANK CASS AND COMPANY LIMITED
c/o International Scholarly Book Services, Inc.
P.O. Box 4347, Portland, Oregon 97208

ISBN 0 7146 3002 0

Library of Congress Catalog Card No. 73-82523

Printed in Great Britain by
Richard Clay (The Chaucer Press), Ltd.,
Bungay, Suffolk

Contents

قُلْ لَا أَمْلِكُ لِنَفْسِي نَفْعًا وَلَا ضَرًّا

إِلَّا مَا شَاءَ اللَّهُ وَلَوْ كُنْتُ أَعْلَمُ

الْغَيْبَ لَاسْتَكْثَرْتُ مِنَ الْخَيْرِ

Say: 'I have no power over any good or harm to myself except as God willeth. If I had knowledge of the unseen, I should have multiplied all good . . .'

The Koran, Sura Al-A'raf v.188

It is important to rely upon authentic contemporary records and the expressions of opinion set down when all was obscure.

Sir Winston Churchill
Preface to *The Second World War*, Volume I (1948)

Foreword

The first Kuwait oil concession agreement, granted on 23rd December 1934 by the late Shaikh Ahmad al Jaber al Subah to the Kuwait Oil Company Ltd., has been of fundamental importance in the history of the State of Kuwait; whose great increase in prosperity and international influence during the last twenty-eight years has directly resulted from its oil production since 1946 under the Agreement and its 1951 successor. Hitherto no full or authoritative account of the very long and complex commercial and political negotiations preceding the Agreement has been compiled; and consequently all descriptions of them so far published (except those merely quoting the Agreement's terms and naming the parties concerned) have contained substantial omissions or errors of fact involving a wrong impression of what actually occurred.

Early in 1970 the Kuwait Government's Minister of Information, as a result of conversations with Mr Yusuf Ahmad Alghanim (whose family and personal connections with the negotiations began in 1925), decided that an authoritative record of the negotiations culminating in the 1934 Agreement should be prepared and published as a source of accurate information for students and historians of Kuwait affairs; and that I, now the only survivor of the Agreement's signatories and negotiators, should be invited to compile it.

In May 1970, after discussions with H.H. the Amir of Kuwait, H.H. the Crown Prince and Prime Minister, and the Minister of Information, I was engaged by the Kuwait Ministry of Information to research and compile for them a documented record of the 1934 Agreement's negotiations originating in 1911 and intensively conducted from 1923 onwards. Permission, essential for such a project, to consult and quote from its four major sources of contemporary records and documents, namely the private archives from 1911 onwards of the British Petroleum Co. Ltd. in London and of the Gulf Oil Corporation in Pittsburgh, and the public archives of the same period of the British Government's Foreign and Commonwealth Office in London and the United States Government's National Archives in Washington, was then accorded and is now gratefully acknowledged. The record now published is based on and mainly documented from those sources, supplemented by my personal knowledge of the negotiations and acquaintance with all those principally concerned with them and the 1934 Agreement.

To prevent the record from becoming even longer than its complexity necessitates, such supplementary details as are required of persons, organisations, or events mentioned in it are included, with copies of relevant documents, in its annexed Notes, which are indicated in the text by bold numbers in brackets. A partial exception to this procedure has been made when recording the intervention in the negotiations from April 1934 onwards of Traders Ltd.; because the remarkable facts about that little-known occurrence and its effects on the negotiations have never previously been published. Particularly through its documentation (which has been made as comprehensive as possible because no contemporary records or documents on the subject have survived in the Kuwait Government Archives) I trust that this record will fulfil its object, namely to provide for the first time a source of complete and accurate information of events, hitherto often misunderstood, of fundamental importance for students of the history of Kuwait.

In addition to the acknowledgements made above, I am particularly indebted to

Mr Yusuf Ahmad Alghanim for his very useful recollections of the negotiations, and to Mr Sadun M. Aljassim, Under-Secretary of the Kuwait Ministry of Information, and his staff for their most agreeable collaboration since May 1970.

London, 1975

Archibald H. T. Chisholm

Glossary

Spelling
In accordance with modern practice, the spelling Kuwait, Shaikh, Ahmad, Muhammad, etc. is used in this Record. At the time of the Kuwait oil-concession negotiations, the variants Koweit, Sheikh, Ahmed, Mohammed, etc. were also in common use, and in contemporary texts and documents quoted in the Record and Notes the original spelling has been retained.

Place-names
1. Before 1935 Iran was generally known as Persia and that name is used in this record of negotiations preceding that date.
2. At the time of the Kuwait oil-concession negotiations the Persian Gulf was universally so described, but in accordance with modern practice in Arab countries it is referred to in this Record as the Arabian Gulf. In contemporary texts and documents quoted in the Record and Notes the original text has been retained.

Currency
At the time of the Kuwait oil-concession negotiations, the currency in general use in Kuwait was the Indian rupee, *one* rupee being divided into *sixteen* annas (e.g. rupees $3\frac{1}{2}$ or $3\frac{1}{4}$ were written as Rs. 3/8 or Rs. 3/4). At that time £1 = 20s and 1 shilling = 12d.

The rupee/£ sterling exchange rate was Rs. $13\frac{1}{3}$ = £1; one rupee being equivalent to one shilling and sixpence and one anna being equivalent to $1\frac{1}{8}$d. Rupee/£ equivalents of various amounts mentioned in this Record are:

Rs./Annas	£	*s.*	*d.*
/2			$2\frac{1}{4}$
/4			$4\frac{1}{2}$
/8			9
2/10		3	$11\frac{1}{4}$
3		4	6
3/4		4	$10\frac{1}{2}$
3/8		5	3
1,000	75	0	0
10,000	750	0	0
20,000	1,500	0	0
30,000	2,250	0	0
100,000	7,500	0	0
475,000	35,625	0	0

Names
Names of persons or organisations mentioned frequently in the Record are given in full on first mention and afterwards in shorter form, e.g.:

Shaikh Ahmad or the Shaikh = Shaikh Ahmad Al Jaber
Cadman = Sir John (later Lord) Cadman
Chisholm = Mr A. H. T. Chisholm
Dickson = Lt-Col H. R. P. Dickson
Holmes = Major Frank Holmes
Wilson = Sir Arnold Wilson
APOC = Anglo-Persian Oil Co. Ltd.
EGS = Eastern & General Syndicate Ltd.
Gulf = Gulf Oil Corporation
H.M.G. = His/Her Majesty's Government
IPC = Iraq Petroleum Co. Ltd.
KOC = Kuwait Oil Co. Ltd.
P.A. = Political Agent
P.R. = Political Resident
U.S.G. = United States (of America) Government

Bibliography

The following books are recommended for specialised information regarding the Kuwait oil-concession negotiations:

The Story of the Negotiations for Oil Concessions in Bahrain, El Hasa (Saudi Arabia), the Neutral Zone, Qatar and Kuwait, by Thomas E. Ward (see Note 33). Published in 1965 for private circulation only by Ardlee Services Inc., New York, U.S.A.; Library of Congress catalog card number 65/23760.

Oil in the Middle East, by S. Longrigg. 3rd edition published in 1968 by Oxford University Press for Royal Institute of International Affairs, London, England.

The following books are recommended for information regarding (*a*) Kuwait at the time of the oil-concession negotiations; (*b*) Kuwait and the rest of the Middle East area at the same period:

(*a*)

The Arab of the Desert, by H. R. P. Dickson (see Note 39). Published 1949 (fourth impression 1967) by Allen & Unwin Ltd., London, England.

Kuwait and Her Neighbours, by H. R. P. Dickson. Published 1956 (second impression 1968) by Allen & Unwin Ltd., London, England.

Adventure in Oil, the Story of British Petroleum, by Henry Longhurst (with a foreword by The Rt-Hon. Sir Winston Churchill). Published by Sidgwick & Jackson Ltd., London, England (Chapter 22, pages 228–47, refers to Kuwait).

Forty Years in Kuwait, by Violet Dickson (see Note 39). Published 1971 by Allen & Unwin Ltd., London, England.

(*b*)

Britain's Moment in the Middle East 1914–1956, by Elizabeth Monroe. Published 1963 by Chatto & Windus Ltd., London, England (see especially Chapter 4, pages 95–115, 'The Role of Oil in British Government Policy').

Chronology of the Negotiations

1911

November: Anglo-Persian Oil Company (APOC) wrote from London to the British Political Resident in the Gulf enquiring whether an oil concession was obtainable from the Shaikh of Kuwait (Shaikh Mubarak) and asking him, if this were the case, to apply for a prospecting licence on behalf of the Company. The Resident replied that conditions in the area were then too disturbed for such an application to be put forward.

1913

March: The first geological survey of Kuwait was made by a member of a British Admiralty Commission then visiting the Gulf.
October: The Shaikh of Kuwait (Shaikh Mubarak) and the British Political Resident exchanged letters in which the Shaikh agreed that no Kuwait oil concession would be given except to a person nominated and recommended by the British Government.
November: A British Admiralty Commission visited Kuwait to inspect bitumen seepages at Burgan and elsewhere.

1914

February: An APOC geologist visited Kuwait to examine the oil seepages at Burgan and Bahra.

1917

March: The Shaikh of Kuwait (Shaikh Salim) agreed to APOC making its first geological survey of Kuwait.
October: APOC applied for permission to negotiate for oil rights in Kuwait to the Political Resident who advised postponement because of war conditions then prevailing.

1918

May: APOC, in a letter to the British Foreign Office applying for an oil concession covering 'such part or parts of Mesopotamia' as might come under British control after the war, asked for such a concession to be extended to include Kuwait.

1921

May: APOC reminded the Foreign Office of its 1918 application and, as regards Kuwait, requested that the Political Resident in the Gulf be instructed to apply to the Shaikh (Shaikh Ahmad) for an exclusive prospecting licence.
December: The British Colonial Office (which had recently succeeded the Foreign Office as the British Government department responsible for Kuwait affairs) informed APOC that its negotiations for an oil-concession agreement could be opened with Shaikh Ahmad but, in accordance with Shaikh Mubarak's letter of October 1913, these had to be conducted through the Political Resident in the Gulf.

1922

October: The British Colonial Office informed APOC that the Political Resident had been instructed to assist the Company's representatives in negotiating a Kuwait oil-concession agreement with Shaikh Ahmad.

1923

January: The Political Agent in Kuwait initiated APOC's negotiations for an oil-concession agreement with Shaikh Ahmad.
May: Shaikh Ahmad received a telegram from Major Holmes of Eastern and General Syndicate Ltd. (EGS) in Bahrain, saying that he had just obtained King Ibn Saud's oil concession and advising the Shaikh not to grant any oil concession before considering his own company's terms; which favourably impressed the Shaikh when described to

him later in the month by Holmes before leaving for London to get British Government approval of his Company's negotiations.
June: APOC presented draft concession terms to Shaikh Ahmad.
July: Shaikh Ahmad informed the Political Resident that APOC's terms as presented in June were unacceptable.
September: The Colonial Office wrote to the Political Resident, and to APOC and EGS, defining British Government policy regarding negotiations for oil concessions in the Gulf generally and, as regards Kuwait, according priority of consideration to APOC's application.

1924

January: Shaikh Ahmad informed the Political Agent in Kuwait that his views on APOC's draft concession terms would not be ready for at least six months.
April: Dr Heim arrived in Kuwait to make a geological survey of Kuwait (and the Neutral Zone and Hasa) for EGS.
May: Shaikh Ahmad and King Ibn Saud jointly granted to EGS an oil-concession option on the Kuwait/Saudi Arabian Neutral Zone.

APOC informed by the Colonial Office that its negotiations for Kuwait oil-concession agreement should not be unnecessarily prolonged and that its priority for consideration would terminate on 31st March 1925.
July/August: Shaikh Ahmad visited APOC's oilfields and Abadan Refinery in Persia.

1925

March: APOC's prior negotiating rights for Kuwait oil-concession agreement lapsed.
December: The Shaikh of Bahrain granted the Bahrain oil concession to EGS.

1926

January/February: Shaikh Ahmad agreed to APOC geologists making a survey of Kuwait.
February/April: An EGS offer to sell their Bahrain, Hasa, Neutral Zone and Kuwait interests considered but eventually refused by APOC.
September/October: Holmes (EGS) visited the USA and tried unsuccessfully to interest major American oil companies, among them Gulf Oil Corporation (Gulf), in EGS's operations in the Gulf.

1927

November: Gulf, after protracted negotiations, made option agreements with EGS acquiring its Bahrain, Hasa, Neutral Zone and Kuwait interests.

1928

April: Holmes (EGS) informed Shaikh Ahmad that he would shortly offer improved terms for an oil agreement and that his company had American backing.
June/August: After protracted discussions Shaikh Ahmad rejected Holmes (EGS/Gulf) terms as inadequate.
November: Holmes (EGS/Gulf), when reopening discussions in Kuwait with Shaikh Ahmad after visiting the USA and obtaining Gulf agreement to a further increase in their terms, was informed by the Political Agent that the British Government now required, in any oil-concession agreement with the Shaikh, a British Nationality clause (which would in effect exclude American participation). A similar British Nationality clause was also now required in the Bahrain Concession (which had just been taken over from EGS/Gulf by the Standard Oil Co. of California).

1929

May: After representations to the British Government by the American Government, the Colonial Office informed EGS that the British Government agreed in principle to American participation in the Bahrain oil concession.

1930

January: Agreement reached between EGS and the Colonial Office over the terms of the Bahrain concession, which was accordingly assigned to Bahrain Petroleum Company, a subsidiary of Standard Oil Company of California.
August: Holmes (EGS/Gulf) resumed negotiations with Shaikh Ahmad in Kuwait but was informed by the Political Agent (Dickson) that, regardless of the EGS/Colonial Office agreement regarding Bahrain, any Kuwait oil-concession agreement must include the British Nationality clause as specified in November 1928.

1931

January: EGS informed by the Colonial Office that it had ascertained that Shaikh Ahmad refused to grant an oil concession to any concern not entirely British and would insist on inclusion of a British Nationality clause in any concession agreement.
July: Holmes (EGS/Gulf) obtained a letter from Shaikh Ahmad which he claimed showed that the Shaikh did not insist on a British Nationality clause in his concession agreement.
August/October: EGS/Gulf discussed significance of Shaikh Ahmad's letter of July with the Colonial Office which referred it to other British Government departments concerned and to the Political Resident in the Gulf.
October: Shaikh Ahmad agreed an APOC request to send geologists in December to survey Kuwait's oil prospects.
December: Following representations made by Gulf in November to the State Department in Washington, the American Government instructed its Ambassador (Dawes) in London to request the British Foreign Office to accord the same freedom of negotiation for American companies in Kuwait as already accorded in May 1929 in Bahrain.

1932

February: APOC geological survey party in Kuwait followed their preliminary reconnaissance with a test-drilling programme.
March: The Foreign Office in London informed the American Embassy that a reply to the latter's representations since December regarding freedom of negotiation for American companies in Kuwait was not yet ready owing to the necessity for departmental discussions; also that APOC was considering a renewal of its application for a Kuwait concession. On 30th March the American Government formally requested the British Government that their 'open door' policy for American oil companies, already existing in Iraq and Bahrain, should also apply in Kuwait.
April: The American Ambassador (Mellon) in London informed the State Department in Washington that the British Government had agreed the American Government's request for an 'open door' policy for American oil in Kuwait.
May: Shaikh Ahmad received from Holmes (EGS/Gulf) a new draft-concession agreement for consideration, and is also informed by the Political Resident and Mr Chisholm (APOC) that APOC were preparing a new draft-concession agreement for his consideration.
June: Oil in commercial quantities discovered in Bahrain.

Shaikh Ahmad refused an APOC application for an oil prospecting licence and requested them to submit comprehensive concession proposals so that he could compare them with EGS/Gulf's and decide between them.
August: Shaikh Ahmad received APOC's draft oil-concession proposals from Chisholm.
September: Shaikh Ahmad informed by the Political Resident that no further negotiations with either EGS/Gulf (Holmes) or APOC (Chisholm) could take place until the British Government's views, which were being prepared in London, on the merits of both their concession proposals were communicated to the Shaikh.
October/December: The Chairman of Gulf and the Chairman of APOC discussed in London and Pittsburgh the possibility of joint action in obtaining the Kuwait oil concession. Although no agreement was reached discussions between the two companies were to continue.

1933

January: Shaikh Ahmad received from the Political Resident the British Government's views on EGS/Gulf's and APOC's concession proposals. Holmes (EGS/Gulf) and Chisholm (APOC) reopened negotiations with the Shaikh.
March: Shaikh Ahmad, after receiving improved proposals for his oil concession from Holmes on behalf of EGS/Gulf, requested similar action from Chisholm on behalf of APOC. The Chairman of APOC (Cadman) visited Shaikh Ahmad and offered very substantially improved terms.
April: Shaikh Ahmad informed Chisholm that Holmes had offered terms on behalf of EGS/Gulf exceeding those recently offered by Cadman for APOC.
May: On 14th Shaikh Ahmad informed Holmes (EGS/Gulf) and Chisholm (APOC) that he had decided to suspend all negotiations with them until further notice.

On 23rd APOC and Gulf decide, and instruct Chisholm and Holmes accordingly, to standstill their negotiations for the Kuwait concession as their discussions of possibly combining to obtain the concession were within sight of succeeding.

On 29th King Ibn Saud granted the Saudi Arabia oil concession to Standard Oil Co. of California.

December: APOC and Gulf agree to become joint and equal partners in negotiating a Kuwait oil-concession agreement through a jointly owned Company (Kuwait Oil Co. Ltd.) to be formed for that purpose.

1934

February: Chisholm and Holmes, who had been appointed joint negotiators for Kuwait Oil Co. Ltd. (KOC) presented the Company's first draft-concession proposals to Shaikh Ahmad; who rejected them as unsatisfactory.

March: Shaikh Ahmad informed the joint KOC negotiators of the terms which he required for his oil concession.

April: On 11th the joint KOC negotiators presented Shaikh Ahmad with revised concession proposals, which the Shaikh rejected as not fully meeting his required terms. He also informed the negotiators that another company '100% British' had recently asked him to negotiate an oil concession with them, offering better terms than KOC's proposals.

On 23rd the Political Agent reminded Shaikh Ahmad of his treaty obligation not to negotiate any oil concession without the British Government's prior consent.

May: Shaikh Ahmad informed the joint KOC negotiators that his required terms could not be reduced and requested an early reply from their Company; he again rejected their subsequent compromise offer made after referring his request to their London principals.

June: Shaikh Ahmad instructed the joint KOC negotiators that unless their Company completely accepted his terms, he would not negotiate further with them until September and they should return to London forthwith for discussions with their principals: he undertook not to entertain or discuss any offer from other parties in their absence.

The joint negotiators therefore left Kuwait for London after briefly interviewing the Shaikh on 14th June to confirm their understanding of his instructions.

September: Shaikh Ahmad wrote to the representative of Traders Ltd. (this was the 100% British Company whose approach he had mentioned to the joint KOC negotiators on 11th April) in Basrah that, having accepted the conditions of their proposed concession agreement, he would sign it if the Company obtained the British Government's assent and he was informed to that effect through the Political Agent in Kuwait, and provided also that the Company agreed certain points marked in the text of the copy of their proposed concession agreement which he enclosed.

October: On 11th, when the joint negotiators were in Basrah *en route* to Kuwait, Holmes heard rumours of a British company having been negotiating during their absence from Kuwait with Shaikh Ahmad for his oil concession and offering better terms than KOC. When Shaikh Ahmad confirmed this to him in Kuwait on 16th, Holmes adopted the (wholly baseless) theory, and similarly convinced the Shaikh, both that the company, Traders Ltd., represented an APOC scheme to double-cross their Gulf partners in KOC and obtain the Kuwait concession for a 100% British company controlled by themselves, and also that the Shaikh and Holmes must work together, keeping Chisholm and APOC in the dark, to thwart this APOC plot and secure the concession for KOC.

On 17th, the joint KOC negotiators presented KOC's revised concession proposals to Shaikh Ahmad, and thereafter informed KOC in London in their jointly signed weekly reports on 17th, 24th, and 31st, that they had not yet received any indication of the Shaikh's reaction to their proposals; although in fact on 25th October Holmes, without the knowledge of Chisholm, had after several private discussions with Shaikh Ahmad from 16th October onwards reached a satisfactory compromise with the Shaikh on his two major requirements as yet unsatisfied in KOC's proposals.

November: On 3rd, Shaikh Ahmad expressed his disappointment to the joint KOC negotiators with their proposals as presented on 17th October and his intention to send them shortly written details of his complete requirements. These (embodying the secret compromise reached between Shaikh Ahmad and Holmes on 25th October) were received by the joint negotiators on 7th November and referred to their London principals for decision.

On 19th the British Government in London received indirect but authoritative information that Shaikh Ahmad, without informing them in accordance with his Treaty obligation, had been negotiating for some time past with Traders Ltd., which information was confirmed to them as correct by the Political Resident in the Gulf on 23rd in response to their enquiry.

On 22nd, the joint KOC negotiators informed Shaikh Ahmad that KOC in London had authorised acceptance of all his requirements of 7th November subject to some minor modifications; the Shaikh replied that he would give them his decision and specify all his final demands by letter as soon as possible.

On 28th, the Chairman and Secretary of Traders Ltd. visited the India Office (the British Government Department then responsible for matters concerning Kuwait) in London, and informed them that their Company had been negotiating with Shaikh Ahmad for an oil concession since April; and that in September the Shaikh had written to their Basrah representative granting them the concession subject to British Government approval which they now requested in a letter to the India Office dated 28th, describing their Company's negotiations with Shaikh Ahmad, giving details of the various British interests controlling or associated with Traders Ltd., and enclosing a copy of their proposed concession agreement with the Shaikh.

December: On 4th, the India Office in London requested the Political Resident in the Gulf to enquire from Shaikh Ahmad whether Traders Ltd.'s account of their dealings with him since April were correct, and, if so, why such concession negotiations had been conducted without the Shaikh informing the British Government in accordance with his treaty obligation, of which he had been reminded on 23rd April.

On 9th, the joint KOC negotiators received from Shaikh Ahmad his finally amended draft of their concession proposals as discussed on 22nd November, informing them that if this amended draft was completely accepted by KOC he would grant them the concession, which important information was telegraphed to KOC London by the joint negotiators on 10th December.

On 13th, KOC in London having learnt from the India Office on 12th December that another Company had applied for permission to negotiate an oil concession with Shaikh Ahmad, telegraphed to the joint negotiators in Kuwait to inform the Shaikh that his amended draft of 9th was completely accepted by KOC and the negotiators were authorised to sign the concession accordingly.

On 15th December, the joint negotiators having given this information to Shaikh Ahmad, he said that he would grant KOC the concession and sign the agreement document as soon as it could be prepared. On the same day he informed Traders Ltd.'s representative in Basrah that he had granted his concession to KOC.

On 16th, the Political Agent in Kuwait telegraphed to the India Office in London the explanations which he had received from Shaikh Ahmad, in three letters dated 15th/16th December replying to the India Office's enquiry of 4th December, of his dealings with Traders Ltd. since April.

On 21st, the British Government approved Shaikh Ahmad's grant of the Kuwait concession to KOC and the Political Agent was instructed to inform the Shaikh accordingly. On the same day the India Office confirmed the decision to KOC and also informed the Chairman of Traders Ltd. that, as Shaikh Ahmad had granted his concession to KOC on 15th December, they could not act as requested in his letter and discussion of 28th November.

On 23rd December, the Kuwait Oil Concession Agreement between Shaikh Ahmad and KOC was signed by the Shaikh, and by Holmes and Chisholm jointly on behalf of KOC.

RECORD

NOTE: Bold numbers bracketed in the text refer to the Notes on pages 83–250

CHAPTER 1

November 1911 to June 1923

The first recorded expression of interest in the oil possibilities of Kuwait was in November 1911. Shaikh Mubarak al Subah, then in the fifteenth year of his reign, had secured the independence of Kuwait twelve years previously from the designs of the Turkish Empire by signing an Agreement with Britain in 1899. The Agreement, signed on behalf of the British Government by its Political Resident in the Gulf **(1)**, included an undertaking that the Shaikh, his heirs and successors would not 'cede, sell, lease, mortgage or give for occupation or for any other purpose any portion of his territory to the Government or subjects of any other power' without the previous consent of the British Government. This proviso was destined many years later to cause complications in the negotiations for Kuwait's oil concession from 1928 onwards.

On 3rd November 1911 Mr Greenway, Managing Director of Anglo-Persian Oil Company Ltd. **(2)** wrote from London to the British Political Resident **(3)** in Bushire requesting his opinion as to whether 'a valid concession for working oil in Kuwait' was obtainable from the Shaikh;

> if so, I should like you to put forward an application on behalf of the Anglo-Persian Oil Co. for a Prospecting License, the draft terms of which I will send you later on if you think there is any chance of a Concession being obtainable. The question of whether or not there are oil deposits of any value in Koweit is, of course entirely problematic, and consequently we should only in the first place be prepared to take out a Prospecting License, which in view of the difficulties of prospecting in this place should be for a fairly long period, say 2–3 years.

British Admiralty Commission and first geological survey of Kuwait, 1913

At that time conditions in Arabia, where Kuwait's Arab and Turkish neighbours were struggling for supremacy, were too disturbed for this application for an oil concession to be put forward. But two years later a British Admiralty Commission headed by Admiral Slade, which had arrived in Persia to examine and report on the operations and prospects of the Anglo-Persian Oil Co. in 1912 and 1913, visited Kuwait in November 1913 **(4)** to discuss its oil potentialities with Shaikh Mubarak; who had also permitted them earlier in that year to carry out the first geological survey of Kuwait. This took place in March 1913 and its findings, issued in May **(5)**, assessed the area's oil chances as highly speculative though on the whole 'not unfavourable', recommending the central plain of Burgan as the best site for test drilling.

In October 1913, in connection with the Commission's November visit, Shaikh Mubarak and the British Political Resident exchanged letters **(6)** of importance for the subsequent history of Kuwait's oil. Writing to the Shaikh for his agreement to this visit by Admiral Slade to Kuwait 'in order that he may inspect the places showing traces of bitumen at Burgan and elsewhere' the Political Resident also asked that, if hopes of finding oil resulted, the Shaikh would agree not to give a concession in this regard to anyone other than a person nominated and recommended by the British Government. Replying the same day (the letters were confirming their discussion of the subject the day before), Shaikh Mubarak agreed both requests.

Early next year Anglo-Persian Oil Co. sent a geologist to Kuwait to inspect the seepages at Burgan and Bahra, but owing to the outbreak of the 1914–18 War it was not until 1917 that the Company was granted permission by Shaikh Salim (Shaikh Mubarak had died in 1915, being succeeded by Shaikh Jaber, who died early in 1917) to make its first geological survey in Kuwait. This conformed generally with its predecessor of 1913, recommending test operations at Burgan, and resulted in the Company's asking the Political Resident in October 1917 for permission to negotiate for oil rights in Kuwait. Owing to the difficult war conditions then prevailing throughout Arabia and Turkish Mesopotamia (now Iraq), the Political Resident advised postponement. It was not until May 1918 that the Chairman (see Note 2) of APOC in London formally applied to the British Foreign Office **(7)**, in a letter whose main purpose was to apply for a concession 'covering all oil deposits which may be contained in such part or parts of Mesopotamia' as might come under British control after the war, for the scope of any such concession to be also extended 'so as to cover the territories of the Sultan of Koweit in which we have already carried out a considerable amount of geological investigation in anticipation of obtaining the concession'.

Shaikh Ahmad, Ruler of Kuwait 1921–50

The final stages of the war in 1918, and Middle East peace treaties and resettlement in 1919 and 1920, prevented any progress with this application until May 1921. Three months previously Shaikh Salim had died, being succeeded by Shaikh Ahmad **(8)**, whose long rule was to cover not only the thirteen years of negotiations culminating in the first Kuwait oil-concession agreement of December 1934 but also, before his death in 1950, the first phase of his State's huge oil developments.

In May 1921 APOC reminded the British Foreign Office (see Note 7) of their 1918 application, requesting as regards Kuwait that the Political Resident should be instructed to apply to the Shaikh for 'an exclusive prospecting licence' for oil in favour of the Company in his territory, with a concession to follow if oil were discovered. The Company offered to send their own representative to negotiate with the Shaikh if this course were preferred by the political authorities.

This letter received a brief acknowledgement in June on behalf of Mr (later Sir) Winston Churchill **(9)**, then head of the Colonial Office (the British Government department to which principal responsibility for Kuwait affairs had recently been transferred from the Foreign Office), stating that after consulting the India Office and the Petroleum Department of the Board of Trade (see Note 7), 'a further communication' would follow 'in due course'. But it needed a further reminder by APOC in October to elicit on 24th December a full reply. The Colonial Office then stated that there were no objections to negotiations for such an oil agreement being opened immediately with the Shaikh of Kuwait, but pointed out that, following Shaikh Mubarak's exchange of letters with the Political Resident in October 1913, such negotiations had to be conducted by the Political Resident acting on British Government instructions.

Mr Winston Churchill authorises APOC negotiations, October 1922

The Company replied on 6th January 1922 agreeing and after considerable further correspondence regarding negotiating points and procedures required by the British Government (which were all agreed, though some reluctantly **(10)**, by the Company) they were at last informed on 16th October 1922 that

> Mr Secretary Churchill has caused the India Office to be requested to instruct the Political Resident in the Gulf to render assistance to the representatives of your company in negotiating an agreement with the Shaikh of Kuwait of the nature indicated in the correspondence ending with your letter under reply.

Early in 1922 the APOC had received a telegram from their Resident Director in Persia, Sir Arnold Wilson **(11)** to whom, on receipt of the Colonial Office's letter of 24th December 1921, they had countermanded previous instructions to start direct negotiations with Shaikh Ahmad. Sir A. Wilson, who was on terms of personal friendship with the Shaikh, had telegraphed that 'the Shaikh of Kuwait has informed me privately that he will readily accept any arrangement desired by the Company and approved by Government'. As soon as Sir A. Wilson had been informed of the British Government's letter of 16th October 1922, he telegraphed to the Company in London on 7th November proposing to go immediately from Abadan to Kuwait to open negotiations which, as indicated in that letter's last paragraph, no longer had to be conducted solely by the Political Resident. In reply he was told that the opening of negotiations should be delayed, owing to the Company's heavy programme of engagements and developments in progress elsewhere. Meanwhile, drafts of possible proposals to be submitted to Shaikh Ahmad, embodying conditions required by the British Government departments concerned in conformity with current concession procedures and precedents and agreed by the Company before the Colonial Office had sanctioned their negotiations, were the subject of much correspondence both between APOC and H.M.G. in London and between the Political Resident and the Company in Persia.

In January 1923 the Political Agent **(12)** in Kuwait without informing APOC initiated their negotiations with Shaikh Ahmad, describing to him the Company's proposed terms **(13)** for exploration and prospecting licenses and a subsequent mining lease (this being then the form required by old Colonial Office precedents for overseas mineral concessions) for his consideration, as a basis for negotiation with the Company's representatives as soon as he wished to discuss the matter with them. The Shaikh's initial reaction was favourable to the exploration and prospecting licenses but not to the mining lease, whose proposed Rupees 3 Annas 8 per ton royalty on crude oil he did not like, suggesting instead '25% royalty on net crude oil'.

During the next two months these proposed terms continued under discussion between APOC and H.M.G. and the Political Resident in the Gulf, and also between the Political Agent and Shaikh Ahmad in Kuwait. In March the Company provided the Political Agent with a draft concession agreement **(14)** as a basis for further negotiation with the Shaikh, including all his proposed modifications except his suggested alteration in the royalty basis. It also combined the two licences and the mining lease as originally discussed into one 'Concession Agreement' in what the Company considered a more modern and convenient form.

Major Holmes (EGS) concession proposals, May 1923

But, while this revised draft was being considered by the Political Resident and before he had agreed its presentation to Shaikh Ahmad, an unexpected development occurred which, by introducing a rival candidate for Kuwait's oil concession, was to involve the Shaikh in ten years bargaining with the two rivals in competition, with one more year to follow before he eventually granted to them jointly the Kuwait oil concession in December 1934. This was the following telegram received by Shaikh Ahmad on 9th May 1923 from Bahrain:

> I have most important letters from Ameen Rihani who has made enquiries concerning myself and Company advising Your Excellency not to grant oil concessions to any other Company without first seeing the terms offered by my Company. My representative will bring by next steamer the letters and terms to present to Your Excellency. Am pleased to inform you that I have secured the approval of His Highness Ibn Saud of my Company and secured the concessions against all other Companies who negotiated with His Highness.
>
> Major Holmes

Major Frank Holmes **(15)** was to play such an important part in Kuwait's oil history from 1923 onwards that his career to that date must be briefly described (for full details see Note 15). Born in New Zealand in 1874, he was a mining and metallurgical engineer of world-wide practical and managerial experience, having worked for various companies, mainly gold and tin mining enterprises, in Africa, Australia, Malaya, New Zealand, Mexico, Russia, and South America before the 1914–18 War. During the war he served as a major in the British forces in France, Egypt, and Mesopotamia (now Iraq). In 1918, when engaged in arranging supplies from Abyssinia through Aden for the British Army in Mesopotamia, he first visited the Arabian Gulf at Basrah. He had previously had no acquaintance with the oil industry, but as a mining engineer was interested in what he then saw of APOC's Abadan refinery and the oil industry in Persia, and what he heard of oil seepages and water problems on the Arabian Gulf coast. Demobilised from the Army in 1919, Holmes rejoined his pre-war mining associates in London, who in August 1920 formed a small company, Eastern & General Syndicate Ltd. **(16)**, with the object of obtaining concessions and investigating business opportunities in Arabia. Holmes was employed by this company, for whom he went in 1921 to Aden and in 1922 to Bahrain, which thereafter became the headquarters of his Arabian activities. Having visited King Ibn Saud at Ojair (see Note 22) in November 1922 for inconclusive discussions of an oil concession, he eventually succeeded on 6th May 1923, in obtaining from the King for the EGS an exclusive option for exploration of oil and mining rights in his province of El Hasa. One of Holmes's friends and supporters in his negotiations with Ibn Saud was Amin Rihani, a well-known naturalised-American writer and historian of Syrian origin, and during those negotiations he had obtained first-hand information about Kuwait from a Kuwaiti then in the King's entourage. Such was the background of Holmes's telegram from Bahrain to Shaikh Ahmad in Kuwait on 9th May 1923, three days after signing his El Hasa oil option with King Ibn Saud.

Holmes followed up his telegram by sending his interpreter-assistant (Holmes spoke no Arabic), Mohammed Yatim **(17)**, to Kuwait by the mail-boat from Bahrain with introductory letters both to the Shaikh and to several prominent Kuwaitis. These included the Shaikh's secretary Mullah Saleh (see Note 17), who was already on friendly terms with Yatim, a member of one of the most important merchant families of Bahrain. Holmes made his first visit to Kuwait a few days later. With such introductions and as a friend of King Ibn Saud he was well received by the Shaikh, with whom before leaving for Baghdad and London on 25th May, he had outlined suggested terms (on more generous lines than those hitherto proposed to the Shaikh by APOC; see below) for an oil concession for future negotiation after discussion with his EGS principals in London and also laid the foundations for what was to prove a life-long personal friendship with Shaikh Ahmad.

Holmes's intervention had immediate repercussions both on the British political authorities in the Gulf and on the APOC who were already negotiating through them with the Shaikh. The latter was reminded by the Political Agent, while Holmes was in Kuwait, of his obligation (in accordance with Shaikh Mubarak's undertaking in 1913) not to negotiate oil concessions except with persons nominated and recommended by the British Government which had not so far been approached by either Holmes or his London principals. To this the Shaikh rejoined on 23rd May that as Holmes was a British subject and had suggested terms more favourable than those hitherto proposed by APOC, and was going to London to pursue the matter with his principals, a British company which would doubtless make the necessary arrangements with the British Government, his actions to date had been wholly correct; as regards negotiations with Holmes he would not undertake these until the British Government had approved the EGS application in London.

The APOC in London, on hearing of Holmes's visit to Kuwait, telegraphed their Abadan management on 17th May to make every effort to get their latest draft Kuwait concession, which H.M.G. in London had by now approved for presentation, signed by the Shaikh; which the Political Resident had been instructed to try to get him to do. They added that Eastern & General Syndicate Ltd. had already asked them to join them in obtaining a Kuwait Concession, but they had declined.

Sir Arnold Wilson's (APOC) concession proposals, June 1923

APOC's General Manager (see Note 11) at Abadan, Sir A. Wilson, arrived in Kuwait on 31st May, a few days after Holmes's departure to London. The Political Resident arrived from Bushire the next day, when the APOC's draft concession was formally presented by Sir A. Wilson to the Shaikh and commended to the latter's favourable consideration by the Political Resident. On 2nd June, Sir A. Wilson accompanied by Mirza Muhammad (see Note 119) met the Shaikh **(18)** to discuss the draft and, he hoped, open negotiations with a view to its early agreement. The Shaikh however said he was unable as yet to discuss the terms of the draft concession, though he would have many observations and suggestions for its modification in due course. As the matter involved the vital interests of his country and of its people, it would be unfair for him to expedite its conclusion without due consideration and consultation. On the other hand, it would be unjust to the Company to reject any of their proposals off-hand. 'This affair is of a commercial nature, and it needs considerable deliberation in order that full justice may be done to both parties.' He would read one article of the agreement after another, suggest such alterations as he considered might be required by the interests of his country, and submit such new articles as he thought fit. It would take some time to do so, but he assured Wilson 'that I will not enter into any agreement whatever with any Company except the one of which you are the head without informing you beforehand'.

In reporting this interview to London, Wilson, who at that time had no knowledge of what monetary terms had been proposed by Holmes to Shaikh Ahmad ten days before, surmised that they were more favourable than those hitherto suggested by APOC; consequently the Shaikh, pending the authorisation of his negotiations with Holmes's Company which he expected to receive following their application to the British Government in London, would not want to involve himself in negotiations with APOC before he had improved his bargaining position by having another competitor also negotiating with him.

The monetary terms (the 'operating' terms were broadly similar) which Holmes had proposed on behalf of EGS were in fact considerably more favourable than those of the APOC draft **(19)**. His initial payment, for instance, was to be £2,000 (against APOC's Rs. 10,000 or £750) with a minimum annually thereafter of £3,000 (against APOC's £2,300). By themselves they justified Shaikh Ahmad's preference for Holmes's offer to APOC's. Furthermore the Shaikh was impressed both by Holmes's personality, by his strong recommendation by various friends, and especially by his success with King Ibn Saud, who had followed up his granting of his El Hasa concession to Holmes on 6th May by a letter to Shaikh Ahmad proposing that they should jointly grant to the EGS a concession for the Neutral Zone **(20)** between El Hasa and Kuwait (a proposal which a year later resulted in the EGS being granted an option over that area on 17th May 1924).

On the other hand, the Shaikh had had long dealings with APOC and friendship for Sir A. Wilson, and was aware both of the APOC's strong position and that it was favoured by the British Government, which at the end of June informed him that they had not yet been approached by, and consequently could not approve of his dealings with the EGS.

In the history of Kuwait's oil-concession negotiations June 1923 marked the end of the first phase, during which the APOC alone, from 1911 until the EGS/Holmes intervention of May 1923, had attempted to secure oil rights over the territory. In the second phase, to continue for the next ten years, the APOC and EGS (the latter from November 1927 onwards with the powerful American backing of the Gulf Oil Corporation of Pennsylvania) were to compete for the Shaikh's favour until May 1933. Thereafter APOC and Gulf Oil Corporation were to join forces and, in the third and final negotiation phase beginning in February 1934, despite complications caused by the last-minute appearance of a formidable competitor, bring the negotiations to a successful conclusion when the Shaikh granted the concession in December 1934 to their jointly owned Kuwait Oil Company.

State of negotiations, June 1923

The position at the end of June 1923 can be summarised as follows:

Shaikh Ahmad was much more impressed by Holmes's EGS proposals and much more inclined to prefer them to those of APOC than either the British political authorities or APOC believed at the time. This was due to several cogent reasons besides Holmes's better monetary terms. A principal reason was Holmes's success in having obtained Ibn Saud's concession, and the fact that he was in the King's favour and recommended by his friends; also the King had made the grant of his concession conditional on Holmes's undertaking not to sell the whole or part of it to APOC **(21)** whose H.M.G. connection (see Note 2) might, he believed, entail political complications. Another powerful factor in the Shaikh's reasoning, despite his respect for and obligations to the British Government, was his resentment at their action in November–December 1922 at their Ojair conference **(22)** with Ibn Saud in fixing the south boundary of Kuwait, without his knowledge or consent, so that a very large territory, previously undefined, had been lost to Kuwait. Besides these, Holmes's arguments in favour of his company's (see Note 16) purely commercial backers (including a world-famous mining concern) ready and willing to exploit any oil found in Kuwait, as compared with the British Government-controlled APOC, whose major commitments in Persia must modify their interest in developing Kuwait, must have appealed as much to the Shaikh and his council (see Note 8) as they already had in Holmes's negotiations with Ibn Saud. After Shaikh Ahmad's first interviews with Holmes and Sir A. Wilson in May and June, during which terms better than those on offer from APOC were mentioned by Holmes, the Shaikh had called a meeting of his Council and laid the two offers before them, coupled with the fact that negotiating with Holmes was not yet approved by the British Government. The meeting decided that the Shaikh must avoid accepting the APOC offer, unless obliged by the British

Government to do so. Holmes, who had left for England at the end of May with that knowledge, had assured the Shaikh that his EGS principals in London would soon secure British Government permission for his Kuwait negotiations, citing in support of this their approval of his recent successful negotiations in competition with APOC for King Ibn Saud's concession.

CHAPTER 2

July 1923 to May 1932

In July and August 1923, both APOC and EGS in London were advocating their cases to the British Government, the former urging the merits of their latest draft concession and the latter seeking official approval for their negotiations. The only development in Kuwait was that Shaikh Ahmad on 13th July informed the Political Resident that the terms of the APOC draft concession, as presented to him in June, were 'entirely unacceptable to himself and his people'.

Colonial Office defines British policy

On 6th September 1923 the situation and future procedure as between the two companies in Kuwait (and also in Bahrain for whose oil concession APOC and EGS were then also in competition), were the subject of a long letter **(23)** from the Colonial Office in London to the Political Resident in Bushire (which is so important both for their subsequent negotiations and also for its definition of British policy at that time that it is quoted in full in the notes). On 21st September letters conveying similar information were sent by the Colonial Office to both APOC and EGS in London.

The Colonial Office pointed out that the British Government had many years previously obtained from the Shaikh of Kuwait, and other Arab rulers in the Gulf, formal undertakings not to grant oil concessions to their territories to any persons not approved by them. The object of obtaining these undertakings were twofold. Firstly to protect the rulers against exploitation by unscrupulous concession-hunters, and secondly to restrict the grant of concessions to reputable British firms. In order to avoid advantage being taken of the rulers' inexperience in such matters, it was laid down that applicants for concessions should not negotiate direct with the Shaikhs, but should in the first instance seek British Government permission to apply for a concession on approved lines and, after obtaining permission and agreeing terms which the British Government would be prepared to recommend to the Shaikhs as acceptable, should then open negotiations through the Political Resident or Political Agent concerned. As Holmes 'avowedly, and probably actually' in ignorance of these requirements had made his initial approach direct to the Shaikh of Kuwait in May proposing his grant of an oil concession to EGS on terms requiring substantial modification before the British Government could recommend them, the Shaikh had been informed in June that the British Government had not been approached by EGS and Holmes's proposal was not approved. This did not, however, mean that the British Government had any objection in principle to EGS except for the irregular manner in which their proposal was made. So far as they were aware, EGS although possessing neither the wide experience of oil production nor the efficient organisation and financial strength of the APOC were a substantial and reputable firm which had now volunteered to modify their concession proposals so as to conform with the British Government's wishes. In view of APOC's prior application and correct procedure, and if they were prepared to offer terms at least as favourable as those submitted by EGS, the British Government would not recognise the EGS proposed concession until the Shaikh had explained his objections to the APOC proposals, as revised according to British Government requirements, and stated how he wished them modified. If thereafter his proposals were not accepted by APOC, the British Government would be prepared to approve his granting a concession to EGS, provided that their terms were recast on approved lines, and were as favourable to the Shaikh as those rejected by APOC.

Competitive negotiations authorised by H.M.G.

These letters, officially authorising competitive negotiations by EGS and APOC as soon as the Shaikh had expressed his views on the latter's draft terms as revised to meet H.M.G.'s requirements, gave great satisfaction to the Shaikh and to the EGS and Holmes. The APOC were correspondingly disappointed but remained confident because of their long and close connection with Kuwait and the Shaikh's views as expressed to Sir Arnold Wilson on 2nd June, of his decision in their favour before too long. Although still underrating the strength of Holmes's bargaining position with the Shaikh and of King Ibn Saud's influence in his favour, and Shaikh Ahmad's resentment at British treatment of his interests at the Ojair Conference ten months before, they instructed their friends and connections in Kuwait to be active in countering the pro-Holmes propaganda being conducted in Kuwait by Mohammed Yatim, whom Holmes had left in Kuwait while he went to England. Their main and strongest argument was, as also indicated in H.M.G.'s letter of 6th September to the Political Resident, their own position as a powerful and established oil company, compared with which the EGS, inexperienced and unknown in the oil industry, was unlikely to be able to exploit a Kuwait concession unless it could secure the backing of some oil company strong enough to withstand the powerful competition of the APOC's dominant position in Persia only a few miles across the Gulf.

This argument, whose validity had already been shown in May by the EGS's unsuccessful offer of collaboration with APOC, whom they were to approach again in 1926 (see Note 28) before eventually succeeding in obtaining the Gulf Oil Corporation's backing, was countered by Holmes's argument that such backing would certainly be forthcoming; and that, in any case, APOC's established position and large prior commitment in Persia would incline them against the sort of vigorous exploitation of any Kuwait concession, if they obtained it, which his own Company intended. While Mohammed Yatim had the support of his own friends, including the very powerful assistance of the Shaikh's secretary Mulla Saleh, and the pro-Ibn Saud party in Kuwait, in his efforts to sway public opinion, the APOC had equally strong support from their own established connections there (led by their oil-sales agent in Kuwait, the highly regarded head of the leading Al-Ghanim family; see Note 27), and the pro-British Government party.

By the end of 1923, while the Shaikh himself had given no indication of his intentions either to the Political Resident or to either of the Companies concerned, reports received by the latter from their local representatives were that local opinion was about evenly balanced. On the 28th November APOC in London had replied to the Colonial Office's letter of the 21st September, agreeing its proposed procedure for the Kuwait negotiations including the revision of its existing draft, and the Political Resident in Bushire was informed accordingly by the British Government and requested to report progress.

On 2nd January 1924 the Political Agent in Kuwait wrote to Shaikh Ahmad asking when he might receive his views on the APOC's draft concession. The Shaikh replied that it would be impossible to give a reply for six months, and in subsequent conversation said that, as he was going into every clause in detail, and getting the opinion of various people on different aspects, at least that time would be required. While this was undoubtedly correct, the Shaikh must have been disinclined, as an astute bargainer, to hurry his decision at a time when the two candidates for his concession through their local representatives were pressing their claims in a way which should result in improving their offers when the competitive stage of negotiating began. The APOC, for instance, wishing to demonstrate the extent and efficiency of their operations were encouraging him to visit their oilfields in Persia; Holmes, from London and Bahrain, was keeping him well informed as to his Company's continued interest in Kuwait, and also in a concession over the Kuwait/Hasa Neutral Zone (see Note 20), as well as his negotiations for a concession in Bahrain.

Dr Heim's (EGS) geological report, September 1924

During the rest of 1924, although no move was made by Shaikh Ahmad to start oil-concession negotiations, various developments of eventual importance to them took place. On 31st March (during which month Holmes paid his second visit to Kuwait) the Shaikh asked the Political Resident if there was any British objection to his granting jointly with King Ibn Saud oil rights for the Neutral Zone to the EGS and was informed that as King Ibn Saud had already given EGS a concession for his own territory, there was no objection. Consequently, on 17th May 1924 an option on the Neutral Zone **(24)** was granted to the Syndicate by the Shaikh and the King jointly. This favourably affected the EGS status in Kuwait, which was also improved by the arrival in April of Dr Heim (see Note 24), a well-known Swiss geologist, with three assistants, sent to Kuwait by the EGS, with the Shaikh's permission, *en route* to examine the Neutral Zone and Hasa territories on which they held options. In his report **(25)** dated 5th September 1924 Heim stated that they had found some earth impregnated with oil at the Bahra seepage. They had subsequently visited Burgan, but failed to find the seepage there. Heim's conclusion was that Kuwait was a country of some oil possibility, but not of high promise. As to the Neutral Zone, he stated that there was no seepage there, and that he could not recommend it, and that in his view drilling anywhere on the Arabian side of the Gulf would be a pure gamble. While this report did not discourage EGS in their efforts in Kuwait, it did make it difficult for it to interest other parties in oil possibilities there.

From the APOC's point of view, 1924 was marked by two main developments. On 27th May they were informed by the Colonial Office that it was undesirable for their negotiations for the Kuwait Concession to be prolonged longer than was absolutely necessary, especially in view of the fact that an application for it had been received from another party. While the Company's claims to priority would be recognised up to 31st March 1925, if their object had not been achieved by then H.M.G. would feel bound to allow other applicants to institute negotiations. This information, when Shaikh Ahmad heard of it from EGS, strengthened his policy of procrastination in opening discussion with APOC until their competitor was also in the field. And it caused APOC to intensify efforts to accelerate his decision in their favour, especially when, anticipating that their stalemate in Kuwait might continue beyond 31st March 1925, they suggested to the Colonial Office that their priority might be extended beyond that date and received a blunt refusal on 28th June. In July, therefore, APOC decided both to outmatch the terms proposed to the Shaikh by Holmes as soon as their negotiations could begin, and meanwhile renewed their invitation to the Shaikh to visit their oil operations in Persia; believing that his first sight of such operations in actual progress, and their very impressive scale of development, should influence him in their favour.

Shaikh Ahmad visits APOC oilfields and Abadan refinery

Shaikh Ahmad accepted the invitation, and subsequently visited the Persian oilfields from 25th to 29th July, and Abadan Refinery on 14th August, expressing great interest in all he saw during talks with the APOC officials who conducted him.

During his final interview on 14th August 1924 with APOC's General Manager, Mr T. L. Jacks (see Note 11) at Mohammerah the Shaikh described the position as regards his own oil concession and his recent agreement to grant equally with King Ibn Saud a concession to Major Holmes to search for oil in the Neutral Zone. As regards the latter, his own agreement was politically necessary because of Holmes's previous agreement with Ibn Saud. As regards the former he said that he and his advisers were studying both APOC and EGS proposals; that on his return he expected to be able to advise the Political Agent of such clauses in the APOC draft which it would be essential to modify; and that after his most impressive visit he was convinced of the APOC's strong position as a candidate for his concession. But when the General Manager suggested that he should indicate what clauses needed revision as this would facilitate negotiations, he replied that he preferred to leave this until his return to Kuwait, where his Council was in process of considering the Company's and Major Holmes's proposals. As public opinion in his territory was a factor in the negotiations which he could not disregard, he was anxious not to afford the Council any opportunity for criticising his actions.

The Company concluded that, though the visit had certainly much impressed the Shaikh, he was unlikely to take any action to prejudice his position before March 1925, when he was aware that with EGS free to negotiate his bargaining position would be strengthened. This proved to be the case, for after his return to Kuwait and during the rest of the year, Shaikh Ahmad made no move to express any opinion on the APOC draft, despite reminders by the Political Agent of the desirability of his doing so, on the lines discussed by him in Abadan in August. Besides his own inclination to postpone any action until after the following March, another factor making for delay (in addition to pressure maintained by the local supporters of both EGS and APOC) was opposition expressed by many merchants in Kuwait to any oil concession being granted, on the grounds that resultant employment would weaken their economic control over labour employed in boat-building and pearling, then the key industries in Kuwait.

In February 1925 APOC again requested the Colonial Office to consider an extension beyond 31st March of their 'prior rights' in the Kuwait negotiations, and also to withdraw their insistence on limiting to ten years the customs exemption proposed in APOC's draft terms for imports of machinery, etc. required for operation in Kuwait. In the latter connection APOC pointed out that no such limitation had been required by the Colonial Office in their then current negotiations for an oil concession in Muscat (this concession was subsequently agreed and signed on 18th May 1925) but the Colonial Office replied on 24th March refusing both requests.

APOC's prior rights of negotiation lapse

Although from 31st March 1925 onwards APOC's prior negotiating rights therefore lapsed, it was not until the following year that either they or EGS took any action in Kuwait apart from keeping public opinion warm, while Shaikh Ahmad himself made no move to approach them. Several reasons contributed to this pause in Kuwait. The Shaikh was preoccupied by his dealings with Ibn Saud and EGS over the Neutral Zone and by observing developments in Bahrain, where Holmes was drilling water-wells for Shaikh Hamad and on their successful conclusion was granted the Bahrain oil concession on 2nd December 1925. In Kuwait public opinion regarding the oil concession remained divided and controversial, Mohammed Yatim's efforts on behalf of Holmes being countered by those of Haji Abdullah Williamson **(26)**, who had been sent to Kuwait by APOC to add his influence to that of the Al Ghanim family **(27)** and their other friends and supporters. On behalf of EGS, Holmes was preoccupied in Bahrain and with the Neutral Zone arrangements. He was also having difficulty in fulfilling the obligations, both monetary and operational, of his Hasa Concession of 1923 with Ibn Saud (early in 1926 the EGS was to offer, unsuccessfully, to sell their Bahrain, Hasa, and Neutral Zone concessions and their anticipated Kuwait Concession to the APOC) **(28)**. This and the failure of Holmes's hopes of getting an EGS interest in the Iraq Concession which was granted in 1925 to the Turkish (later Iraq) Petroleum Company **(29)**, were adverse factors offsetting his success in Bahrain. The APOC remained unwilling to accept the Colonial Office stipulation of a ten-year limit to the customs-exemption on imported materials provided for in their draft Kuwait Concession, and representations that permanent exemption had been agreed in their Muscat Concession of May 1925 were unsuccessful in altering the Colonial Office's point of view. APOC, considering this to be unreasonable, and also that any weakening of their attitude over the question of customs exemption would establish a dangerous precedent in connection with their major concession in Persia which provided for no such time limit, were therefore in no hurry to begin negotiations in Kuwait while that difficulty was unresolved **(30)**. But at the end of 1925 (by which time APOC were aware that EGS had obtained Colonial Office approval of their draft Kuwait Concession for discussion with the Shaikh when he reopened negotiations), in order to maintain the favourable relations established with Shaikh Ahmad since his visit to Persia in 1924 (when the geological connection between oil possibilities on the Arab and Persian side of the Gulf had been among the subjects discussed with him), the Colonial Office agreed, provided that the Shaikh's agreement was also obtained, an APOC request that they should be permitted to make a geological survey of Kuwait. The Company had pointed out that, as EGS had sent Dr Heim's party of geologists to Kuwait in 1924, there should be no objection to APOC now doing likewise. Yet, when the Shaikh was approached by the Political Agent on the matter early in January 1926, it was only after long deliberation, and after receiving assurance that it had no connection with the question of oil-concession negotiations, that he gave his approval. Major Holmes was at that time in Kuwait *en route* to Bahrain, and had no doubt informed the Shaikh that EGS had recently received Colonial Office approval of their draft concession for discussion when the Shaikh was ready, so the Shaikh's reluctance in approving the APOC's geological survey was evidently due to a desire not to appear to be doing a favour to either party.

APOC geological survey of Kuwait

The APOC geologists arrived in January 1926, and completed their reconnaissance by early February; their report dated 13th February **(31)** reached the same discouraging conclusion as previous reports on Kuwait oil prospects.

When reporting to London in mid-February 1926 the result of their geological survey and also that Holmes had rented a house in Kuwait and was to arrive there shortly for discussion and possibly negotiations with the Shaikh, APOC in Abadan asked their London office if they too should take similar steps in Kuwait, but were instructed in March to suspend any such action 'in view of negotiations at present taking place with the EGS in London'. EGS had approached APOC on 23rd February, offering to transfer to the latter their Bahrain, Hasa, and Neutral Zone concessions together with any Kuwait Concession they might obtain, in return for reimbursement of all their expenditure on obtaining and validating them to date, amounting to some £50,000.

EGS offer to sell their Bahrain, Hasa, Neutral Zone, and Kuwait interests rejected by APOC

As a result of subsequent discussions and correspondence **(32)** throughout March, although APOC were ready to accept the transfer of all four concessions, they proposed monetary terms and other conditions which the EGS found difficulty in accepting. And on 7th April 1926 the APOC, in the light of their geologists' unfavourable opinion of the various territories, broke off the discussions.

With neither EGS nor APOC inclined to take any initiative, the former until they could ally themselves with an oil company able to supply the finance and expertise necessary for exploiting an oil concession, the latter because of their objection to the Colonial Office's limited customs-exemption proviso and their discouraging geological reports on Kuwait, and with Shaikh Ahmad also disinclined to make any move in the matter, negotiations for an oil concession in Kuwait remained at a standstill for the next two years. It was not until June 1928, following EGS's success in making an agreement with the Gulf Oil Corporation of Pennsylvania in November 1927 (see below) to take over their concession interests in the Arabian Gulf, that Holmes reopened his discussions in Kuwait with Shaikh Ahmad, this in turn spurring the APOC to similar action.

Meanwhile, EGS's geological surveys in Bahrain in 1925 during their water-drilling operations and in 1926 subsequent to obtaining the Bahrain oil concession in December 1925, had disclosed encouraging oil possibilities there which might also denote similar possibilities on the mainland in Hasa and Kuwait. Holmes therefore was sent by EGS in September 1926 for a month's visit to America, where he contacted various major oil companies and tried to get them interested in EGS's operations and prospects in the Arabian Gulf area, and also in the Red Sea where they were negotiating for a concession in the Farsan Islands. But he had no success (mainly because the majority of geological opinion was so adverse, though the remoteness and small areas of the territories concerned were contributory factors) except that the Gulf Oil Corporation expressed some interest in the Red Sea area. This, however, came to nothing a few days later when the news was received that EGS had failed to obtain the Farsan Island Concession.

Despite Holmes's failure to get any American backing by the time he left for London on 27th October 1926 it was decided that EGS's New York representative, Mr T. E. Ward **(33)**, should continue attempting to do so, especially with Gulf Oil Corporation, and he was fully briefed by Holmes with the latest information, particularly regarding APOC's opposition in Kuwait and recent refusal to take over EGS's Gulf interests. Holmes also informed Ward that he was drilling for water in Kuwait, in the hope of repeating his success in Bahrain both as regards obtaining useful geological evidence and as a further inducement to Shaikh Ahmad to grant him the oil concession.

Gulf Oil Corporation take over EGS's Bahrain, Hasa, Neutral Zone, and Kuwait interests, November 1927

In 1926 and 1927 Holmes completed four water-wells in Kuwait under a water-drilling contract which he had arranged with the Shaikh in March 1926, and although no water was found, one well showed slight traces of oil. This closely guarded secret greatly encouraged EGS and also Gulf Oil Corporation of Pennsylvania, who in November 1927, after protracted negotiations in New York and London and reconsideration of the latest geological information, took over EGS's Bahrain, Hasa, Neutral Zone, and Kuwait interests in two separate agreements with EGS both dated 30th November 1927.

Under one of these agreements Gulf Oil Corporation secured an option to acquire from EGS their Bahrain Concession of December 1925. Under the other agreement Gulf secured options to acquire from EGS their Hasa Concession of May 1923, Neutral Zone Concession of May 1924 and 'Kuwait Concession' as regards which, although no concession had been obtained at the time the option agreements were made, EGS was regarded as likely to secure it and undertook to use their best efforts to do so. (About a year later Gulf Oil Corporation, because of their participation **(34)** in the Turkish (later Iraq) Petroleum Company, had to transfer their interest in the Bahrain Concession to Standard Oil Company of California, but retained their Kuwait, Hasa, and Neutral Zone interests; they eventually renounced the latter two in April 1932.)

In April 1928 Holmes again visited Kuwait when *en route* to Bahrain, and was well received by Shaikh Ahmad and his advisers, whom he told that his company, having been approved by the British Government, wished to negotiate for an oil concession for which he would propose new terms in a draft concession which he would come and discuss in a month's time. In reporting this in a letter to Gulf Oil Corporation from Bahrain on 6th May Holmes expressed surprise that the Shaikh had been so ready to discuss the question of a concession, as he had refused to do so with an APOC representative who had visited him the day before. Holmes believed that the reason was threefold. Firstly, Shaikh Ahmad had opened their meeting by telling his advisers 'that I have proved a friend of the Arabs and would be more likely to help them than any other party, and that I had no political aspirations, being purely commercial as had been proved in Bahrain'. Such a remark clearly referred to the British Government's advocacy of, and their 51 per cent share interest in, APOC. Secondly, wrote Holmes 'it is interesting to observe the greedy attitude the Shaikh exhibited when I hinted to him that the Company I was representing had an American tang about it. The Arab Rulers certainly suspect every move of the British, this suspicion is largely removed when American money is suggested'. And thirdly, 'the Shaikh has been in constant communication with Ibn Saud during the past two months, and I feel sure that Ibn Saud has told him to ask plenty for his territory'. Although Holmes's reasoning showed his usual shrewdness, in the light of subsequent events and Shaikh Ahmad's adroit bargaining over the next six years, a further very powerful factor may well have contributed to his reasoning. While public opinion in his State was clearly deeply and about equally divided on the merits of the two contestants for his oil concession, both of whom were now pressing him to open negotiations, both of them with influential *political* backers (Ibn Saud and the Shaikh of Bahrain for EGS, the British Government for APOC), now, with their new American backing, EGS could be regarded for the first time as being *commercially* as able as APOC to exploit his oil concession. On such reasoning, this occasion may have been the first indication of the strategy to be so successfully followed by the Shaikh during all his subsequent negotiations, namely commercially to get the benefit of the best possible terms with two powerful competitors bidding against each other and politically to strengthen his position by forcing them to combine, thus adding an American element to his hitherto solely British dependence, on the same pattern as had recently occurred in his northern neighbour Iraq with the inclusion of an American group in the internationally owned oil company exploiting its concession (see Note 34).

Shaikh Ahmad receives and rejects Holmes's (EGS/Gulf) draft concession proposals, June 1928

Holmes revisited Kuwait in June and presented his draft concession document to whose terms the Political Agent advised the Shaikh that the British Government had no objection. But it proved unacceptable to Shaikh Ahmad and his advisers, despite increases in the monetary terms conceded by Holmes in the course of discussions which continued until August. This was unexpected by Holmes, who on 3rd July had telegraphed that his revised draft **(35)** being submitted that day had been amended as agreed with the Shaikh's 'State Adviser' (Mullah Saleh, Mohammed Yatim's friend who was the EGS party's strongest Kuwaiti supporter) and that he had high hopes of its approval when considered the following day by Shaikh Ahmad and his Council of State. It was equally disappointing to Gulf Oil Corporation, who had approved the increased payments in his proposed draft with some reluctance **(36)** (expecting they would clinch the matter) but informing him 'that one of the tenets of the religion of the oil people is to take what is obtainable as and when it can be had, because experience had taught that the future is always uncertain, particularly when political questions or viewpoints become involved'. Furthermore Holmes had informed them early in July that 'opposition have been most virulent and are still very active; their attitude has been more to prevent the signing of our draft rather than endeavouring to secure concession at this moment for themselves'.

In August 1928 the Shaikh and his Council definitely rejected Holmes's draft terms, demanding both larger royalties and also a provision that the concession was granted to Holmes personally and could be invalidated if transferred to any other company or interest. These revisions were so far-reaching that it was agreed to suspend the negotiations to enable Holmes to discuss them with his principals in London, which he reached in late August 1928 going on to New York a month later for discussions with Gulf Oil Corporation. Holmes's unexpected failure to reach agreement with Shaikh Ahmad in August had been attributed to the strength of public opinion in Kuwait and among his Council in favour of negotiating also with APOC. While that was undoubtedly one factor, coupled with the Shaikh's strategy of developing a competitive situation and keeping the position open meanwhile by hard bargaining, the implication of his 'non-transfer' proviso seems so significant in view of a major development about to occur in November that the Shaikh's advisers and especially his Bahrain connections may well have anticipated it.

H.M.G. requires a British Nationality clause in Kuwait and Bahrain Concessions, November 1928

That major development, described below, occurred on 29th November 1928 when the Political Agent in Kuwait notified Holmes (who had returned to Kuwait after his American visit, during which he had obtained Gulf Oil Corporation's approval for continuing his negotiations with the Shaikh on the lines proposed by the latter in August, and believed he was on the verge of an agreement **(37)**) that the British Government would require the insertion in any oil concession which the Shaikh of Kuwait might grant of a British 'Nationality Clause' which (see below) would in effect exclude American participation in it. Simultaneously, EGS in London were informed by the Colonial Office that the same clause was now required in their Bahrain Concession. It was to take EGS and their American backers in Bahrain (Standard Oil of California) until January 1930 to get this clause sufficiently modified for their concessionary operations in Bahrain to proceed and thereafter it was not until April 1932 that EGS and Gulf Oil Corporation, with U.S. Government support, secured the British Government's agreement to omit a British Nationality clause in their draft concession agreement for negotiation with Shaikh Ahmad. Meanwhile no further negotiations for a Kuwait concession could be undertaken by Holmes, although in frequent visits he maintained his very cordial relations with the Shaikh. The latter both sympathised with Holmes's frustration and himself disliked being thus prevented from opening negotiations with him, in competition with APOC, for an oil concession whose proceeds were becoming increasingly necessary for Kuwait from 1929 onwards, as successive annual failures of the pearling season aggravated the economic distress caused in Kuwait by Ibn Saud's blockade of Kuwait trade with the interior of Arabia.

APOC, although as a British company they were unaffected by the stipulated Nationality Clause, rather welcomed its immobilising effect on the Kuwait negotiations at a time when, in the world trade depression of 1930/1931 following the New York Stock Exchange slump of 1929, all oil companies were cutting down on expenditure. But they too continued to maintain good relations with the Shaikh and his community through their local representatives, against the time when EGS/Gulf had resolved their 'Nationality Clause' problem. Until then they saw no reason to offer to reopen negotiations with Shaikh Ahmad calculating that he would not himself desire it until their competitor was again in the field. Furthermore they remained unwilling to accept the Colonial Office's requirements in their draft Kuwait Concession terms for a ten-year limit to customs exemption for imported plant and material, and for a royalty of Rs. 3.8 per ton of oil produced.

Terms of proposed 'British Nationality' clause

The 'Nationality Clause' required by the Colonial Office on 29th November 1928 to be inserted in any oil concession to be granted by the Shaikh of Kuwait and also in EGS's existing Bahrain Concession of December 1925, read as follows:

The Company shall at all times be and remain a British company, registered in Great Britain or a British Colony, and having its principal place of business within His Majesty's Dominions, the Chairman and Managing Director (if any) and the majority of other Directors of which shall at all times be British subjects. Neither the Company nor the premises, liberties, powers and privileges granted and demised nor any lands occupied for any of the purposes of this lease, shall at any time be or become directly or indirectly controlled or managed by a foreigner or foreigners or any foreign corporation or corporations, and the local General Manager of the Company, and as large percentage of the local staff employed by them as circumstances may permit, should at all time be British subjects or subjects of the Shaikh.

In this clause the expression 'foreigner' means any person who is neither a British subject nor a subject of the Shaikh, and the expression 'foreign corporation' means any corporation other than a corporation established under and subject to the laws of some part of His Majesty's Dominions and having its principal place of business in those Dominions.

Hitherto, the only relevant clause in both the EGS's Bahrain concession and its latest draft Kuwait Concession (which in that and most other respects, but not its monetary terms, was similar to the Bahrain Concession) as submitted to the Shaikh with Colonial Office approval earlier in November, read as follows:

The rights conveyed by this lease shall not be conveyed to a third party without the consent of the Shaikh acting under the advice of the Political Resident in the Persian Gulf. Such consent shall not be unreasonably withheld.

This new restriction to British firms only of third parties to whom these concessions could be conveyed was in general accord with the Colonial Office policy as defined in its letter of September 1923 (see Note 23) whereby such concessions should be restricted to 'reputable British firms'. The Colonial Office had only become aware in October 1928, when Gulf Oil Corporation had unsuccessfully offered their option on the Bahrain concession to the British and other groups in Iraq Petroleum Company before successfully transferring it to Standard Oil of California, of Gulf's acquisition of EGS's Bahrain and Kuwait options a year earlier. Hence its action in November 1928 to impose British nationality on concessions in both those territories in accordance with that 1923 ruling, regardless, as both EGS and Gulf were well aware and hastened to point out, of the British Government's action since that date in Iraq in 1924, after strong pressure from the United States Government, to agree to an 'Open Door' policy for American oil interests in the Middle East.

Representations were made in December 1928 and January 1929 by EGS to the Colonial Office in London as to the inequity and damaging nature of a 'Nationality Clause', as now proposed for its Bahrain Concession, when no such stipulation as to the British character of any assignee had been imposed in the concession as originally approved by themselves; but with no success. Consequently, on 20th March 1929, the attitude of the British Government in respect of the Bahrain Concession was discussed in Washington by representatives of both Gulf Oil Corporation and Standard Oil of California with the United States Secretary of State, who subsequently had representations on the subject made to the British Government by the American Embassy in London. As a result the British Government receded from its position as regards Bahrain, EGS being informed as follows in a letter from the Colonial Office dated 30th May 1929:

As a result of further consideration, His Majesty's Government are prepared in principle to consent to the participation of United States interests in the oil concession in the Bahrein Islands, subject to their being satisfied as to the conditions on which United States capital will participate.

I am to suggest that the conditions governing the admission of United States capital should form the subject of discussion between representatives of your Syndicate and the Colonial Office, and to say that a further letter in this regard will be addressed to you in due course.

Modification of British Nationality Clause for Bahrain Concession agreed between H.M.G. and EGS

Subsequent discussions and correspondence culminated in a letter dated 16th September to EGS from the Colonial Office imposing various conditions required by H.M.G.

for insertion in the Bahrain Concession in lieu of the 'Nationality Clause'. As these conditions were not entirely satisfactory from the American point of view, modifications were suggested and eventually a version acceptable to both sides was settled in January 1930 **(38)**. The Colonial Office then advised the Shaikh of Bahrain to agree to the assignment of his concession, on such conditions, to a Company controlled by American interests, and the concession was accordingly assigned to the Bahrain Petroleum Company, a Canadian subsidiary of the Standard Oil Company of California.

H.M.G. insists on a British Nationality Clause in Kuwait Concession

The British Government, in a note in May 1929 to the American Embassy in London consenting to participation by American interests in the Bahrain Concession, had 'found themselves unable to make any general statement of their policy on this question such as the United States Government desire'. But representatives of EGS and of Gulf Oil Corporation on various occasions in 1929 and 1930, had understood from officials of the Colonial Office that the conditions agreed with respect to the Nationality Clause in Bahrain would serve as a precedent in the similar matter of the Kuwait Concession negotiations. This however proved not to be the case when Holmes, after returning to Bahrain from London early in 1930 following the agreement reached by EGS with the Colonial Office, arrived on 4th August in Kuwait, and had a cordial discussion the next day with Shaikh Ahmad who, reported Holmes, 'is friendly and quite willing to give us the Concession'. On 6th August 1930, however, Holmes was informed by the Political Agent in Kuwait (Lt-Col H. R. P. Dickson **(39)**) that he was under instructions from the Political Resident to inform Shaikh Ahmad that the British 'Nationality Clause' would have to be inserted in any oil concession granted by him. Holmes, leaving Kuwait on the 20th August after suspending his talks with the Shaikh, reported this problem to EGS in London, who took it up with the Colonial Office expecting its early settlement on Bahrain lines.

But in fact it was not to be until nineteen months later, in April 1932, that the British Government agreed to renounce their insistence on the Nationality clause in Kuwait thus enabling Holmes to return there in May 1932 and renew his negotiations on behalf of EGS/Gulf with Shaikh Ahmad. And they only did so then after strong representations made to them by the U.S. Government from December 1931 onwards, Gulf Oil Corporation having requested this Government support when joint EGS/Gulf representations in London had had no success during the previous fifteen months.

Events during that period had been as follows:

In August 1930 EGS sent to the Colonial Office the draft Kuwait Concession which Holmes had been discussing with the Shaikh and the Political Agent in Kuwait earlier that month. They pointed out that it contained all the clauses which the Colonial Office had accepted in January 1930 in the EGS Bahrain concession in substitution for the original 'Nationality Clause', and which EGS were equally willing to accept in connection with a Kuwait Concession. Colonial Office replied that before discussing the matter they had to obtain the comments of other interested British Government departments in London and of the Political Resident in Bushire. It was not until 13th January 1931, after various enquiries and reminders, that EGS was informed by Colonial Office that all these comments had been received and that the Syndicate was to expect a letter on the subject within a few days. On 31st January the Colonial Office wrote to EGS as follows:

> I am directed by Lord Passfield [Secretary of State, Colonial Office] to state that it has now been ascertained that the Sheikh of Kuwait has definitely refused to grant a concession to any concern that is not entirely British, and that he will insist on the inclusion of a British 'Control Clause' in any concession which he might grant in respect of his territory. In the light of this information your Syndicate's application has been carefully considered in consultation with the other interested Departments and the Government of India, and I am now to inform you that His Majesty's Government are not prepared to advise the Sheikh to reconsider his attitude in this matter.

Holmes (EGS/Gulf) disbelieves H.M.G.'s statement that Shaikh Ahmad requires British Nationality Clause

This letter greatly surprised EGS and particularly Holmes, who was then in London and expressed disbelief that the Shaikh's attitude was correctly stated. After further discussions the Colonial Office informed Holmes and his EGS colleagues in March 1931, while reiterating that Shaikh Ahmad insisted upon the 'Nationality Clause', that he might be induced to change his attitude eventually through negotiations. The Colonial Office had no objection to Holmes returning to Kuwait for such negotiations, and in fact recommended him to do so, so Holmes left London for Kuwait on 26th March, resuming his negotiations with the Shaikh shortly after his arrival there on 16th April 1931.

Gulf Oil Corporation had written to Holmes in London impressing him with the importance of his mission and of securing the Shaikh's agreement to having no British Nationality Clause. They had been encouraged as to the oil potentiality of Kuwait by traces of oil found in one of the water-wells which Holmes had drilled for the Shaikh, and they suspected that news of this had also reached APOC whose representatives had unsuccessfully tried to investigate the well early in March; action which might mean that APOC were contemplating reopening their own negotiations which had now been at a standstill since 1928.

APOC were not in fact considering any such action at that time. In October 1930, the British Government had written **(40)** to APOC enquiring if, in view of EGS's active interest in Kuwait, they too were prepared to revive their interest in its oil prospects. APOC's reply (see Note 40), after mentioning that APOC's previous attempts to negotiate a Kuwait concession had broken down because of the Colonial Office's insistence on unacceptable terms, was discouraging and in January 1931 **(41)**, APOC's disinterest in Kuwait was confirmed again to the British Government.

Holmes's discussions with Shaikh Ahmad were protracted until early July 1931. He emphasised to his American principals the undesirability of hurrying the Shaikh in what was evidently a difficult situation for him, which a visit by the Political Resident on 28th April made no easier. Mullah Saleh reported to Holmes that Shaikh Ahmad had been advised by the Political Resident not to grant an oil concession to Holmes's company, and that the Shaikh had replied that 'he had promised a concession to Major Holmes and desired to carry out his promise'.

Shaikh Ahmad's letter of 2nd July 1931 to Holmes

Finally, after further discussions with Shaikh Ahmad Holmes at the end of June wrote to him asking for his decision regarding the concession. To which the Shaikh replied **(42)** on 2nd July 1931:

> I am in receipt of your letter dated 28th June 1931 contents of which are noted. The draft of the conditions which you have presented for an oil concession in the territories of Kuwait has been considered by us.
>
> We have mentioned to you in the course of our conversations regarding the clauses which have been indicated to you by H.B.M.'s Government that their incorporation in any oil concession is necessary. Because I believe that H.B.M.'s Government is my own sincere Government which is always devoting its care to the welfare of my country and the safeguard of my rights, we must not ignore what it considers to us and our country.
>
> Therefore if you and your company agree with H.B.M.'s Government on the said clauses mentioned to you and it allows you to omit them, then we shall have another opportunity of discussing matters with you.

This letter (for facsimile and Holmes's translation see Note 42), which was to prove an important factor in solving the problem of the 'Nationality Clause', was taken by Holmes and EGS/Gulf as clearly showing that Shaikh Ahmad did *not* insist on the inclusion of such a clause in his concession, as had been stated in the Colonial Office letter of 31st January, and that if the British Government agreed with EGS on its omission (on the pattern which he knew had been adopted in the Bahrain Concession) he would continue negotiating with Holmes accordingly. But it was not taken in that sense by APOC's General Manager in Abadan, who received a copy of it from his Kuwait agent and sent it to APOC in London on 9th July with the comment that it showed 'that the Shaikh at the present time shows no disposition to come to terms with Major Holmes and his group regarding a Concession for the Kuwait area'.

The fact is that the letter was not intended by Shaikh Ahmad to be a formal definition of his position or to be quoted as such. As will be seen, in May 1932 he was to express great displeasure with Holmes (it is the one recorded instance of their long friendship being even temporarily interrupted) for, in Holmes's own words, 'misleading him and using the letter to get decisions made behind his back'. The letter was, however, taken seriously enough by APOC in London to alert them to the imminent possibility of the Shaikh giving Holmes a concession for EGS/Gulf if the British Government could be persuaded by EGS to waive, as in Bahrain, the Nationality Clause; on whose impregnability they were so relying to thwart both EGS's protests in London and Holmes's negotiations in Kuwait that they had believed there was little urgency about any approach by themselves to the Shaikh. APOC therefore on 25th August informed the British Government **(43)** that, after obtaining Shaikh Ahmad's permission, they were proposing to send a small party of geologists to Kuwait later that year to make a thorough examination of its surface geology; a proposal which the British Government welcomed as being in line with its suggestion in the previous October (see Note 40) that APOC should revive its interest in Kuwait.

Holmes had arrived back in London from Kuwait on 14th July 1931 with the Shaikh's letter of 2nd July, and on 4th August EGS sent a copy of it to the Colonial Office asking whether they would now notify the Political Resident that the British Government had no objection to the Shaikh granting them an oil concession with the 'Nationality Clause' omitted. They also informed the Colonial Office that Shaikh Ahmad was fully aware that their concession was to be exploited by American interests and that the Syndicate would include in the Kuwait Concession similar conditions as regards nationality as those approved by the Colonial Office in their Bahrain Concession. By the time Holmes returned to Kuwait in October no reply had been received from the Colonial Office except that they would have to 'consult the authorities in the Persian Gulf' before doing so, and despite reminders by EGS no reply, except that the Political Resident had not yet sent his views, had been received by the end of November 1931.

Holmes, who telegraphed from Kuwait that Shaikh Ahmad told him on 21st October that he stood firmly by his letter of 2nd July although the Political Agent 'had done nothing but badger him during the past few days', reported on 1st November that the Political Resident, visiting Kuwait during the previous week, had had many talks with the Shaikh. The latter, wrote Holmes, told him that the Political Resident tried to persuade him to withdraw his letter, kept impressing upon him that there was no need to hurry granting a concession to Holmes or anyone else, and also informed him that APOC would like his permission to send geologists to Kuwait.

Shaikh Ahmad agrees APOC geologists' visit to Kuwait

As regards the latter Shaikh Ahmad, who two days previously had learned from Holmes (whose London office had cabled him) of APOC's intention, said he had agreed. Holmes also reported a week later what 'the Secretary of the Council of State' (Mullah Saleh) had told him on the subject. According to Saleh, the Shaikh and Council of State believed that APOC was primarily seeking political power for the Colonial Office and Shaikh Ahmad had told the Political Resident and Political Agent that he would not permit himself to be made the cockpit of trouble between EGS/Gulf and the Colonial Office. On being asked why he had given Holmes the letter of 2nd July, Shaikh Ahmad had replied by asking 'where does it state in any treaty between the Kuwait Government and the British Government that the Sheikh is not permitted to write to persons like Holmes regarding concessions?' Shaikh Ahmad also told the Political Resident that he had shown the letter to the Acting Political Agent in Kuwait Dr Greenway **(44)** before its delivery to Holmes, and that the Acting Political Agent had told the Shaikh that as the letter was only putting the Nationality Clause point to the Colonial Office, he saw no harm in it.

Shaikh Ahmad's anxiety to have Kuwait's oil resources developed

Holmes also reported that Shaikh Ahmad, when dining with him on 30th October, had said that his letter made it clear that the Nationality Clause was a question to be settled between Holmes's Company and the British Government.

> If they agree on the Clause question and their decision is transmitted to the local political officials in order that their opposition may be withdrawn, I am still willing to grant the Kuwait Concession to Major Holmes's Company. I shall not, under any circumstances withdraw my letter to Major Holmes as it expresses my feelings as I consider this question is one between Major Holmes's Company and the British Government. In common with our people, I am anxious that the oil resources should be developed with as little delay as possible. It was a stab to my heart when I observed the oil work at Bahrein and nothing here.

In retrospect, that last sentence contains the heart of the matter. In the then very bad state of Kuwait's finances and general economy, the revenue from an oil concession was badly needed and if oil were found it would mean economic salvation. That this prospect had been denied him for more than a year because the British Government insisted on a proviso which they had waived in Bahrain must have been intolerable to the Shaikh. And the fact that APOC, whom he knew were favoured by the British Government and to whom the Nationality Clause was no obstacle, were showing no interest in renewing their offer of a concession but only wished to make yet another geological survey of his territory must have strengthened his resolve to force the British Government's hand, if only to stimulate APOC into action.

In November 1931 Holmes's verdict on the situation **(45)** was that as fifteen months' representations to the Colonial Office since August 1930 had failed to alter their attitude or to elicit any reason for it except that the Shaikh endorsed it, which the Shaikh's letter to him of 2nd July and subsequent discussions with the Political Resident had contradicted, 'I personally consider that we have been patient and considerate to a fault, and that we are now justified in making use of every weapon, diplomatic or otherwise, which we possess.' He was clearly advocating the same request by Gulf Oil Corporation for U.S. Government representations to the British Government which had proved successful in 1929 in removing the 'Nationality Clause' obstacle from the Bahrain Concession negotiations.

Gulf Oil Corporation's discussion with State Department, Washington

Holmes's advice was followed. A memorandum detailing the current impasse in his Kuwait negotiations on their behalf and its back history, was discussed with the State Department in Washington by Gulf Oil executives on 30th November 1931. The final paragraph of the memorandum's covering letter was as follows:

> As the foregoing brief statement of facts indicates, the British Government has for three years prevented our obtaining the Kuwait Concession. During all that period the Syndicate and we have exercised the utmost patience, and have tried in every proper way known to us to secure some satisfaction from the Colonial Office. We feel now that our efforts are exhausted, and unless we have the prompt assistance of the United States Government, the combined pressure of the British Government upon the Sheikh and activities of the Anglo-Persian Oil Company in Kuwait may shortly result in completely destroying any opportunity for us to obtain the Kuwait Concession.

In referring to the activities of the Anglo-Persian Oil Co. in Kuwait, Gulf Oil Corporation had principally in mind not only that Company's forthcoming geological survey as approved by Shaikh Ahmad in discussion with the Political Resident on 27th October, but also what they had heard about increased activity by APOC representatives in Kuwait to influence the Shaikh against EGS through members of his family and Council and influential merchants. Holmes had written to EGS/Gulf from Kuwait on 1st November:

> The APOC's agents have been very active and are still so, both with their money and propaganda. The APOC subscribed liberally to the hospital and schools and managed to secure the help of some of the leading merchants.

American Embassy in London discusses Kuwait situation with British Foreign Office

The State Department took prompt action. With the object of securing freedom of negotiation for American companies in Kuwait on the precedent they had established with the Bahrain Concession in May 1929, on 3rd December 1931 they instructed the American Embassy in London to take the matter up with the British Foreign Office. By the end of December it had reached the highest level there, having been twice discussed by the American Ambassador, General Dawes, with the British Secretary of State for Foreign Affairs, Sir John Simon **(46)**, when doubts were expressed by the latter as to the correct interpretation and significance of Shaikh Ahmad's letter of 2nd July to Holmes. In January 1932 with no decision reached and the news from Kuwait of APOC's geological survey party's arrival there in December and activity thereafter, the Secretary of State in Washington, at Gulf's request, instructed the American Embassy in London to press the Foreign Office for a decision, to be told on 3rd February that the latter were still awaiting the India Office's views. At the end of February, Gulf's Kuwait news from Holmes that APOC's geologists had followed their preliminary reconnaissance with a test-drilling programme and that Shaikh Ahmad was under strong pressure from their supporters in Kuwait, caused the

Embassy to renew its representations to the Foreign Office, only to be informed on 15th March both that further delay was inevitable, as divergent views received from various Departments had to be considered, and also that APOC was considering a renewal of its application for a concession in Kuwait.

Hitherto the Embassy–Foreign Office exchanges, though conducted at top official level, had not been the subject of a formal request from the American Government to the British Government for their 'open door' policy for American Oil Companies, already existing in Bahrain and Iraq, to apply to Kuwait. On 26th March, no definite reply having then been received as a result of increasingly urgent representations since early December, the Embassy in London was instructed by the State Department **(47)** to take this formal action, and on 30th March delivered a note on the subject to the Foreign Office; in reply to which they were informed that the matter was to be discussed by the British Cabinet on 6th April.

H.M.G. letter to American Embassy, London, agreeing to 'open-door' policy in Kuwait

On 8th April the U.S. Ambassador, Mr Andrew Mellon (who had recently succeeded General Dawes; see Note 47), informed the State Department in Washington that he understood a favourable reply was being drafted, and on 11th April 1932 he reported that the matter had at last been settled in a letter from the British Secretary of State for Foreign Affairs, Sir John Simon, dated 9th April 1932 (see Note 47) agreeing the American request; which good news was slightly dimmed for Gulf Oil Corporation by the letter also including the information that the Shaikh did not consider himself as in any way committed by his letter to Holmes to grant the EGS his concession.

Shaikh Ahmad's quarrel with Holmes

As soon as he received this news Holmes left Bahrain for Kuwait where he arrived on 27th April, at first reporting a very cordial reception by Shaikh Ahmad but, a few days later, a very serious setback on 3rd May **(48)**. On that day, in discussion with Shaikh Ahmad, Holmes had been describing the events leading up to his Company's recent success in getting the British Government to withdraw its insistence on the 'Nationality Clause' for Kuwait and mentioned the key role played in them by the Shaikh's letter to him of 2nd July. Whereupon Shaikh Ahmad expressed great anger, accused Holmes of having misled him and used the letter to get decisions made behind his back, and broke off the conversation. It was not to be until ten days later, Holmes's Kuwaiti friends having meanwhile exerted all their influence with the Shaikh, that the latter consented to see him again, when cordial relations were re-established on 14th May in a two-hour conversation during which Holmes succeeded in explaining matters to Shaikh Ahmad's satisfaction. This episode (the only one of its kind in the long history of Holmes's close friendship, from 1923 until his death in 1947, with the Shaikh) was inexplicable to Holmes. In fact it was the result of the following sequence of events resulting from dealings which APOC had been having with the Shaikh over the same period.

Two weeks before Holmes had returned on 27th April 1932 to Kuwait APOC, considering that the geological evidence **(49)** in Kuwait was not yet sufficient to justify their seeking a concession there, had abandoned their geological operations then in progress and written to the Shaikh accordingly on 13th April **(50)**. This decision came as a great disappointment to Shaikh Ahmad, which he freely expressed to APOC representatives and the Political Agent, because he had been counting on APOC's renewed interest to stimulate the profitable competition for his concession essential to alleviate Kuwait's current economic straits. When therefore Holmes on 3rd May mentioned his letter of 2nd July 1931 as having been helpful in securing the withdrawal of the Nationality Clause which could be considered a setback to APOC, the Shaikh felt that it could also be a mark of British Government displeasure and was furious at the thought that Holmes's use of his letter could have helped to cause it. But a week later, on 11th May, Shaikh Ahmad had received from the Political Agent the good news that APOC (who on learning that the 'Nationality Clause', on which they had been relying to block EGS/Gulf's progress, had been waived by the British Government had decided that they must act to protect their position in Kuwait) wished to reopen their negotiations with him for a concession. Although Shaikh Ahmad refused an APOC offer, also then conveyed to him by the Political Agent, of £2,000 for a two-year exclusive prospecting licence because he already had a full concession on offer from Holmes, he asked the Political Agent to tell APOC that he was ready, as always, to consider a similar full offer from them, whereafter he would compare the two and eventually grant the concession on the best terms available, subject to His Majesty's Government's approval. It was against that background, with the competitive position which he had always worked for and had so recently seemed to have lost, unexpectedly restored, that the Shaikh resumed cordial relations with Holmes three days later on 14th May.

By the end of that month all the parties to be concerned in the intensely competitive phase of the negotiations about to open, with any previous obstacles removed, had completed their arrangements. On 26th May Shaikh Ahmad received from Holmes his latest draft concession **(51)**, which was under consideration by his Council of State. He knew that the APOC's draft concession, as requested by him to the Political Agent on 11th May, was being prepared and would be submitted to him as soon as possible for similar consideration. He had been informed of this both by the Political Resident, who had visited him on 23rd May (when he also informed him that both Holmes's Company and APOC were approved by His Majesty's Government as negotiators for his concession) and also through Mr A. H. T. Chisholm **(52)** of APOC who visited Kuwait from Abadan on 31st May. Holmes was reported on 6th June by Gulf in London to New York as having cabled **(53)** from Kuwait that the Shaikh, in discussion on 27th May, had complained of strong British political pressure on him to favour APOC but had added:

> No matter what terms APOC offers I will not discuss them, nor will I upon any consideration grant to APOC the Kuwait concession, I have promised the Kuwait concession to you and shall stand by my word.

On 30th May APOC in London telegraphed to Abadan, 'Consider it essential to use all methods in our power to prevent Sheikh giving concession to Syndicate'.

The position of the negotiations at the end of May 1932 can be summarised as follows.

Shaikh Ahmad had at last achieved the strong bargaining position which he had hoped for and, according to all available evidence, consistently and skilfully manoeuvred for during the nine years since Holmes had first arrived in May 1923 to compete with APOC's offer of five months earlier. Both the rival parties now about to open formal negotiations for his oil concession had shed, or removed, such commercial or political handicaps as had hitherto been restraining them. Holmes's company, originally insignificant in comparison with the APOC, now represented the equally powerful Gulf Oil Corporation and had removed British Government objections to this American connection. APOC, whose preoccupations in Persia combined with successive discouraging geological reports and a belief that British Government policy **(54)** would long delay American competition in Kuwait had induced a stalling attitude in Kuwait, had had to abandon that attitude to prevent a powerful competitor commencing operations so near their own. APOC's urgency for that reason was matched by Gulf Oil Corporation's for an equally powerful one, namely their growing belief, about to be proved correct, that the Bahrain Concession which they had unwillingly had to transfer in 1928 to Standard Oil of California was shortly to prove an oil-winner; which would enormously enhance Kuwait's oil potential, regarding which they already had encouraging evidence on the spot from one of Holmes's water-wells.

Mr A. H. T. Chisholm's (APOC) report on the Kuwait negotiations position, June 1932

Writing on 2nd June 1932 to APOC in London from Abadan, after spending the previous three days visiting the Political Resident in Bushire and the Political Agent in Kuwait, Mr A. H. T. Chisholm (see Note 52) analysed the position as follows from the APOC point of view:

> The Sheikh is very impressed by Holmes, and especially by the fact that he was offering good money down for a concession document expressly stipulating that he would immediately set to work to prove Kuwait territory as he had done in Bahrein. The Sheikh was equally disgruntled at the way in which he considered that he had been treated by the APOC, namely (*a*) that during many years of discussion he had never been made any definite offer (*b*) that the only apparent effort made to prove his territories had been abruptly broken off in April with no reason given to him (*c*) that now similarly with no reason except that APOC are frightened into it by Holmes's competition, an offer of £2000 has been made apparently to

lock the door on the search for oil in Kuwait for two years and then even that offer was not direct from the APOC but semi-officially through the Political Agent.

Two of his grievances are genuine, namely (*a*) that he has been played with by us for so long to no purpose (*b*) that even with Holmes in the field with a plain offer for a concession we have as yet made no counter-offer. We must come down quickly with a proposition in the shape of an offer for a concession involving an attempt to see what Kuwait can produce in the way of oil if we want to prevent Holmes getting Kuwait as he did Bahrein.

As regards Holmes's offer, the position is this, as soon as we left Kuwait in April, Holmes was there, fresh from his triumphs in Bahrein. He made the Sheikh an offer for an exclusive concession; the concession document was on the same lines as that which he had obtained in Bahrein, except that he doubled the price, offering Rs. 20,000 p.a. to start with instead of Rs. 10,000 (the Bahrein document is an agreement with three annexes; the first being a prospecting licence involving payment of Rs. 10,000 p.a. If work under this proved satisfactory the second annexe, an exploration licence, came into play, with a further payment of Rs. 10,000 p.a., while if commercial oil, over 100 tons in quantity, was found, a royalty of Rs. 3/8 per ton produced was promised, with a minimum of Rs. 30,000 p.a.). This document was unsatisfactory when shown to the Political Resident ten days ago, so Holmes took it back and remodelled it. It has now been despatched to the Colonial Office.

Although the Sheikh made no bones of his admiration of Holmes's offer and methods, he has informed the Political Resident that, other things being equal, he would rather see the Concession go to a British concern in view of his relations with Britain. This view was encouraged by the Political Resident, but in view of the fact that we had made no counter-offer to Holmes he could not press the point. We must come down with a counter-offer if we want to keep Holmes out. The Sheikh needs money badly as his state finances are in a very bad way; he is being hard pressed by what *vox populi* there is in Kuwait to raise money from oil, or at least from people who want to look for it; he is being constantly urged by Holmes to grant him a concession on terms which mean money down, more money spent in the country finding oil, and much more if oil is found; although Holmes represents U.S.A. the Sheikh knows that the Foreign Office has written to the U.S. State Department that they have no objection to his group operating in Kuwait; and no other offer than Holmes's has yet been made.

I called on the Political Agent in Kuwait on my way back from Bushire yesterday. He told me of the effect which the Foreign Office's waiving of the Nationality Clause had on the Sheikh, and the play which Holmes had made with it, saying that it was the result of his backer's direct appeal to the American Ambassador in London, also that it was due to this appeal that we withdrew our geologists so hurriedly in April **(55)**. Holmes in fact has *vis-à-vis* the Sheikh a very strong position, what with his success in Bahrein, his free way with money, and his apparent faith in Kuwait oil. He has put up a straight-forward offer for a concession which the British Government departments concerned will have no reason for refusing in the absence of other offers.

On such an informal visit I did not see the Sheikh, but the Political Agent will inform him that an APOC representative passed through, and that there is no doubt that an official presentation of our proposals will shortly be made.

Three points in this early analysis of the situation are worth remarking as relevant to subsequent developments.

Firstly, the Shaikh's statement to the Political Resident that 'other things being equal' he would prefer the British competitor, directly contradicts Holmes's simultaneous report (see page 22) of Shaikh Ahmad's statement that he would upon no consideration grant APOC the concession. There is a wealth of evidence that the Shaikh's attitude both then and later was to tell both competitors that whichever of them gave him the better terms would succeed, and also that Holmes, both then and thereafter, consistently conveyed to his principals a too favourable impression of the Shaikh's attitude towards them **(56)**.

Secondly, the amounts quoted as having been paid for the Bahrain Concession and offered by both Holmes and APOC in Kuwait, though seemingly trivial in view of later developments, may be compared with the fact that the entire expenses of Shaikh Ahmad's household at that date were budgeted within Rs. 18,000 p.a.; it is also relevant that up to that time not a drop of oil had been found anywhere on the Arab side of the Gulf and all the geological surveys carried out there during the previous twenty years had been discouraging to the prospects of any being ever found there. Furthermore, what with the heavily depressed state of world trade at that time following the American Stock Exchange slump of 1929 and the simultaneous world glut of oil which had led some of the major international oil companies in 1928 to mutually agreed action **(57)** to avoid further overproduction, there was reluctance throughout the world oil industry to embark on potentially expensive new ventures unless in extraordinarily favourable conditions.

Thirdly, the 'success in Bahrein and apparent faith in Kuwait Oil' mentioned as one of Holmes's chief assets *vis-à-vis* the Shaikh, was undoubtedly the main reason for his enduring hold on the Shaikh's respect. Holmes was, and always had been from 1921 onwards, as optimistic **(58)** regarding the oil potentialities of the Arab side of the Gulf as the whole weight of expert opinion throughout the international oil industry was pessimistic. From June 1932 onwards he was to be proved right to an increasingly extraordinary degree. But even by May 1932, despite no hard evidence of what was shortly about to happen, his own unflagging optimism and his Company's willingness to back his opinions with their money, had made an enduringly favourable impression with Shaikh Ahmad.

CHAPTER 3

June 1932 to February 1934

On the night of 31st May 1932 oil was struck in Bahrain and, when the well was tested next day, proved to be in commercial quantities.

Effect of Bahrain oil-discovery on Kuwait negotiations

This event revolutionised previous geological theories of the oil possibilities of Arabia, **(59)** confounding the pessimists and vindicating Holmes's unique optimism, and immediately accelerated the tempo of negotiations in Kuwait. It greatly increased Holmes's status and reputation, and Gulf Oil Corporation urged the EGS to take every advantage of it in pressing Shaikh Ahmad for an early and favourable decision. APOC at Abadan, who when telegraphing to London on 1st June the results of Chisholm's visit to Bushire and Kuwait had requested permission to reopen negotiations with the Shaikh and offer up to Rs. 20,000 p.a. telegraphed again three days later that 'the Bahrain news must have great influence on Shaikh of Kuwait' and in reply were told by London to go ahead 'up to Rs. 25,000 p.a.'. That Shaikh Ahmad was fully aware of the favourable effect on his own position of the Bahrain discovery, and took full advantage of it, is shown by records of his conversations on 10th and 11th June with APOC's Deputy General Manager in Persia, Mr N. A. Gass **(60)**. The latter had arrived in Kuwait from Abadan to pave the way for his Company's negotiations for a concession, as proposed on 11th May by the Shaikh when he had rejected their suggested option, and as now agreed by London. In contrast with the close and friendly relations maintained by Holmes over the previous nine years, this was the first occasion since 1923 (when Sir A. Wilson had visited him; see Note 18) that Shaikh Ahmad had been visited in connection with his oil concession by any official of APOC, whose last correspondence (see Note 50) with him had concerned the abrupt termination of their geological survey on 13th April. Consequently Gass did not expect a very cordial reception by the Shaikh and in fact reported that it was 'dignified but extremely frigid. His attitude was one of detachment and it was obvious that under no circumstances would he broach the subject upon which he knew quite well I had come to see him'. Eventually, after recounting the geological and other reasons based on their long experience in Persia and Iraq that had inhibited his Company hitherto from an active interest in Kuwait, and arguing from them that a full concession Agreement as requested by the Shaikh might not be in his best interests pending further exploration of his territory, Gass proposed that his Company should negotiate with the Shaikh for a prospecting licence for three years, in the course of which it would advance their research work in the way which they knew to be to the best advantage of his territory. The Company was prepared to pay generously for permission to go ahead on these lines, and during this prospecting period it would be able to formulate its proposals for a concession.

Shaikh Ahmad replied that for reasons of state he could not in any circumstances agree such a proposal. His State was in dire need of funds and employment. The pearl trade was a fraction only of its past activities, and commerce with the interior had practically disappeared. He was now a seller of the right to develop the oil of his territory and he was in possession of a concession offer in very comprehensive terms from a rival company; APOC's offer gave him no security and, if he accepted it, the rival concern would be lost to him. He was only prepared to grant a concession of sixty or seventy years' duration and he desired from the APOC, if they were interested, a document as comprehensive as their EGS rival's in order that he could compare the two, consult the Council and decide which it was in the best interests of his State to accept. He expected the APOC to take a risk as the EGS had done in their draft and guarantee regular payments throughout the concession. He had only granted free permission for APOC's recent geological survey in the face of the strongest opposition from his relatives and councillors, because he had been genuinely anxious to welcome the Company to his territory. He had been accused of giving their birthright for nothing when generous offers were waiting from others who might withdraw and be lost for ever and doubts had been expressed as to the Company's sincerity, as their past actions were not encouraging. The Shaikh had told them all to go to hell [*sic*], and when he had received the Company's notice of withdrawal it had been a terrible blow to his prestige.

In view of Shaikh Ahmad's uncompromising attitude (he also rejected a proposal that the APOC should be granted a one-year exploration licence against immediate payment of Rs. 20,000) Gass had to agree that the Company would prepare, for presentation to him as soon as possible, a full concession document on the lines he had requested. The Shaikh asked how long this would take, and on hearing that it would be some months replied that, although he was prepared to leave the matter open for a reasonable period, more than this he could not do, with a definite alternative in his hands and in view of the very straitened circumstances of his State. Great pressure was being brought to bear on him on all sides and he could not keep Holmes waiting indefinitely for an answer.

From the APOC point of view this interview was satisfactory, both because they now had time to prepare a counter-offer to EGS's and also because the atmosphere of frigid indifference and injured prestige at first displayed by the Shaikh (as they had expected) had been converted into one of friendliness and cordiality. Shaikh Ahmad, after the first day's discussion, had changed his plans so that Gass could be his guest at dinner the following night. A subsequent exchange of letters **(61)** confirmed this new relationship and also that a comprehensive concession, as desired by the Shaikh, was being prepared for discussion with him as soon as possible. Although both Holmes and Yatim reported to EGS/Gulf that the Shaikh had assured them, after his discussion with Gass, that his preference for them was as firm as ever, the actual position was evidently as expressed by Shaikh Ahmad on 28th June in conversation with the Political Agent. When the latter expressed the view that the results of Gass's visit must have been satisfactory to the Shaikh, he replied, 'Yes, I have now two bidders and from the point of view of a seller that is all to the good'.

Chisholm presents APOC's draft concession proposals to Shaikh Ahmad, August 1932

It took two months for APOC's draft concession document to be finalised for presentation to Shaikh Ahmad. Its operating clauses had to be cleared with and approved by the Colonial Office and other British Government departments in London in the same way as Holmes's first and later draft concessions had been, and its monetary terms had to be settled at a level sufficient to outbid what could be learnt about the EGS offer. Meanwhile Holmes left Kuwait on 2nd July for London, where he and his principals (Gulf Oil had sent a representative to London to support EGS's efforts) were similarly engaged in getting the approval of British Government departments to his latest concession document which from the Shaikh's point of view had already been agreed to be satisfactory during Holmes's discussions with him in May. During July APOC's agents in Kuwait leased a house there for occupation by Mr Chisholm, who was to bring the draft concession for presentation to Shaikh Ahmad as soon as it was ready and conduct negotiations with him thereafter on behalf of his Company. With a letter of introduction **(62)** to the Shaikh, Mr Chisholm arrived in Kuwait from Abadan on 14th August 1932, was courteously received by Shaikh Ahmad on 15th, and on 16th had a first business discussion with him, during which he presented the APOC's draft concession document **(63)**, at the same time securing the Shaikh's agreement for a first discussion of it with him three days later together with Gass who was to visit Kuwait again for that purpose.

These events caused considerable concern to Mohammed Yatim, who in telegraphing regarding them to Holmes in London advised his early return. But Shaikh Ahmad had evidently decided on delaying tactics, for though informing Messrs Chisholm and Gass, during the latter's visit to Kuwait on 20th August, of his general approval of the form of the draft concession agreement (he told the Political Agent that he was impressed with its proposed initial payment of Rs. 50,000 being so much larger than Holmes's Rs. 30,000) he declined to discuss any aspects of it in detail until he had first received the Political Resident's formal approval of it as a basis for discussion. This he had not yet received and he recalled that it had been necessary before he discussed Holmes's draft. When he received it, he said, he would examine the Company's proposals with his advisers and discuss any modification required with Chisholm. He also laid stress on the importance he attached to dissipating any impression among his Council and subjects that he had been rushed into discussing an APOC concession without affording them the opportunity of expressing their views.

The Shaikh's attitude was to result in postponing both APOC's and Holmes's negotiations with him for nearly five months, as the Political Resident declined **(64)** to approve his opening negotiations with APOC until he had received H.M.G.'s authorisation for him to do so, and in writing for the latter to the Colonial Office in London he enclosed not only the APOC draft concession agreement but also various comments on it resulting from his and the Political Agent's discussions of it with the Shaikh. It was consequently not to be until January 1933 that the British Government departments concerned in London, who already had under review Holmes's draft concession proposals of May 1932, had completed their examination both of his and APOC's proposals, and also a comparison of their merits from the Shaikh's point of view which they decided to provide for his consideration when choosing between them. While advising Shaikh Ahmad that all this information was being prepared for him the Political Resident also informed him in September that no further negotiations with either Company could take place pending their receipt.

This long delay was disagreeable to EGS and Gulf Oil Corporation, both of whom did their best to accelerate matters in London, while Gulf in America again obtained State Department intervention (see page 29) with the Foreign Office in London on their behalf. But it is unlikely to have been equally objectionable to the Shaikh. That time was now on his side must have been apparent to him, with two powerful oil companies, encouraged by the Bahrain oil discovery, ready to bid against each other for his concession. And the fact that one of them was American could be turned to his advantage in extracting stronger British Government support, hitherto rather reluctant, in some current claims of his against Iraq if, in his oil concession negotiations, he appeared lukewarm towards their favoured APOC. Also the possibility of an eventual Anglo-American link-up between the two rivals may, as already suggested, have been among his calculations.

Gulf/APOC correspondence and discussion regarding their competition in Kuwait

Certainly, whether it was due to luck or good judgement, two events which occurred during this five months' interval pointed in that direction. Firstly, following a rather contentious exchange of letters in July/October 1932 **(65)** between APOC's New York office and Gulf regarding their competition in Kuwait, in October 1932 the Chairman, Colonel Drake, of Gulf Oil Corporation, during a visit to London, discussed the Kuwait position with the Chairman of APOC, Sir John Cadman (see Note 77); and the discussion was resumed in Gulf's Pittsburgh head office, when Cadman was in America in November with Mr Leovy, Gulf's Deputy Chairman. Drake in October had expressed surprise at APOC's renewed competition for a Kuwait Concession, which they had turned down EGS's offer to sell them in 1926, but Cadman said that they had refused it then mainly on geological grounds and also with a view to doing a better direct deal with the Shaikh; now (meaning, Gulf thought, since the Bahrain discovery) he did not think Gulf would ever get the concession 'except over his dead body'. In Pittsburgh a month later considerable discussion of possible combined action in Kuwait ensued but ended with neither side in any way committed and no suggestion that either should relax its current efforts. However, each side's strong attitude had so impressed the other with the desirability of joint action if a satisfactory arrangement could be reached, that from December onwards correspondence and discussion aimed at a mutually satisfactory understanding continued between the two companies. Secondly, right at the end of the five months' pause in the negotiations when, on 8th January 1933, the Political Agent at last presented to Shaikh Ahmad the British Government's views on both the rival concession documents, and their comparison between them, he also impressed on him that if the EGS obtained the concession, although technically a British concern, this would mean the exploitation of Kuwait's oil resources by definitely American interests. This reminder, coupled with what he had probably already heard from Holmes of talks having taken place between APOC and Gulf of possible co-operation, may well have crystallised the Shaikh's determination to manoeuvre the two competitors into coming together to his own advantage. Writing from Kuwait on 16th January 1933, when negotiations were about to be resumed, Mr Chisholm included that aspect in his following appreciation of the situation:

> While the intrinsic merits of our APOC offer, and the better relations and realisations of our interest and intentions which have been fostered during the past five months are causes for optimism now that direct negotiations are about to begin, there is no doubt that Holmes will now make more play than ever (unless during his visit to London his principals have given him some new line of policy consequential on our own principals' discussions with them) with his pro-American and anti-British argument, and at this particular juncture such an argument unfortunately must have some appeal to the Sheikh. It appears that the Sheikh's inclination towards us or the EGS may be dictated by his solution of a larger problem; namely whether he is to continue to put all his eggs in the British basket, or whether, at this stage of the development of international feeling towards the Middle East and small nationalities generally, it is safe to give his British protectors a reminder that he cannot be considered as entirely in their pocket by deliberately giving his concession to an American concern when a British concern is also in the field. The Political Resident states that the, at best, over-cautious attitude of the British Government over the Sheikh's problems looms very large in the latter's mind.

Five months' pause in negotiations

During the enforced five months' pause in their business dealings with the Shaikh from August 1932 to January 1933, both Holmes and Chisholm continued in friendly contact with him. It was during this period that the rivals first met each other, by chance on the first occasion in the Political Agent's house on 27th September and on frequent social occasions thereafter, when talk on oil matters was avoided, although each knew that the other was doing his best, through their friends and informants, to discover and counter his intentions and plans. It is perhaps significant, and was suspected of being so by Chisholm at the time, that at their first meal alone together, on 1st October, Holmes remarked that from what he knew of the Shaikh's character they would probably have to partition the concession eventually. Holmes had left London for Kuwait on 27th August, but after meeting the Shaikh in Baghdad (where the latter was visiting King Feisal in early September) went on to Bahrain, whence he reported to Gulf that there was no reason to believe that his Kuwait position had been unfavourably affected by APOC's pressure on Shaikh Ahmad. He returned to Kuwait at the end of September, reporting shortly afterwards such extensive entertaining and other social activities by the APOC party ('APOC is feeding all and sundry') that he needed further funds to counter it, although he was constantly assured both by the Shaikh, and by his friends on the Council of State, that they remained firm in his favour. Chisholm, whose first meeting with the Shaikh since the Gass discussion was on 19th September, reported both then and after subsequent meetings in October and November that the Shaikh refused to be drawn into any talk of oil matters. In September Shaikh Ahmad told him that since a preliminary perusal he had hardly looked at the APOC draft and that he did not intend to do so until he got the approval to deal with it which he had requested from H.M.G. As by that time he expected also to have H.M.G.'s views on Holmes's proposal, he expected then to have no difficulty in deciding the matter.

Meanwhile both EGS and APOC had been doing their best to accelerate action on their drafts with the Colonial Office and other Government Departments (Foreign Office, India Office, Admiralty and Petroleum Department of the Board of Trade; see note 7) in London, where in December both Holmes and Chisholm were recalled from Kuwait by their companies to be briefed with the latest information. By that time APOC was so preoccupied with the difficult position caused by the Shah of Persia's abrogation on 27th November 1932 of its Persion Concession **(66)** that the prolonged delay in Kuwait was not unwelcome to it. Gulf Oil, however, objected to its Kuwait plans, apparently so near success in May, having been first blocked by APOC's intervention in June and now kept at a standstill on British Government orders until both APOC's proposals and its own could be simultaneously considered by the Shaikh. They were also so apprehensive, from Holmes's reports of anti-American pressure being exerted on Shaikh Ahmad by the Political Resident and Political Agent, that the Shaikh might have to decide in favour of APOC, that they had obtained State Department action in August for representations to the British Government to ensure equal favour for both parties. Although on 16th September the British Foreign Office assured the State Department in Washington that no political bias was being shown and that both parties, APOC and EGS, were being treated *pari passu*, the American Ambassador (see Note 47) in London continued during the next three months to press the Foreign Office for an early end to the British Government's embargo since August on the Kuwait negotiations.

On 3rd January 1933 Gulf were reassured to hear from Holmes in London (he had left Kuwait on 29th December) that the Shaikh (and his secretary Mullah Saleh) had promised to telegraph him as soon as H.M.G.'s comments and permission to resume negotiations were received, 'and would not grant the concession to APOC under any circumstances'. On 11th January EGS in London received a telegram from Muhammad Yatim in Kuwait that the Shaikh had received the British Government's comments from the Political Agent on 8th January, that the Political Resident was visiting Kuwait shortly for discussions, and that the Shaikh requested Holmes to return immediately. Leaving London on 14th January, he arrived in Kuwait on 20th to find that Chisholm, who had returned to Kuwait on 3rd January after his London visit, had already reopened APOC's negotiations with the Shaikh on the previous day.

EGS/Gulf and APOC negotiations with Shaikh Ahmad authorised by H.M.G.

The reopening of negotiations with the Shaikh by both EGS and APOC had been authorised by the Political Resident on 14th January 1933, which date marked the beginning of the last and keenest phase in their competition against each other for the Kuwait Concession. On 29th December the Political Agent had received from the Political Resident in Bushire the documents containing the British Government's advice to Shaikh Ahmad as to the merits of both Companies' offers to him, with instructions to have them translated and presented to the Shaikh on 8th January, at the same time informing him that the Political Resident would visit him a few days later to discuss the matter and authorise reopening of the two companies' negotiations. The Political Resident arrived in Kuwait on 11th January and, after two days' discussion with the Shaikh, on 14th gave Chisholm a letter **(67)** (a similar letter was to be given to Holmes on his return from London where meanwhile the Political Resident telegraphed to the India Office to inform EGS similarly) saying that he had given the British Government's comments on APOC's proposal for an oil concession to the Shaikh, and that the Company could now commence negotiations.

The Political Resident also told Chisholm on 14th January that he had gone through the British Government's comments on APOC's proposals with the Shaikh (at no time then or later did Chisholm or APOC see those comments, although Holmes and Gulf Oil were more fortunate; see below); that the latter had given no indication as to which party he favoured; and that he had informed the Shaikh that, as to future procedure, he should now on the basis of the British Government's views strike the best bargain that he could with whichever party he favoured although, whichever party that proved to be, the eventual concessionary document before final signature must be re-submitted to the British Government for the inclusion of such final safeguards as might be necessary. Chisholm, who soon after his return from London had a long interview with the Shaikh on 6th January **(68)** at which, among other news of interest from London, he had said that his company had agreed, when negotiations were reopened, that various payments in APOC's draft would be revised upwards from those last proposed in August, after his talk with the Political Resident requested an interview with the Shaikh in order to present this revised draft and commence negotiations. Shaikh Ahmad granted this interview on 19th January when, on receiving APOC's revised proposals **(69)**, he said that he would study them closely in comparison with their competitors with a view to further discussions with both parties before coming to any decision. Stating that now 'armed with H.M.G.'s views, he regarded himself simply as a dealer with something to sell as favourably as he could', he emphasised that any agreement come to by him was to be binding on his heirs and successors for seventy years, and consequently he must make every effort to secure the best terms possible in the interests of his State; there could be no hurry in such an important matter.

Holmes arrived in Kuwait from London on 20th January and had several interviews with Shaikh Ahmad before leaving for Bahrain on 27th, having expressed his opinion to Chisholm that the Shaikh would take several months before taking any decision. Soon after returning to Kuwait on 2nd February he advised EGS to expect slow progress, as the Shaikh was consulting his Iraqi lawyer, Mr Gabriel **(70)** on the merits of the competing proposals and H.M.G.'s comments upon them; but assured them that his position with Shaikh Ahmad was as firm as ever.

Shaikh Ahmad gives Holmes a copy of H.M.G.'s comments on APOC and EGS proposals

The Shaikh had, he said, given him a copy not only of the British Government's comments (see Note 72) on APOC's and EGS proposals, but also APOC's latest revised draft concession as presented by Chisholm on 19th January. In passing these documents on to EGS and Gulf Holmes stressed that all of them, particularly the APOC draft concession, should be regarded as strictly confidential, the Shaikh having told him that he had always been asked by APOC not to divulge their terms to their competitor.

There may well have been extenuating reasons for this apparent breach of confidence in Shaikh Ahmad's dealing with APOC. Firstly, it may not have been the Shaikh but his Secretary Mullah Saleh, Holmes's close friend and confidant, who divulged these key documents. Secondly, it was at about that time that Yusuf al Ghanim **(71)** told Chisholm that the Shaikh had recently told a friend that he was not bothering over comparisons between APOC and EGS offers as 'they would fight it out and arrange matters between themselves', a remark which was taken by Chisholm to be one more indication that on his return from London Holmes had informed the Shaikh that there were prospects of a combination between APOC and EGS/Gulf which would disembarrass him of the necessity of choosing between them. Against such a background Shaikh Ahmad may well have reasoned that the terms of either party need not be concealed from the other.

The copy of the British Government's comments on APOC and EGS's proposals **(72)**, and comparisons between them, was received and carefully analysed by Gulf in February. Detailed and non-committal, they provide good evidence of the meticulous way in which such matters were handled in London by the various British Government departments interested (in February 1933 the Colonial Office **(73)**, hitherto the principal Department concerned with Kuwait affairs, ceased to have any interest in them, leaving the India Office principally concerned, together with the Foreign Office, Admiralty, and the Petroleum Department of the Board of Trade). Gulf Oil Corporation took the view that they were unfairly biased in favour of APOC and, after discussions with the State Department, the latter took steps to have the matter taken up by the American Embassy with the Foreign Office in London at the end of February **(74)**. But by that time the discussions between APOC and the Gulf for joining forces over Kuwait were so far forward that no action was taken. Both companies however remained determined to continue their efforts to get the concession for themselves until a joint effort was actually agreed.

Meantime in Kuwait Shaikh Ahmad made no move towards either party. On 9th February the Political Agent conveyed to him an enquiry from H.M.G. as to whether he had yet decided to whom he would give his oil concession. This was badly received by Shaikh Ahmad who replied that he deprecated such an enquiry less than a month after the British Government had given him their views which had taken them five months to deliberate. He had not yet made his decision, it would very likely be weeks or months before he did so, and he had not yet even received the opinion on the two documents for which he had asked his legal adviser in Basrah. The Shaikh also reminded the Political Agent that the British Government had recently required him not to commit himself to granting any railway rights or township areas to either party at this stage, and that all such matters were to be regarded as reserved. What with that, he said, and the Political Agent having impressed on him on 8th January that the EGS represented definitely American interests, together with the information that no final agreement could be made by him with either party before certain safeguards had been included by the British Government, his freedom of action was being reduced to a farce.

On 11th February Mr E. H. O. Elkington **(75)**, APOC's General Manager in Persia, visited Kuwait from Abadan. In the afternoon he and Chisholm had an interview with Shaikh Ahmad and were his guests at dinner the same evening. This had been arranged a week previously by Chisholm with the Shaikh who, on learning that Elkington would appreciate an opportunity of visiting Kuwait and of making his acquaintance, cordially invited him to do so. Elkington was able to give Shaikh Ahmad the latest information on APOC's dispute with the Persian Government regarding its Persian oil concession and mentioned the possibility of Sir John Cadman (see Note 77), the Company's Chairman, visiting Persia from London in that connection in the near future, when he might also visit Kuwait if the Shaikh so desired. As regards the Kuwait concession negotiations, Elkington said that although Shaikh Ahmad's resumption of discussions with Chisholm was awaited with interest, APOC well realised that his decision in so important a matter could not be reached rapidly. Both during the interview and subsequent dinner, to which the Political Agent was invited by the Shaikh, a very friendly atmosphere prevailed and when reporting to London on his visit and subsequent exchange of letters **(76)** with Shaikh Ahmad thanking him for his hospitality, Elkington expressed reasonable optimism regarding APOC's prospects in the Kuwait negotiations.

Both Holmes and Chisholm had interviews with the Shaikh about a week later but no progress with the oil negotiations was made. They were both informed by the Political Agent that Shaikh Ahmad had told him that their rival claims were being closely studied and discussed with his advisers and family. On 22nd February, at which time Holmes was making arrangements for one of his EGS directors from London, Mr Janson, to visit the Shaikh in the following month, Chisholm was unexpectedly visited by Mullah Saleh, who stayed for over an hour talking around the topic of the oil negotiations. He appeared anxious to assure Chisholm that Holmes's personal ties with Shaikh Ahmad would not weigh with the latter in taking his decision, and Chisholm anticipated that during Janson's visit EGS might make a new offer to the Shaikh. Writing to Abadan on 25th February he expressed the belief that the Shaikh had decided on the line of action which he thought best to adopt in the negotiations and had informed his family advisers to that effect.

Sir John Cadman (APOC) visits Shaikh Ahmad, March 1933

On 1st March, Chisholm told Shaikh Ahmad that APOC's Chairman, Sir John Cadman **(77)**, would be arriving from London in Abadan later in the month on the way to Tehran, and when the Shaikh enquired if he meant to visit Kuwait, said he would certainly do so if the Shaikh wished it. Though the matter was left open, Shaikh Ahmad seemed impressed by Chisholm's argument both then and at another meeting on 12th March that Sir John's arrival (he was due in Ahadan on 22nd March) made the time particularly opportune to discuss APOC's proposals if he wished to do so; and as the Shaikh had also hinted that in Janson's forthcoming visit he expected a new EGS offer, Chisholm informed Abadan of the desirability of a visit by Cadman.

Mr Janson of EGS arrived in Kuwait from Bahrain with Holmes on 14th March and they both stayed with Shaikh Ahmad until 23rd when they left for Cairo and Jeddah **(78)**. On the morning of 20th March the Shaikh requested Chisholm to visit him, and informed him that he had been thinking over their recent conversation about Sir John Cadman's arrival in Persia providing an opportune occasion for discussing APOC's terms for his concession. He was consequently going to send his Secretary, Mullah Saleh, to Chisholm that afternoon to ask an important question regarding APOC's monetary terms, which were the main matter of interest for him in their draft concession proposals. When Mullah Saleh visited Chisholm he informed him that the Shaikh had recently had discussions with Holmes and Janson at which he had obtained from them new terms with which he was completely satisfied. For comparison with these, he now required to know APOC's final cash offer for:

(*a*) Payment on signature of the Agreement.
(*b*) Guaranteed minimum annual payments for the first three years after 'declaration of Commercial production'.

When Chisholm asked if this was the only substantial point remaining for decision, and if the EGS had also been similarly asked for a final figure, the Secretary replied that he did not know. So Chisholm asked him to reply to the Shaikh that he would like to come and discuss the matter with him on 22nd March and meantime to inform Sir John Cadman who, he was sure, would be pleased to visit Kuwait within the next few days and make a final offer satisfactory to the Shaikh. The following day the Shaikh sent a message that he was pleased with Chisholm's reply, and looked forward to seeing him next morning and to Sir John Cadman's suggested visit, which Chisholm requested Abadan to arrange accordingly.

Before interviewing Shaikh Ahmad on 22nd March, Chisholm had found out the main outlines of the new terms obtained by him from EGS. Shaikh Ahmad had requested EGS to redraft their proposals so that instead of the 1,600 square miles concessionary area previously asked for, to be disposed of as they wished (which the Shaikh knew was to be to Gulf Oil Corporation) they were to make two separate offers for areas of 1,200 and 400 square miles, the 400 square miles to be disposed of to American interests only to exploit, and the 1,200 square miles to be disposed of to British interests only. If, however, no British interests would take up the latter area, EGS were to be free to offer it as well to American interests. The Shaikh had said that this was entirely his own idea (though Janson, in reporting it to Gulf, attributed it to the advice of his lawyer); it would please the British Government; it would bring in American as well as British oil interests; it would fulfil his long-standing friendly obligations to Holmes; and it would result in his getting more money, as he had also told EGS that they must offer Rs. 90,000 (instead of their previous Rs. 50,000 or the Rs. 65,000 offered by APOC) as payment on signature of the agreement. Janson and Holmes had agreed the Rs. 90,000 payment, and had also produced outlines of the two new proposals satisfactory to the Shaikh although these were subject to agreement by their London colleagues as soon as they could consult them after leaving Kuwait on 23rd March.

When Chisholm visited Shaikh Ahmad on 22nd March and informed him that Sir John Cadman proposed to fly from Abadan to visit him on the morning of the 25th, the Shaikh said he would be pleased to see him then and for lunch afterwards. As regards Mullah Saleh's message to him, Chisholm said that in the circumstances Sir John Cadman would like to answer it himself so long as it was not inconvenient for the Shaikh to wait until then, to which Shaikh Ahmad replied that this was perfectly satisfactory.

Sir John Cadman flew from Abadan to Kuwait on the morning of 25th March 1933 accompanied by Elkington and Gass and, with Chisholm, was cordially received by the Shaikh at an interview (of which a detailed account written on the following day is given in the notes **(79)**) lasting two and a half hours, whereafter they were his guests for lunch at Dasman Palace before returning to Abadan.

Sir John Cadman, a highly experienced and persuasive negotiator with a frank and easy manner, at once established friendly relations with Shaikh Ahmad by recalling that it was exactly twenty years since he was last in Kuwait as a member of the Slade Commission (see Note 4) visiting Shaikh Mubarak in 1913.

Sir John Cadman's (APOC) offer to Shaikh Ahmad

After describing the position so far reached, and APOC's belief that they were better able than anyone else to exploit Kuwait's oil possibilities to Shaikh Ahmad's satisfaction, although if he would like American participation with them that could easily be arranged, Cadman said that if the Shaikh wished he could at once give him APOC's improved final offer of cash terms which he had asked for from Chisholm. In reply, Shaikh Ahmad explained the procedure which he proposed to follow. He had recently received an offer from EGS which was better than APOC's. If APOC were prepared to raise their offer, and it was better than EGS's, he would then give EGS another opportunity to raise their offer. If they then offered less than APOC had, the matter would be closed, but if on the contrary they offered more, then APOC would be given another opportunity to outbid them and this procedure would continue until he was able to tell one or the other party, whose final bid was largest, that the concession would be granted to them. When that stage was reached, he would discuss the other (i.e. non-monetary) terms of the concession document, and so far as APOC's was concerned he stated that he would certainly ask for some modification of their document if they were the party ultimately successful in the cash bidding. The Shaikh gave his assurance that in no circumstances would the offer of one party be disclosed to the other. Cadman accepted the Shaikh's proposed procedure and made the following offer of cash terms:

On signature of the agreement Rs. 200,000 (compared with Rs. 65,000 previously offered)

On each subsequent Agreement date after declaration of commercial production, minimum for 1st year Rs. 100,000 (compared with Rs. 75,000)
2nd year Rs. 150,000 (compared with Rs. 90,000)
3rd year Rs. 200,000 (compared with Rs. 120,000)
and subsequently.

Cadman, in making this offer, also said that he would double the amounts payable both on signature and for the third and subsequent years after commercial production was declared (making them each Rs. 400,000) if the Shaikh was prepared to sign an Agreement with APOC forthwith; but, if not, he could not leave this higher offer open.

Shaikh Ahmad expressed his appreciation of Cadman's offer, but said he could not consider the suggested doubled amounts as he had promised the EGS, in accordance with his procedure as already stated, that he would not close with APOC without giving them an opportunity to revise their offer. After lunching with the Shaikh, who wished him success on his important mission to Teheran, Cadman and his party returned to Abadan, having promised that his officials would keep him closely informed on Kuwait developments.

Although APOC were very satisfied with Shaikh Ahmad's reception of their offer they were apprehensive, in spite of his personal assurance, that it might somehow be disclosed to EGS. Consequently nothing in writing was given to the Shaikh on the subject, and no information about the amounts offered were disclosed to anybody else, the Political Agent and Political Resident and Government Departments in London being merely informed that higher figures than before had been proposed to the Shaikh. Shaikh Ahmad himself did not mention the figures even to the Political Agent.

Shaikh Ahmad informs Chisholm that EGS has made a bigger offer than Cadman's

So when Holmes returned to Kuwait on 19th April, the Shaikh's next move was awaited with interest, and it was a considerable disappointment to APOC when Chisholm received only ten days later a letter from Shaikh Ahmad dated 29th April as follows:

> Referring to previous verbal negotiations regarding an oil Concession in the Kuwait territory we would state for your information that Major Holmes, representative of Eastern Syndicate has made a bigger offer than yours.
>
> Please let us have your views.

This development was the result of Holmes's success soon after he returned to Kuwait in discovering, with considerable difficulty according to his own account, what Cadman's offer had been. On 23rd April he had cabled EGS that the situation was much more serious than he had anticipated owing to the largeness of the APOC offer made by Sir John Cadman, and he later reported having had great difficulty in ascertaining its amount, Shaikh Ahmad having told him of his promise to Cadman not to disclose one party's offer to the other. Eventually, however, the Shaikh, he said, had referred him to his Council of State and it was Mullah Saleh by whom he was informed on 28th April of APOC's figure, which Saleh said was Rs. 240,000 (Cadman in fact had offered Rs. 200,000). When cabling this information to EGS on 30th April Holmes said he had, at the Shaikh's request, made him a counter-offer on 29th April of Rs. 250,000 (in two payments of Rs. 125,000 each in respect of the two halves of the concession area, see below) and the Shaikh had then informed Chisholm that the Cadman offer had been exceeded.

There were two features of this episode which may be regarded as mitigating Shaikh Ahmad's apparent disregard of his assurance of non-disclosure of APOC's offer. One was that according to Holmes on 30th April the Shaikh had, during the previous day's discussion, repeated his assurance that he would make no agreement with the APOC. The other was that the Shaikh, when the Political Agent visited him on 29th April to enquire what action he had taken with Major Holmes since his return to obtain a new offer to counter Sir John Cadman, said that Holmes had topped that offer by what he described as 'a fair amount', only on condition that he would now be free to divide the concessionary area 50/50 between British and American interests **(80)** instead of 75 per cent British and 25 per cent American as previously stipulated (see page 32), to which the Shaikh had agreed. These two facts taken together (and also the subsequent course of the negotiations) may indicate that the untypical disregard by Shaikh Ahmad of his undertaking not to disclose APOC's figure was the result of his having been persuaded by Holmes that a 50/50 Anglo-American bid for his concession by APOC/Gulf jointly was so nearly arranged (as in fact it was to be only three weeks later) that it was legitimate to do so. The Shaikh informed the Political Agent on 29th April that he was very pleased with Holmes's action, of which he had informed APOC so that the bidding up might continue, a prospect at which he evinced much satisfaction.

Chisholm, who visited Shaikh Ahmad on 4th May to discuss his letter of 29th April, had had an interesting talk with Holmes on the previous day. Although as usual there was no mention by either of the state of their negotiations with the Shaikh, Holmes expressed the view that his decision was a long way off, again hinting that the concession would probably have to be divided between them. He was leaving for Bahrain the following day and would be back in Kuwait on 17th May, then leaving for London where he was to meet Mr Wallace **(81)** of Gulf Oil early in June. He hoped not to have to return to Kuwait until after the summer, though that would depend on the outcome of his London discussions and APOC's activities in Kuwait. From all of which Chisholm inferred, correctly as it was to be proved, that Holmes was well aware of and closely concerned in Gulf's and APOC's latest negotiations for a link-up over Kuwait which were then being actively pursued in both London and Pittsburgh.

When the Shaikh received Chisholm on 4th May he discussed the situation of the negotiations at unusual length, saying that there was no urgency about replying to his letter of 29th April, which Chisholm said had been referred to Sir John Cadman then (after successfully concluding a new APOC Agreement with the Persian Government **(82)**) in Baghdad on his way back to London from Teheran, for instructions which would reach him shortly.

Shaikh Ahmad 'after laying stress on the fact that he had given Holmes no information as to our figures' (reported Chisholm to Abadan on 5th May) said that after some consideration Holmes had made an offer which exceeded APOC's (he would not say by how much) in all respects, i.e. payment on signature, annual payments, and guaranteed minimum. Holmes had also promised to give, as from signature, free supplies of petrol and kerosene for him and his family. In return for his big offer Holmes now required the whole of Kuwait territory **(83)** for his concession and not the 1,600 square miles for which he had previously asked, in this respect coming into line with APOC's proposals.

The Shaikh also said that Holmes had been offering a Royalty of Rs. 3.8 per ton, compared with APOC's offer of Rs. 2.10; but when Chisholm expressed surprise at this because both the Persian and Iraqi Governments had agreed their royalties at the equivalent of Rs. 2.10, Shaikh Ahmad said that Holmes had now reduced his royalty offer from Rs. 3.8 to only a little more than APOC's Rs. 2.10. In view of the alterations mentioned by the Shaikh as having been made in Holmes's general proposals, Chisholm asked if Holmes was preparing an entirely new EGS draft, in which case the Shaikh might like to discuss possible modifications in the APOC draft proposals also. But Shaikh Ahmad replied that no redrafting was required by either party until the main question, namely the cash terms, was decided according to the procedure he had described and agreed with Sir John Cadman on 25th March.

Shaikh Ahmad receives Elkington/Chisholm (APOC) and suspends negotiations as from 14th May

Following this interview Mr Elkington, after discussions with Cadman in Baghdad, arrived in Kuwait from Abadan on 10th May, the Shaikh having told Chisholm on 9th May that he would be pleased to see him again. On 11th and 14th May he and Chisholm had two long interviews with Shaikh Ahmad. Every aspect of the negotiations situation was discussed and the Shaikh was strongly pressed, on the same lines as those advocated by Cadman on 25th March, to make an immediate decision in favour of APOC. On 13th May the Political Resident, who had arrived in Kuwait on his way back to Bushire from London, also discussed the situation with the Shaikh stressing the desirability, in coming to a decision in a matter of such importance to his State and people, of not overlooking the undoubted strength of the APOC position; especially as Cadman had even offered, if the Shaikh thought it desirable, to bring in American partners. This point, according to the Political Resident, made no apparent impression on Shaikh Ahmad who had merely asked him again to confirm that he possessed an absolutely free hand from the British Government to make whatever decision he thought best; which the Political Resident did, being left after the interview with the firm impression on balance that the Shaikh's inclination was to decide in favour of Holmes.

It was with this in mind that Elkington and Chisholm, at their second interview with the Shaikh on 14th May, represented to him that as APOC's position in the oil world and especially in the Arabian Gulf was obviously so much stronger than EGS's, and as their terms for his concession could be improved as Cadman had told him in any way he wished, there must, in their opinion, be some barrier between themselves and the Shaikh, based perhaps on some misinformation given him by Holmes, or some special commitments to the latter, for his continuing to regard EGS as APOC's equal in every way or even, as they heard widely but they hoped wrongly rumoured in Kuwait, for his having already decided to grant his concession to Holmes. They requested Shaikh Ahmad to tell them if there was any such factor for Holmes or against themselves affecting his decision before they asked for further instructions as to how they should proceed from Sir John Cadman, who had told them to convey his best greetings to the Shaikh, and to do their best to correct any misunderstanding which might have resulted in the unexpectedly disappointing outcome of his own recent visit to Kuwait.

In reply Shaikh Ahmad strongly denied that any such factor as they had suggested, or barrier erected by Holmes in APOC's disfavour, existed and stated emphatically that he did not allow his personal friendship with Holmes or anyone else to interfere with the business of his State. He was, and would continue to be, absolutely impartial between the two parties, taking into consideration only what he and his advisers considered to be in the best interest of his State and people. He had not yet come to any decision, and was following exactly the negotiating procedure which he had described and agreed with Cadman. He went on to say that he had now decided that to continue, and still more to conclude, his oil negotiations at that juncture was not in the interests of his State. He had therefore telegraphed to Holmes in Bahrain, and now wished to inform Elkington and Chisholm, that he had decided to suspend all negotiations for the time being until such time as he notified both parties that he was prepared to reopen them. He stated that he was sure that in taking this action he would be serving the best interests of all concerned, and concluded by sending messages of good wishes and congratulation on the successful conclusion of his Teheran negotiations to Sir John Cadman **(84)**. These messages, and further assurances by the Shaikh that they should not be disappointed at this decision to suspend negotiations as it would certainly be to APOC's ultimate advantage, were repeated to Elkington and Chisholm when he dined at Chisholm's house the following night. At his request, Mr Elkington had written to Shaikh Ahmad **(85)** the previous evening acknowledging his letter of 29th April to Chisholm and confirming their conversation on 14th May regarding his suspension of negotiations.

Shaikh Ahmad's decision to suspend negotiations was believed by APOC to be due to a combination of three factors; the strong pressure being put on him by both competitors and their supporters in Kuwait; his belief based on information received by him from Holmes that the competitors would shortly combine and save him from the difficulty and odium of deciding between them; and, particularly at that moment, the fact that the competitive negotiations in Jeddah for King Ibn Saud's oil concession were approaching finality and he was anxious to know, as relevant to his own decision, what terms would be agreed and whether the partly British Iraq Petroleum Company or the American Standard Oil of California would be favoured by Ibn Saud (see Note 88).

Holmes's adverse reaction to suspension of negotiations affects his relations with Shaikh Ahmad

Holmes's reaction to the suspension of negotiations when he returned to Kuwait from Bahrain on 17th May was one of severe disappointment, and he cabled to EGS on the same day **(86)** that rumours of secret deals being done in London over his affairs had antagonised Shaikh Ahmad. Uncertainty surrounded, and still surrounds, what were the real reasons for what was evidently a bad upset in Holmes's very close and cordial relations with the Shaikh, but the facts are as follows.

On the afternoon of 17th May Holmes visited the Shaikh who returned his visit on the morning of the 18th. They spent the morning of 20th May together, and had tea at Holmes's house on 21st and dinner on the following night. On the morning of 22nd May the Political Agent, Lt-Col Dickson had a three-hour visit from Holmes of a very stormy nature.

After a preliminary talk about his prospects in Bahrain, Holmes turned the conversation to Kuwait affairs and particularly to the visits in his absence of the Political Resident and Mr Elkington of APOC. While talking on this subject, Dickson asked what Holmes's plans were following the Shaikh's suspension of all negotiations with him and APOC for an indefinite period. Holmes showed astonishment and complete surprise at the suggestion that the negotiations were in any way suspended and asked what Dickson meant. Dickson replied that all he knew about the matter was that Shaikh Ahmad had informed Elkington and Chisholm on 14th May that he had decided to suspend negotiations indefinitely both with APOC and Holmes, and that APOC had written the same day to the Shaikh confirming their understanding of this information regarding themselves and Holmes, sending a copy of their letter to Dickson (see Note 85) for his information. Holmes protested most vehemently against this having been done, saying that he had never heard a word from the Shaikh regarding any suspension of negotiations, that if APOC liked to have their negotiations suspended they could, but that his negotiations would go on; that 'it looked like a ramp engineered by APOC and the British Political authorities to keep him out of Kuwait', that he had the American Government behind him and that he might have to request them to send a representative to Kuwait to secure fair play for his negotiations; and he demanded a copy of APOC's letter in which they had so misrepresented his position.

Dickson had some days previously asked, and on 21st May received, from Shaikh Ahmad a letter for his records confirming that he had informed both APOC and Holmes that negotiations with them both with regard to an oil concession were suspended for the time being. Dickson therefore produced this letter from the Shaikh and asked Holmes if, in the face of it, he still denied all knowledge of what the Shaikh stated in writing that he had told him. When Holmes still strenuously denied knowing anything about it, Dickson suggested that, as Holmes had already mentioned that the Shaikh was his guest for dinner that evening, he should discuss and settle the matter with him as soon as possible. Otherwise, as he had standing instructions to report all developments in the oil negotiations to the Political Resident, he would have to report next day this information from Holmes, which was in direct conflict with written information which he had already reported as received from both the Shaikh and APOC. Holmes said that he would do so, adding before he left further protestations at the behaviour of 'APOC and its political supporters' and asseverations of the power and intention of the American Government to support him against the unjustifiable methods to which he was being subjected.

Dickson was astonished at what had happened. He did not believe that the Shaikh had not yet informed Holmes of the suspension of negotiations, but felt there was a ring of sincerity about Holmes's protestations and could not see what his motive was if they were a bluff unless possibly, resenting as he obviously did that negotiations should be suspended, he had been trying to find out if it had been a genuine move by the Shaikh or if it had been dictated or inspired by the Political Resident and APOC in his absence. In that case Holmes could have made considerable political capital out of a move which was otherwise unexpected by himself and his American backers, on whom he had long impressed the fact that his relationship with the Shaikh was so intimate as to preclude the possibility of such a surprising development occurring without notice.

Dickson was even more astonished when he received a short letter from Holmes that evening at 7.30 p.m., at which time Shaikh Ahmad would still have been with him or just left after dinner, saying that Holmes had now learned that Mohammad Yatim, during his own absence in Bahrain, having received from the Shaikh the information regarding suspension of negotiations, had written accordingly to Holmes in Bahrain. But the letter had not reached him there and consequently he had only just then, on questioning Mohammad Yatim, learnt that what Dickson had told him was correct. This explanation, in Dickson's opinion, could not be true. The Shaikh had given Mohammad Yatim on 13th May the information for Holmes, who was known to be leaving Bahrain on the mailboat for Kuwait on 15th. The first mail to Bahrain which a letter could have caught was the airmail on 18th, before which Holmes was due to arrive in Kuwait. Furthermore, not only had Shaikh Ahmad told Elkington and Chisholm on 14th May that the news of suspension had been telegraphed to Holmes in Bahrain, but it was also incredible that Holmes, having arrived in Kuwait on 17th and since been in everyday contact with Yatim as well as having had almost daily discussions with the Shaikh, should not have been told of such an important development until the evening of 22nd May, as he now asserted.

Altogether the position was so unsatisfactory that Dickson visited Shaikh Ahmad on 23rd, described what had happened, and asked him if he could throw any light on it.

The Shaikh said that Holmes had, for some inexplicable reason, misinformed Dickson. When his host at dinner the previous night, Holmes had told him of his interview with Dickson, and that he had told the latter that the Shaikh had as yet said nothing to him about suspension of his negotiations. At this the Shaikh had protested strongly, especially as Mohammad Yatim, who was present, confirmed that a long telegram on the subject had been sent to and received by Holmes in Bahrain. Holmes then admitted that he had lost his temper and made a mistake, and immediately wrote the note to Dickson to try and retrieve the position. Shaikh Ahmad informed Dickson that he had told Holmes that he had made a big mistake; also that as the oil negotiations were definitely suspended for three or perhaps six months there was nothing for him to remain in Kuwait for at present. Holmes therefore would be leaving Kuwait for London almost immediately.

In fact Holmes left Kuwait on 24th May for London, arriving there on 2nd June. Before leaving he wrote to Dickson notifying his departure, but making no reference to their recent interchange about the suspension of negotiations and the episode was never again mentioned between them. In retrospect it may have been due more to Holmes's bad-tempered reception by Shaikh Ahmad on his return from Bahrain (as reported in his cable to EGS of 17th May; see Note 86) than to annoyance at the unexpected news of suspension of negotiations which he had received there from Yatim. The Shaikh's sudden decision to suspend negotiations was probably due partly to a realisation, after the combined visits of the Political Resident and Elkington, that his dependence on Holmes's advice, recently culminating in divulging to him APOC's confidential offer, might be putting at risk the confidence in himself of the British protectors of his State and régime in a way which, in the then difficult state of his relations with his neighbours in Saudi Arabia, Iraq, and Persia (which at that time did not even acknowledge the existence of his State) he could ill afford. Reconsideration also of the prospect of APOC and Holmes's backers, Gulf Oil Corporation, being about to combine for his oil concession, of which there is every reason to believe he had been informed by Holmes and for which the latter had probably claimed credit as good business for the Shaikh, may have brought belated recognition that it could, if he showed too easy an acquiescence in it, mean an end to the profitable bidding-up process which he had been enjoying; and which he knew was then also in process in Jeddah, where the Iraq Petroleum Company was competing for King Ibn Saud's concession with Standard Oil Company of California. Some such background to Holmes's bad reception by Shaikh Ahmad on 17th May and the necessity to re-establish himself in his confidence, could explain both the unusual frequency of Holmes's meetings with the latter over the following few days, and also his interview with Dickson. That interview for a man of Holmes's great negotiating ability, was apparently conducted deliberately in such a way that an account would reach the Shaikh, and was probably intended to show that he and his backers were as good as APOC and theirs and to draw, if possible, some indication from Dickson that the suspension of negotiations had been more the consequence of APOC and political pressure than of the Shaikh's own volition. When Holmes's profession of ignorance of Shaikh Ahmad's suspension order was countered by Dickson's production of the Shaikh's own letter to him confirming it, of which Holmes had probably been hitherto unaware, his bluff was so obviously called that, as he said, he lost both his temper and the objective of his visit. That the episode had no lasting adverse reaction on the Shaikh's lifelong respect and friendship for Holmes is an indication of the latter's negotiating finesse and dominating personality, and possibly also that the Shaikh was aware of the purpose underlying Holmes's visit to Dickson and could therefore condone the 'mistake' involved in it.

APOC and Gulf agree on a standstill of their competitive negotiations in Kuwait

Shaikh Ahmad's suspension of negotiations on 14th May 1933 was soon followed, on 23rd May, by an agreement reached in London between APOC and Gulf (whose discussions since the previous December of the possibility of combining to obtain a Kuwait Concession were by then within sight of succeeding; see Note 77) that, for the next three weeks no efforts were to be made by the representatives of either party in Kuwait or elsewhere to get the concession for themselves alone. The initial three weeks of this 'standstill', as it was called, were subsequently to be continually prolonged to cover the seven months which it eventually took before the formal document of agreement between the two Companies to combine their efforts to get a Kuwait Concession was finally completed and signed on 14th December 1933. Whereafter, following the formation of their jointly owned Kuwait Oil Company in January 1934, the return of Chisholm and Holmes to Kuwait in February 1934 as that Company's joint representatives and negotiators was to mark the end of both 'suspension' and 'standstill' and the beginning of the final phase of the concession negotiations.

Such events of relevance to the negotiations as occurred in the eight months from June 1933 to February 1934, during which Shaikh Ahmad's 'suspension' and the APOC/Gulf 'standstill' combined to enforce an armistice in the battle for the Kuwait Concession which had been going on since May 1923 between APOC and EGS/Gulf, were as follows:

As instructed by their principals, Holmes and Chisholm and their respective supporters in Kuwait from 23rd May 1933 onwards abandoned their previous efforts either to obtain the concession from Shaikh Ahmad or to persuade him to reopen negotiations. But as, until the actual conclusion and signature on 14th December of APOC/Gulf's agreement to combine, there was always a contingency, however unlikely, of agreement not being reached and their competition being resumed, their policy of cultivating the best possible relations with the Shaikh was actively continued. There were also two other reasons for such a policy. There was the possibility, then considered exceedingly remote by all concerned though it was in fact to occur just a year later, of another competitor for the Kuwait Concession making proposals to Shaikh Ahmad. Also, as until the APOC/Gulf agreement was concluded each party felt that it might weaken its own position **(87)** to inform the Shaikh—although he was probably aware of it being under arrangement from May onwards, and it was common gossip in Kuwait by July, it is a curious fact that his first official notification of it was not to be until the Political Resident informed him on 15th February 1934, the day on which Chisholm and Holmes arrived in Kuwait to begin their joint negotiations on behalf of the Kuwait Oil Company—it was always possible that either to accelerate matters or in order to have the terms already on offer to him further improved, Shaikh Ahmad might revoke his suspension order and request both APOC and EGS to reopen their negotiations. The fact that the Shaikh made no move of that sort, but was content to bide his time can have only have been due to his faith in his old friend Holmes's willingness and ability to get him the best possible terms in any circumstances.

Effect on Kuwait negotiations of Standard Oil of California obtaining Saudi Arabian concession on 29th May 1933

Holmes, after leaving Kuwait on 24th May 1933 for London, remained there until August when he returned to Bahrain. By then he and his supporters in Kuwait had been encouraged, and Chisholm and the pro-APOC party correspondingly disappointed, by the news which reached Kuwait from Jeddah in early June that King Ibn Saud had granted the Saudi Arabian Oil Concession **(88)** on 29th May to the Standard Oil Company of California; whose agreement to pay £30,000 (approx. Rs. 400,000) on signature with a further £20,000 eighteen months later had far outbid the £10,000 offered by Iraq Petroleum Company (in which APOC was a partner). Holmes, when in Kuwait in March accompanied by Mr Janson of EGS and Mr Lombardi of Standard Oil of California, had forecast the latter's success with Ibn Saud to the Shaikh, and had renewed his prediction with even greater certainty in May after himself visiting Jeddah in April in an unsuccessful attempt to interest the Saudi Government in again granting to EGS oil rights in the Kuwait/Saudi Neutral Zone. The correctness of this prediction, as well as the fact that King Ibn Saud had awarded his concession to an American company, whose payments were on a scale unprecedented in Arabian oil negotiations, in preference to the partly British Iraq Petroleum Company, were calculated to raise Holmes's and his American backers' prestige with Shaikh Ahmad and to encourage the latter to expect a much higher payment on signature of his own concession than the Rs. 250,000 proposed in April by EGS, then the best offer in his own possession.

Holmes, after arriving back from London in Bahrain on 3rd August (he had a short talk *en route* with Shaikh Ahmad on Kuwait aerodrome the same day) remained there until late September. While in London he had been assisting his EGS/Gulf colleagues—towards the end of June a senior Gulf Official Mr Guy Stevens **(89)** had arrived in London to carry on negotiations with APOC which Mr Wallace (see Note 81) had had there six weeks previously resulting in the 'standstill' on 23rd May—in their discussions of the Kuwait situation and arrangements to combine with APOC. During his absence Mohammed Yatim represented him in Kuwait, and when the Shaikh entertained Chisholm and his assistant Haji Abdullah Williamson (see Note 26) to dinner on 30th June the fact that he also invited as the only other guest Yatim whom he well knew was extremely disliked by Williamson, was regarded by Chisholm as the most conclusive indication yet that the Shaikh was aware of the impending combine between APOC and Gulf.

APOC complains to Gulf that Yatim (EGS) in Kuwait is contravening 'standstill'

By the end of July, APOC in Kuwait had heard from so many trustworthy sources that Yatim had informed Shaikh Ahmad that Holmes in London 'had been approached by APOC with a view to combining with him to acquire the Kuwait Concession, but was refusing to do anything detrimental to the Shaikh', that they complained to London about such action as being contrary to the 'standstill' and likely to benefit no one except Holmes and Yatim in their personal relations with the Shaikh. When APOC complained to Gulf Oil Corporation the latter instructed Holmes to query Yatim, who cabled him on 31st July denying that there was any truth in the story.

There was to be an interesting sequel nearly two months later when Holmes, again passing through Kuwait on 21st September *en route* back to London from Bahrain, was met on the aerodrome there by Mullah Saleh with whom he had a long conversation. When Holmes described this in London to his EGS/Gulf colleagues, he reported Mullah Saleh as having said that Chisholm, when visiting the Shaikh that morning, had asked him if he was now ready to discuss a concession with APOC as Holmes was just leaving the Gulf (Holmes had just been obliged to give up his appointment in Bahrain with the Bahrain Petroleum Company); that the Shaikh replied that he would not do so and could not believe Holmes would not return; that Chisholm then told the Shaikh that EGS/Gulf were negotiating with APOC without informing the Shaikh; that the Shaikh replied to Chisholm that, on the contrary, he had been informed and knew 'EGS/Gulf would not throw him to the wolves'. This account of Chisholm's conversation, evidently meant as a riposte to APOC's recent accusation against Yatim (Mullah Saleh's close friend and confidant) was completely untrue, and the actual conversation during Shaikh Ahmad's talk with Chisholm on 21st September was reported by the latter at the time, in a letter dated 27th September to London from Abadan, as follows:

> The Sheikh continues extremely cordial, while showing no signs of interest or renewing activity in his oil negotiations which have now been in suspense since 14th May. He mentioned that he had received news that Major Holmes was being 'exiled' from Bahrein and was not to be permitted to return. He considered that this move is the natural result of the Sheikh of Bahrein's dissatisfaction with the way his concession is being worked. He stated that he had not heard from Major Holmes for a considerable time, apart from a telegram received on 20th September stating that he was leaving Bahrein by air and could not stop to visit Kuwait. Mr Chisholm enquired whether the Sheikh would soon be reopening his own negotiations now that Major Holmes was apparently unemployed. The Sheikh replied that he did not propose to take any action for the present adding, as usual, that the Company's interests would not suffer thereby.
>
> Major Holmes was met on Kuwait aerodrome on 21st September by Mullah Saleh, the Sheikh's Secretary, with whom he had a long conversation.
>
> We continue to believe that the Sheikh remains inactive in these negotiations because he has been informed that, by the good offices of Major Holmes, the two competitors for his concession have joined forces and will shortly offer for a joint concession such as is, in his opinion, desirable for Kuwait from both a commercial and a political point of view.

Holmes was to remain in London from September 1933 until his return to Kuwait, with Chisholm, in the following February. Until December he was engaged in the London discussions between EGS/Gulf and APOC leading up to the Gulf/APOC agreement of 14th December, and thereafter in the discussions between Gulf and APOC of the form and terms of the draft concession to be presented to the Shaikh on behalf of their jointly owned Kuwait Oil Company formed on 2nd February 1934. Holmes himself thus only spent a few hours in Kuwait, on 3rd August and 21st September, between leaving it on 24th May 1933 and returning again on 15th

February 1934, but he was well represented there meanwhile by Mohammed Yatim who kept him closely informed of developments.

Chisholm, although from June 1933 onwards he was based in Abadan leaving his assistant Williamson in occupation of his house in Kuwait, visited there approximately every ten days until he was recalled to London in mid-November for a month's discussion there, jointly with Holmes, with APOC/Gulf on their future negotiating plans in Kuwait. This close touch with Kuwait affairs was at the direct request of Shaikh Ahmad, who at their frequent meetings from 14th May onwards treated Chisholm with a cordiality which on many public occasions he went out of his way to display. On 6th August 1933 when Chisholm, visiting Kuwait in his Company's new aeroplane, had at the Shaikh's request taken him for a flight over his State and his properties in Iraq, he responded by inviting Chisholm to accompany him on his yacht for a visit to Bahrain. Their conversation on 21st September has already been described, and at another meeting in October the same atmosphere of cordiality prevailed. When describing that meeting on 5th November Chisholm stated that anticipation of APOC's combining with the interests represented by Holmes continued to be the subject of conversation in Kuwait business circles, and that Mohammed Yatim was reported as having said on several occasions that his employers were engaged in forming 'a strong combine' which would shortly be ready to exploit the oil resources of Kuwait. In that connection, shortly after Chisholm's return to London in mid-November, an incident occurred proving that, although Holmes had denied in August that either Yatim or he had informed Shaikh Ahmad that EGS/Gulf and APOC were considering joining forces, the Shaikh was by then well aware of it from them.

Holmes's telegrams to Yatim regarded by APOC/Gulf as contravening 'standstill', November 1933

This was a letter dated 20th November 1933 sent by Holmes from EGS's London office to APOC's, enclosing the following telegram which he had received that day from Mohammed Yatim in Kuwait:

> Referring to your telegrams dated 14th and 16th November the Sheikh of Kuwait does not mind which name you call the Company and in which way it be formed as long as you think it would not harm the Sheikh and his people's interests and also that political difficulties would disappear. The Sheikh conveys his greetings to you and Mr Janson and assures both of you that the Sheikh has full confidence and is sure that both of you will preserve him both politically and commercially and adds that he is confident you both feel the same friendly feelings as he has towards you both.

In his letter **(90)** to APOC Holmes explained that this telegram was in reply to one sent by him to Kuwait on 14th November requesting Yatim to find out from the Shaikh whether he had any preference as to what the oil company to operate his territory should be called and whether he minded who the English groups were that would join with the Americans in forming the Company.

> I told Yatim to mention that the Company with whom the Americans would most probably work would be the APOC. You will see from the reply to this telegram that there is no difficulty regarding the name of the Company, nor is there any difficulty regarding the Americans and the APOC forming a Company subject, of course, to the safeguarding regarding the interests, both politically and commercially, of the Sheikh and his people being protected, as mentioned in the cable.

This letter was taken by both APOC and Gulf to be proof that Holmes and Yatim had long since informed the Shaikh of the EGS/Gulf APOC combine being under arrangement.

Although that could no longer do any harm, the merger having by then been arranged and only awaiting the completion of documents for final signature, Holmes's action was considered by APOC/Gulf to be so contrary to his instructions under their 'standstill' understanding since the previous May, that he was asked to come and explain it at a meeting **(91)** at APOC's office on 21st November 1933 presided over by Mr Fraser **(92)**, then Deputy Chairman of APOC, with Chisholm also present. It was pointed out to Holmes that, though he had doubtless acted in good faith, his communication to the Shaikh of matters relating to the proposed combine was contrary to the intention of the 'standstill' agreement; and he was requested to give an assurance that he would not communicate again with the Shaikh regarding the Kuwait oil concession without prior authorisation by APOC/Gulf. Which assurance was given by Holmes, who said that he had only telegraphed to the Shaikh because the latter had telegraphed to him through Yatim asking what progress was being made by him in the matter of his concession negotiations, to which it was difficult not to reply.

The episode was regarded by APOC/Gulf as typical of Holmes's lone-wolf methods of negotiation, which he had been accustomed to use for so long and with which he had been so successful that it was inevitable that, whether or not they accorded with the team spirit necessary in the new combine's plans, he could hardly be expected to change them **(93)**. It was desirable, as both sides had agreed after close consideration **(94)** because of Holmes's unique experience and standing with Shaikh Ahmad, to continue to take full advantage of them in the combine's coming negotiations. It was therefore necessary to condone the sort of idiosyncratic behaviour which he had displayed on this occasion, and also to be on guard against any future examples of it. It was mainly for that reason that very full and comprehensive instructions **(95)** were given to Holmes (and copies sent to Chisholm in Abadan) when he eventually left London in early February 1934 for Basrah. Chisholm, who had returned to Abadan from London in mid-December was to meet him there on 15th February in order that they should proceed together to Kuwait the same day as joint negotiators for the new Kuwait Oil Company.

APOC/Gulf agreements of December 1933: incorporation of Kuwait Oil Company, February 1934

By that time all the essential preliminary arrangements and agreements had been come to between APOC and Gulf, Gulf and EGS, and APOC/Gulf and the British Government, after protracted discussions in London, Pittsburgh, and New York, for the two former rivals' new joint approach to Shaikh Ahmad for the Kuwait Concession.

Two agreements **(96)**, both dated 14th December 1933, were the basis of the partnership in Kuwait between APOC and Gulf. Of these the principal agreement was between APOC and Gulf Exploration Company, the subsidiary of Gulf Oil Corporation of Pennsylvania which dealt with its Kuwait interests. It provided for the formation of a joint company, equally owned and controlled by both parties, to obtain and operate a Kuwait oil concession, and regulated the manner in which its negotiations, expenses, finance, and operation should be conducted. The basic principles of the agreement was equality between partners. The second Agreement, between APOC and Gulf Oil Corporation of Pennsylvania, was entered into for the following reason. Under the principal agreement APOC and all its subsidiaries were bound in various ways, including marketing of any production obtained from Kuwait, and so also was Gulf Exploration Company. The latter, however, was a subsidiary of Gulf Oil Corporation, and APOC's agreement with it did not bind its parent. Consequently the two parent companies bound themselves similarly in the second Agreement.

The joint company, Kuwait Oil Company Ltd., provided for in the principal Agreement of 14th December was incorporated in London on 2nd February 1934, with an initial issued capital of £50,000 owned in equal shares by APOC and Gulf. Of its six directors, three were APOC and three Gulf appointees; the Chairman had no casting vote and was to be appointed annually in rotation by APOC and Gulf. In connection with this Company the principal APOC/Gulf Agreement's provisions as regards negotiations for a Kuwait oil concession included the following:

(1) After the signing of this Agreement neither party shall carry on any negotiations or make any efforts directly or otherwise and/or either alone or jointly with others, to secure a concession or any interest in any concession covering the whole or any part of Kuwait except as may be mutually agreed by the parties or as may be determined by the Company; and any Kuwait oil concession or interest in any Kuwait oil concession heretofore or hereafter obtained by or on behalf of either party shall be considered as held in trust for the benefit of both parties and be transferred forthwith to the Company.

(2) Each party undertakes to employ, in such manner and to such extent as may be mutually agreed or may be determined by the Company, the agencies and facilities at its disposal to secure from the Sheikh a concession in terms satisfactory to the parties hereto.

(3) Each party hereto shall bear all expenses incurred by it prior to the date of this agreement in connection with efforts to obtain a Kuwait Concession.

(4) From the date of this Agreement, all expenses incurred by either party in connection with efforts mutually agreed upon or determined upon by the Company to secure any concession or in carrying out the provisions and purposes of this Agreement shall be for equal joint account of the parties hereto or for account of the Company.

Of the above provisions, the first was to have an important bearing on an extraordinary development (to be described later) which occurred in October and November 1934 in the closing stages of the Company's negotiations in Kuwait. Under the second provision, the services of Holmes continued to be employed by the Kuwait Oil Company in the same way as those of Chisholm. Under the third provision Gulf, under an agreement made with EGS on 6th November 1933, at the end of January 1934 reimbursed them (they, with Holmes, then ceased to have any connection with the Kuwait negotiations except as employed by KOC) for all expenses incurred on its behalf in attempting to obtain the Kuwait Concession prior to that date. Under the fourth provision Gulf then also paid EGS a sum of £36,000 (of which APOC contributed half) in respect of an obligation in their previous contract with EGS to pay the latter a royalty of 1/- per ton on any oil to be found by them in Kuwait, through any concession obtained for them by EGS, in excess of 750 tons per day **(97)**.

H.M.G./Kuwait Oil Company agreement, March 1934

To regulate APOC and Gulf's relations with the British Government in connection with the Kuwait Oil Company's forthcoming negotiations with the Shaikh, the British Government signed an Agreement **(98)** with the Kuwait Oil Company on 5th March 1934. Its preparation was so far advanced by the time of Holmes's departure from London on 7th February for Basrah to meet Chisholm and go on to Kuwait, that it was not felt necessary to delay the start of negotiations any longer on its account. This agreement, on the strength of which the Political Resident in Bushire was instructed to inform Shaikh Ahmad that the British Government had no objections to his negotiating with the new Company in view of its Agreement with the British Government (a copy of which was being sent to the Shaikh for his information), covered the following main points:

(*a*) The Company was to remain a British Company.
(*b*) The concession, if obtained, was not to be transferred without the consent of the British Government.
(*c*) Provisions concerning nationality of employees, appointment of a chief local representative in Kuwait, aircraft landing rights, and wireless facilities.
(*d*) The British Government's right of pre-emption over Kuwait oil and products in the event of war.
(*e*) Should there be any conflict between the terms of the Agreement and those of any concession made between the Company and the Shaikh, the former would prevail.

As will be described later, the necessity of reconciling their Company's concession Agreement with Shaikh Ahmad with its political agreement with the British Government was to cause Chisholm and Holmes some difficulty in their negotiations. But its conclusion meant that the negotiations were otherwise free from the overriding necessity (which had so frequently in the past obstructed and delayed their previous separate negotiations) of having political points referred back from Kuwait to London for clearance as they arose, and would be confined to straightforward commercial issues with the Shaikh concerning the concession's monetary or operating terms. Both Gulf and APOC having already got these issues so far forward with the Shaikh in their previous discussions, and both their joint negotiators (but especially Holmes) being on friendly personal terms with him, the Kuwait Oil Company Board in London anticipated a successful conclusion to the concession negotiations fairly soon after Chisholm and Holmes arrived back in Kuwait on 15th February 1934. They were to take with them, for presentation to the Shaikh, a concession document **(99)** (drafted so as to combine the main points included in the APOC and EGS/Gulf drafts previously discussed with him) which had also been agreed in London with all the H.M.G. departments concerned, and a copy of their joint power of attorney **(100)** from Kuwait Oil Company indicating that they had full powers to negotiate with the Shaikh on its behalf.

Chisholm and Holmes return to Kuwait as joint KOC negotiators, 15th February 1934

The position on 15th February 1934, when Chisholm and Holmes returned to Kuwait as joint KOC negotiators, at the beginning of the final stage in the negotiations, may be summarised as follows:

(1) Shaikh Ahmad could justifiably regard himself as being in a very strong bargaining position, to which his own shrewdness and initiative had contributed on various occasions since he had been first approached for an oil concession by APOC eleven years previously. His early encouragement of Holmes's company as a potential competitor, based originally on nothing more than appreciation of his personal qualities and reputation in Bahrain and Saudi Arabia, had been more than justified by Holmes's success in discovering oil in Bahrain and enlistment of a major American oil company to back his proposals for Kuwait. That the Shaikh was now about to start negotiating, with the approval of the British Government, with an Anglo-American oil combine justified his original opinion (against the advice of many of his family and friends that to encourage American competition was to put at risk Britain's vital goodwill towards his State) that H.M.G.'s attitude to Anglo-American co-operation elsewhere in the Middle East meant that they would in due course agree similarly in Kuwait. Having been personally assured by Sir John Cadman for APOC and by Holmes for Gulf that both their companies were ready to exploit, and pay well for, his concession, he expected that their joint company's offer of monetary terms would be at least as favourable as their highest offer to date. As a good bargainer himself, he must have anticipated that their last word on that subject had by no means yet been said, and, in the light of the monetary terms recently agreed for King Ibn Saud's Hasa concession, the prospects of his further increasing them were obviously good. Both commercially therefore, as well as politically, Shaikh Ahmad's position on the eve of his first meeting with Holmes and Chisholm as joint negotiators for the Kuwait Oil Company must have been one of agreeable anticipation and justifiable self-satisfaction.

(2) From the point of view of both the British Government (and of its local representatives the Political Resident and Political Agent) and the American Government, the situation now reached was at last satisfactory to both. The American State Department must have felt that its interventions on behalf of the Gulf Oil Corporation had successfully contributed to the Anglo-American *entente* now established over Kuwait. The British Government departments concerned were satisfied that their long-term objective (as laid down in the Colonial Office's letter of September 1923) of seeing that Kuwait got proper terms for its oil concession was being achieved, and that the British Government agreement with the Kuwait Oil Company and the latter's draft concession proposals for presentation to the Shaikh adequately safeguarded all aspects of Kuwait's political and commercial interests. That the sometimes acrid interchanges of the past five years had left no bitterness on the subject of Kuwait oil between the American and British Governments is well indicated in a letter from the American Embassy in London to the State Department in Washington at the end of September 1933, at which time the APOC/Gulf Oil Corporation discussions culminating in their Agreements of 14th December 1933 were in full swing. After recounting the current state of those discussions, anticipating their eventual successful conclusion and saying that meanwhile in Kuwait Chisholm and Holmes were 'jockeying for position', the letter emphasised that the American interests concerned believed that the British were as sincere as themselves in their efforts to achieve a true compromise which would avoid the necessity for further diplomatic exchanges.

(3) Both Gulf and APOC, and their Kuwait Oil Company negotiators Holmes and Chisholm, were optimistic that, after a few weeks at most, during which the money terms first suggested would have to be moderately increased and minor alterations made in the operating terms of their draft concession proposals, a Kuwait Concession Agreement should soon be concluded with Shaikh Ahmad. This optimism was to be disappointed.

CHAPTER 4

February 1934 to 23rd December 1934

When Holmes arrived in Basrah from London on the morning of 15th February 1934 he was met by Chisholm, who had come there from Abadan, at the airport and they proceeded together by car later in the day to Kuwait arriving at 5.30 p.m. Holmes had brought with him from Cairo the Arabic translations of the KOC draft concession document which he had had prepared there *en route* from London. Before leaving Basrah they telegraphed to Shaikh Ahmad and the Political Agent in Kuwait announcing their arrival, and when passing through the Customs post on the Iraq/Kuwait frontier at Safwan they received a reply from the latter welcoming them in the joint names of himself and of the Political Resident who was then visiting Kuwait. The wording of the reply 'welcome to the heavenly twins', was a fair indication of the friendly attitude towards them of those officials, who must have been as glad as they were that their long period of business antagonism was now ended.

On the following day (which was a Friday, so they could not visit the Shaikh) they were requested by the Political Resident to visit him as he was leaving Kuwait that afternoon. He informed them that on the previous day he had officially informed Shaikh Ahmad of their expected arrival to negotiate jointly on behalf of a newly formed Kuwait Oil Company, also that the necessary formal permission from the British Government to negotiate with the KOC would very shortly be communicated to the Shaikh and that pending arrival of that permission there was no objection to preliminary discussions taking place with him. That evening Shaikh Ahmad sent word that he wished them to visit him next morning, and they accordingly did so, from 10 to 12.15 p.m. on Saturday, 17th February, accompanied by their interpreter-assistants Haji Abdullah Williamson and Mohammed Yatim for the first time in their joint KOC capacity. The Shaikh was exceedingly cordial, and although the visit was of a purely social nature he evidently realised the implication of their joint visit. On the 18th, when he returned their call and they had dined with him in the evening, they asked him for a first business appointment which he granted for 22nd February, the slight delay being, they believed, because he expected to receive in the meantime the British Government's formal permission to negotiate with them as anticipated by the Political Resident.

At their interview on 22nd February, Chisholm and Holmes described the background and formation of KOC explaining that EGS was no longer concerned with a Kuwait Concession, and APOC and Gulf were now jointly interested in it as equal co-owners of KOC. They informed Shaikh Ahmad that they had a joint power of attorney from KOC to negotiate with him, and also had prepared a draft concession agreement to submit to him as soon as he wished after H.M.G.'s formal permission to do so was received. The Shaikh (according to a report dated 22nd February sent to KOC and signed by Holmes and Chisholm) by his questions and general attitude appeared to appreciate and welcome the formation of the new Company, as 'a combine of two interests both of which were valued by him, but neither of which had he hitherto been willing to favour to the exclusion of the other'. He confirmed that he had not yet received the formal permission to negotiate promised by the Political Resident, but he requested the two negotiators meanwhile to present him with their draft concession in order that he might begin examining it and avoid delay when H.M.G.'s permission was received, and also to send him for record purposes a translation of their power of attorney. These were accordingly sent to him after the interview **(101)**.

Monetary terms of first KOC concession draft

Before sending their concession document to Shaikh Ahmad, Chisholm and Holmes had to decide what monetary figures to insert in the spaces for them which had been so far left blank, for security reasons, in the draft as printed.

The question of the monetary terms to be included in their Kuwait Concession proposals had been exhaustively considered by APOC and Gulf from December onwards, and they had eventually authorised their negotiators to go initially **(102)** to the figures shown in the first column of the table below (which being *maximum* figures would not of course be offered at the start of the negotiation) which also shows in the second and third columns the amounts last offered by APOC and EGS:

		KOC Maximum	*APOC last offer*	*EGS last offer*
Initial Payment	Rs.	200,000	200,000	250,000
Annually Before Commercial Production	Rs.	65,000 *or*	45,000 *or*	50,000 *or*
Royalty		*Rs. 2.12 per ton	*Rs. 2.12 per ton	Rs. 3.8 per ton
Annually After Commercial Production	Rs.	200,000 *or*	200,000 *or*	250,000 *or*
Royalty		*Rs. 2.12 per ton	*Rs. 2.12 per ton	Rs. 3.8 per ton

* Included 2 Annas in respect of taxation exemption.

Both Holmes and Chisholm, when consulted by APOC/Gulf in December, had said that although it was *possible* that they could get the Shaikh to agree to figures in or near the range of such maxima, it was highly probable, especially in view of the terms obtained six months previously by King Ibn Saud (see page 38) for his oil concession, that he would require considerably higher sums. There was also, they felt, the consideration that Sir John Cadman's figure of Rs. 400,000 offered in March 1933 for an immediate deal, would still be in Shaikh Ahmad's mind.

But the KOC Board in London did not accept their view, and instructed them to open their negotiations within the maxima authorised.

With those considerations in mind, the figures inserted by the negotiators in the first KOC draft sent to the Shaikh on 22nd February 1934 were as follows:

Initial payment, Rs. 200,000; annually before commercial production Rs. 50,000 (or royalty Rs. 2.12); annually after commercial production Rs. 150,000 (or royalty Rs. 2.12), very close to the maxima authorised.

It was when awaiting Shaikh Ahmad's reaction to their first draft proposals, that Holmes and Chisholm discussed and agreed on the working procedure which they would adopt for their joint negotiations in accordance with their London instructions (see Note 95). All their formal interviews and conversations with the Shaikh would be conducted together; and all their written communications with him, and with KOC in London, jointly signed. They would send regular written reports by air-mail to KOC in London, weekly if the news justified, otherwise at longer intervals. As Chisholm's house had larger office space, their joint reports and letters would be drafted, and telegrams drafted and coded, by him there, and then agreed or amended with Holmes before being jointly signed and despatched. Holmes's house was on the sea-front about a quarter of a mile to the west of the Sief Palace where all the negotiators' formal talks with the Shaikh took place. Chisholm's house was about a quarter of a mile further to the south-west, inside the then existing town wall near its Jahrah Gate. Shaikh Ahmad frequently entertained the negotiators at his Dasman Palace, and they returned his hospitality at their own houses, either together or singly. On these and other similarly informal occasions they agreed that any references made by the Shaikh to either of them to matters relating to the concession negotiations would be reported as soon as possible to the other, and that no decisions should be taken or offers made without the other's agreement in accordance with their written instructions from KOC. As will be seen, this system worked well until Holmes broke away from it in mid-October 1934.

It was not until 27th February that Shaikh Ahmad requested the negotiators to visit him again. Since their last meeting on 22nd February they had heard from the Political Agent that he had communicated to the Shaikh the British Government's agreement to his opening negotiations with their company, and had also informed him that an Agreement covering political points had been come to between H.M.G. and the Kuwait Oil Company, a copy of which was being forwarded to Kuwait for his information. The Political Agent had asked the negotiators various questions about this Agreement which they had not been able to answer, being themselves without any detailed information on the subject. While they were aware of the existence of this 'political Agreement' (as it was called to differentiate it from the 'commercial' agreement which they were negotiating with the Shaikh), their London principals had, before they left London, told them that, as it was not desirable that they should get involved in any political issues, they were not being informed of its details and should avoid getting into any discussions about it. The negotiators, who had heard that Shaikh Ahmad's lawyer from Basrah had been summoned to Kuwait to assist his scrutiny of their proposals, attributed to that fact the several days which had elapsed since the last meeting and had no reason to believe that this first negotiating session would not go well.

Shaikh Ahmad's unfavourable reaction to KOC proposals

But on arriving at the Sief Palace on the morning of 27th February for their appointment with the Shaikh, they could see from his demeanour that this would not be so. When, after the usual polite overtures, he proceeded to business, he said that, although they were now acting together, he wished first to ask each of them separately some preliminary questions. Firstly, he asked Holmes why the draft as now presented omitted various favourable considerations already offered by him on behalf of the EGS, such as their initial payment offer of Rs. 250,000 and Rs. 3.8 royalty and that Major Holmes would remain resident in Kuwait on behalf of the concessionary company. Secondly, he asked Chisholm how he could reconcile the initial offer of Rs. 200,000 in their KOC draft agreement with the offer of Rs. 400,000 made by Sir John Cadman during his visit to Kuwait in the previous March. Speaking 'with considerable and unusual heat' (according to the negotiators' report of the interview to London) Shaikh Ahmad said that their draft fell so very far short of his expectation that he was inclined to reject it altogether as a basis for discussion. If it was the Kuwait Oil Company's opinion that they alone were interested in his concession he wished them to know that he was aware that the Standard Oil of California and other companies would be very ready to treat with him.

After some general discussion, the negotiators thought it best to adjourn the meeting in view of the Shaikh's mood, having arranged with him to renew the conversation later after having fully considered his attitude towards their terms. They subsequently decided that, as they had been each separately charged with breaking faith with the Shaikh, it would be best to seek separate interviews in which to re-establish themselves. In this way an unwieldy double interview, of a difficult nature for both the Shaikh and themselves in view of his past dealings with each of them when in competition with the other, would be avoided and, they hoped, the air cleared for the future of any misunderstandings over the past. They consequently wrote a joint letter to the Shaikh requesting separate appointments which were accorded by him on 28th February to Chisholm and 1st March to Holmes. They then answered his questions as best they could by reference to the changed basis of negotiations and the changed form of their draft which was claimed to contain new favourable features to counterbalance the omission, if any, of advantages promised in previous negotiations. As regards the Rs. 400,000 initial payment offered by APOC in the previous March, Chisholm pointed out that, as that offer had been definitely conditional on its immediate acceptance, it had on non-acceptance become null and void and could not now be regarded as any criterion for an offer at this stage by the Kuwait Oil Company. These two conversations had the desired effect of dissipating any feeling that the KOC proposals represented any back-tracking from previous proposals of either APOC or EGS, and the Shaikh's cordiality towards both negotiators was fully restored. But he made it very clear to each of them that he saw no reason for their joint terms for his concession being in any way less favourable than the best that he had been offered by either of them separately in the past, and that when they next came to discuss the matter with him jointly, which he requested them to do on 3rd March, he proposed to inform them of the terms which he would consider acceptable as a basis for their future negotiations.

When reporting the situation on 1st March to KOC in London, the negotiators emphasised that Shaikh Ahmad, notwithstanding his vehement attack (which they said could mainly be attributed to the disregard of their own advice that the monetary terms authorised for their presentation to the Shaikh were too low in comparison with their past offers), continued to express very friendly sentiments towards the Kuwait Oil Company, from whom he professed 'willingness to accept tens where from others he would demand twenties'. They believed that his unexpectedly peremptory tone was due to instigation by other parties, feeling that there was more behind it than a natural desire to obtain the best of all terms already offered, whether conditional or not. There was also a possibility that his dignity had been offended by being kept so far in the dark about the Political Agreement, and the fact that an important emissary from King Ibn Saud had arrived in Kuwait on 27th February might also have had a bearing on his attitude. The latter suspicion was confirmed when the negotiators kept their appointment with Shaikh Ahmad on 3rd March. He then informed them that he had not yet been able to complete his examination of their proposals in the light of their recent interviews, and it would be three or four days yet before he could give them his opinion and proposed terms. The concession affected his whole people and was for a long period and therefore needed very careful consideration. He was favourably inclined towards the Kuwait Oil Company, particularly because it was a British Company, but if its terms were too hard he must consider other alternatives in the interest of his people. He added that he had received communications from Ibn Saud extolling the performance of the Standard Oil Company of California, and enquiring his intentions with regard to the Hasa–Kuwait Neutral Zone.

When the negotiators tried to obtain an idea of his general opinion of KOC's proposed terms the Shaikh replied that the monetary terms were his main objection to them, though he might require also a few minor alterations in the operating terms; whereupon the negotiators withdrew, having arranged to resume discussions as soon as they heard again from the Shaikh. This was not until 8th March, when they were requested to meet him two days later.

Clauses referring to H.M.G./KOC political agreement to be included in KOC concession draft

By then they had received telegraphed instructions on 6th March from London, where the British Government/KOC Political Agreement (see Note 98) had been formally signed on the previous day, to add two clauses referring to it to their draft agreement as already submitted to the Shaikh. These clauses were as follows:

> If the Company shall fail to observe any of the Terms of the Agreement between the Company and His Majesty's Government signed in London on 5th March 1934, and, if the matter is referred to Arbitration under Article 18, fail to remedy such failure within the reasonable time which shall be fixed by the Arbitration for so doing . . .
>
> It is hereby declared that should any of the terms of this Agreement be inconsistent or in conflict with the terms of the Agreement between the Company and His Majesty's Government signed in London on 5th March 1934, this Agreement shall, to the extent of any such inconsistency or conflict, be subordinate to and controlled by the terms of that Agreement between the Company and His Majesty's Government.

When discussing these proposed additional clauses with the Political Agent on 7th March, the negotiators asked him whether the British Government/KOC political agreement had yet been communicated to the Shaikh, as until that had been done it was clearly inadvisable for them to discuss the insertion of such clauses into their concession document. The Political Agent said that no copy of the Political Agreement (or details of it) had yet arrived in Kuwait, and he agreed with the negotiators that possibly Shaikh Ahmad was deferring discussion of their draft pending its arrival and his information regarding its contents. If the British Government/KOC agreement was to override the Shaikh's own agreement with KOC he would obviously wish to study the former before coming to any definite decision on the latter.

Shaikh Ahmad's written statement of his required terms, 12th March 1934

When the negotiators kept their appointment on 10th March with the Shaikh he again postponed any discussion of his opinion of their terms, informing them instead that he would send them a written statement on the subject two days later. This duly arrived on 12th March, reading as follows (the figures in brackets are the comparable amounts already offered by the negotiators).

Our remarks in regard to payments:

(1) Payment on signature Rs. 550,000 (200,000).
(2) Paymend during exploitation and drilling Rs. 120,000 (50,000) at the end of each year.
(3) After the Company's declaration of the existence of commercial oil, royalty on each ton of oil Rs. 3/12 (2/10) provided that the yearly payment shall not be less than Rs. 250,000 (150,000).
(4) As duty on exports and that which may be sold in the same country upon which the aforementioned royalty is payable the Company shall pay on each ton of oil Annas 5 (2) at the end of each year.
(5) A. We have the right to appoint one member in the Company's Office in London.
B. We also have the right to appoint a local representative for things connected with this agreement between us and the Company.

We have a few points in some of the texts of the Agreement and will point them out to you if agreement is reached.

When telegraphing these written demands to London, the negotiators said they regarded them as only a basis for bargaining and before reporting further would discuss them further with Shaikh Ahmad.

This they did at an interview on 14th March, their object being to obtain an admission from him that such high terms, notably the initial payment and the royalty, were distinctly figures for negotiation and that they might expect some compromise between his figures and theirs. This the Shaikh readily admitted. When the negotiators pointed out that his royalty demands in particular were surprisingly in excess of anything he had ever mentioned to either of them during their previous separate negotiations, the Shaikh did not demur, but indicated that he would like them to get their London principals' opinion of his demands and also that he considered that they bore comparison with the royalties being paid in the Persian and Iraq concession agreements. On this subject the negotiators felt, though Shaikh Ahmad had not mentioned the subject and they had always avoided it, that he might have in mind the gold/sterling depreciation clauses in those agreements **(103)**. As regards the Shaikh's request to have the right to appoint a Director (this, it became clear in their discussions, was meant by his request for a 'member') of the Company in London as well as a representative in Kuwait, the negotiators expressed surprise that this should be required and doubt as to what useful work such an appointee could do. The Shaikh replied that he was entitled to such an appointment to protect his interests and requested them to get their London Principals' views on this point too; in which connection the negotiators subsequently reported to London that it must be recollected that such an appointment had been agreed in principle by EGS during their former negotiations.

The negotiators were not unduly disturbed by Shaikh Ahmad's demands or his reaction to their arguments and informed their London principals, who telegraphed asking for their urgent assessment of what compromise figures they considered would be acceptable to the Shaikh, that they would seek further meetings with him before doing so. They felt that by going a good way towards the initial payment which he demanded his royalty figure, a much more important one from the Company's point of view, could be reduced to a more reasonable figure.

By 20th March, after two further meetings with Shaikh Ahmad, the negotiators were able to report quite substantial progress towards reducing his monetary demands. To achieve this they had not only to exert all the arts of persuasion but also, at the Shaikh's request to them as old personal friends, to give him their own estimtat (involving figures much higher than the maxima so far authorised to them by their principals) of what they felt he would be justified in asking. On 18th March the Shaikh had had lunch at Chisholm's house with the negotiators, and both then and at an appointment with him on the following morning the amounts, history and backgrounds of his demands were argued strenuously though in the friendliest terms. The negotiators told him that their London office was urgently requesting their own advice on the subject and appealed to him to revise his terms as much as he felt justified in doing, so that they could put proposals to London best calculated to attain a compromise satisfactory to both parties and thus at last honourably end the long argument over Kuwait's oil concession and get it into operation. The Shaikh, on 19th March, at last agreed that he would do so, but only on condition that the negotiators gave him some indication of what they jointly thought might be acceptable to their principals in London. After stressing that they could, in such a matter, only do so as friendly advisers to the Shaikh and not as negotiators for the KOC, they gave as their opinion that if the royalty figures remained as in their previous negotiations (i.e. at or very near Rs. 3—including taxation exemption) it might be possible to approach an initial payment figure of Rs. 400,000 with payments before and after 'declaration' of Rs. 70,000 and Rs. 200,000. Shaikh Ahmad's reception of these suggestions was unenthusiastic, but he stated that he would consider and answer them at a further meeting the next day; when he authorised them to inform London that he would reduce his demands of 12th March to the following:

Initial payment	Rs. 500,000
Before declaration	Rs. 100,000
After declaration	Rs. 250,000
Royalty	Rs. 3/4
Tax exemption	/4

The negotiators, though stressing to the Shaikh their fears that such demands would be regarded as too onerous in London, were encouraged by them to believe that their arguments were beginning to impress him and anticipated further success in the process of reducing his demands.

Reduced monetary terms agreed by Shaikh Ahmad

The next day, 21st March, Shaikh Ahmad agreed after a further discussion, this time at Holmes's house, to reduce his initial payment figure to Rs. 475,000 and Royalty figure to Rs. 3, although at the same time demanding 5,000 gallons of petrol annually before, and 10,000 gallons after, declaration of commercial production.

When reporting these developments to London on 22nd March, the negotiators gave their explanation for the Shaikh being so unexpectedly resistant to their arguments and for his holding out for money terms which were considerably higher even than those which they themselves had advised KOC to expect when in London the previous December. They attributed his attitude to three main causes over and above the normal desire to drive a hard bargain. The intensified activity of the Standard Oil Company of California in Bahrain and Hasa since December might well be causing Shaikh Ahmad to put a higher value than before on the oil prospects of his own territory. They had also learnt that the Political Resident, 'in a well-meaning but unfortunate manner', had referred in conversation with the Shaikh at some time in the previous year to the possibility of such a combine as the KOC being formed 'in order to eliminate competition and squeeze the Shaikh'. Such a remark, they said to a man of his temperament and position, could have been a positive incentive to extract the best possible terms from their company. They also believed that they were 'suffering from the irresponsible advice given to the Shaikh by his Iraqi lawyer, Mr Gabriel'. They were to discover some seven months later that that lawyer's influence had been the biggest factor then working against them.

On 24th March the negotiators received a telegram from their London principals agreeing to cancel the maximum money terms first authorised for their negotiations, and indicating that Shaikh Ahmad's latest demands for the initial and annual payments were negotiable if the royalty figure, to which they attached great importance, could be reduced so as not to exceed Rs. 3 including taxation exemption (comparing with the Shaikh's latest demand for Rs. 3/4). Money terms on this scale must, however, be subject to the assumption that alterations required by the Shaikh in the draft concession's general terms were not of major importance, and the negotiators were requested to inform London as soon as possible their own views as to the money terms the Shaikh was likely to accept.

How best to reply to this request from London presented a problem to the two negotiators, who had come to different conclusions on the subject. Chisholm had visited Shaikh Ahmad on 25th March to discuss his demand to nominate a director to the KOC Board. The negotiators had felt that it would be easier for him to do this than for Holmes, who in his previous negotiations on behalf of EGS had agreed to a similar request by the Shaikh. The main argument used by Chisholm was that in a joint Company like KOC, carefully constructed to secure exact 50/50 participation, it was impossible to have a third party represented on the Board without upsetting the balance, though this would not be so if the Shaikh had a London representative without Board powers but entitled to obtain all necessary information for him from the Company. The Shaikh did not accept this alternative at that time (although he was to do so eventually) and the conversation then turned to his royalty demand, concerning which he indicated to Chisholm that he might consider a figure of Rs. 3 including taxation exemption if all his other monetary and general demands were met. When Chisholm reported this to Holmes, the latter expressed great surprise, forecasting (correctly) that in fact the Shaikh would never lower his Royalty (including taxation exemption) figure below Rs. 3.2 and would probably insist on his then latest demand of Rs. 3.4. Finding that they were not unanimous in their views as to what were the minimum terms which the Shaikh could be expected to accept, they decided to inform London accordingly and give their separate opinions, telegraphing London as follows on 26th March:

> We are not unanimous as to terms acceptable to Sheikh and therefore submit our separate views. Following is from Holmes.
>
> Consider minimum Sheikh will accept is Rs. 450,000 (on signature) Rs. 90,000 (annually before declaration) Rs. 250,000 (annually after declaration) Rs. 3.2 (Royalty plux tax exemption). I believe we could close the agreement on above figures. Policy of increasing yearly payments in order to secure reduction of royalty has its dangers in that we emphasize that we consider low royalty of prime importance and will convey idea to Sheikh that we are prepared to pay handsomely to secure royalty reduction of which fact Sheikh would be ready to take full advantage. It would be unwise to assume at this stage that Sheikh will not insist upon nominating member of Board of Directors, secondly that Sheikh has given up the idea of dealings with other oil concerns. My advice is for the Board either to instruct us to accept agreement on above figures or to cable us definite maximum figures acceptable to Board so that we can use such as basis leaving us to endeavour to secure reduction below maxima if possible. Following is from A. H. T. Chisholm. I estimate minimum terms acceptable to Sheikh as follows: Rs. 425,000 (on signature) Rs. 80,000 (annually before declaration) Rs. 250,000 (annually after declaration) Rs. 2.14 (Royalty plus tax exemption). High initial payment is difficult to avoid owing to past history of these negotiations and developments in neighbouring territories and attempt at appreciable reduction would entail prolonged and difficult negotiations.

For easier reference a table **(104)** shows the position as at 27th March of the terms so far authorised by KOC, submitted, demanded by the Shaikh, and proposed by the negotiators, from the beginning of the KOC's negotiations in February to that date.

Revised terms authorised to negotiators by KOC

KOC's reply to the negotiators was prompt, and as it was to be the basis of their subsequent dealings with the Shaikh, their telegram dated 29th March is quoted in full.

> Glad to note your estimate of terms acceptable to Sheikh of Kuwait are closely approximate. With regard to Sheikh's desire for Director we are willing to give Sheikh letter to the effect that when high production attained he and company shall consider desirability of appointing Sheikh's representative in London with the right to attend special Board meetings when his interests are involved. Subject to agreement on foregoing and on Sheikh's textual amendments we are willing to authorise you to settle on best terms you can secure not exceeding initial payment and annual payments suggested by Holmes if by agreeing to these the Sheikh would accept Royalty plus tax exemption figure stated by A. H. T. Chisholm. We consider royalty much more important question than either initial payment or annual payments therefore greatly favour anything that could be done to keep royalty as low as possible.

Before receiving this telegram the negotiators had had a meeting with Shaikh Ahmad on 29th March at which, as had been forecast by Holmes, he had not confirmed the possibility of reducing his royalty demand from Rs. 3.4 to Rs. 3 as suggested in his conversation with Chisholm on 25th March, although reducing his demand for annual payment before declaration from Rs. 100,000 to Rs. 95,000. As regards his appointment of a Director he said that, once his right had been admitted, further arrangements could be left for later discussion; he must also be consulted regarding the Company's appointment of its Chief Local Representative. Holmes and Chisholm, on that evidence, were pessimistic as to the possibility of negotiating on the level of royalty required by London, but less so as to the matter of his Director's appointment.

When they had communicated the Shaikh's latest attitude on the Royalty to London, the latter on 6th April increased their authorisation for it to Rs. 3 from their previous Rs. 2.14, and it was on that basis that the negotiators next approached Shaikh Ahmad. This was not to be until their interview with him on 11th April, destined to be a crucial date for the future course of the negotiations.

Since their last interview on 29th March, although they had met Shaikh Ahmad several times on social occasions there had been no discussion of the main outstanding question of his monetary demands (regarding which the negotiators were engaged in obtaining London's instruction until 6th April). But as the Shaikh had sent them on 23rd March a long list of proposed textual alterations in the concession draft, the negotiators had been discussing and simplifying these at his request with Mullah Saleh before referring them to London on 5th April with their recommendations for acceptance or amendment. Although none of them were of particular importance or proved eventually to present much difficulty to negotiate, the full list as received from the Shaikh is annexed **(105)** to show the careful and detailed way in which the whole text of the agreement, quite apart from such major aspects of it as the monetary terms and the Shaikh's representation on the KOC Board in London, were handled by Shaikh Ahmad and his advisers and lawyer.

The negotiators were by now beginning to be concerned lest the Shaikh, who they reported on 5th April as by then not having yet been shown the text of the Political Agreement of 5th March between the British Government and KOC, would query them regarding its possible relevance to their own agreement with him. They knew that its text had been received by the Political Agent on 24th March, but the Political Resident had not yet authorised its presentation to the Shaikh. When that authorisation was at last received on 10th April, they had already received from the Shaikh on the previous day a request to be informed if there was to be any reference to the Political Agreement in their own concession document, whose draft in his possession did not mention it.

Negotiators request interview to discuss proposed revised terms and political clauses

In response to his request, and having been assured by the Political Agent that full details of the Political Agreement would be in the Shaikh's possession by the time that he received their answer, they replied on 10th April **(106)** informing him of the two additional clauses which their London principals had asked them to add to their draft concession document in accordance with the terms of the British Government/KOC Agreement; and requested an interview with him on the following day, 11th April, to discuss that matter and also the replies from London which they had by then received regarding the other points raised in their last interview on 29th March, i.e. his monetary demands and the question of his representation in London.

When the Shaikh received the negotiators on 11th April, they therefore requested his consideration of three separate major points:

(1) the proposed additional clauses to their concession document concerning the Political Agreement.
(2) London's proposal to meet his desire to appoint a KOC Director by a letter agreeing to the appointment of a London representative at a later date.
(3) London's reply to his financial terms namely that his other figures could be agreed on condition that he would reduce his demand for Royalty plus taxation exemption of Rs. 3.4 to not more than Rs. 3.

As was Shaikh Ahmad's custom at such interviews, he requested the negotiators to give their fullest reasons and arguments for the points under discussion before giving them his replies, and the two negotiators did so at some length. Whereafter the Shaikh, after a pause for reflection so unusually long that both negotiators, when comparing notes after the interview, found that they had both then realised from their previous knowledge of his ways that a very unfavourable reply was about to be given, gave his views on all three questions with even more than his usual air of dignified determination.

Shaikh refuses proposed terms, and political clauses; and states he has been offered better terms by another British company

As regards the Political Agreement, Shaikh Ahmad said that he saw no reason why the KOC concession agreement should contain any reference to it at all. The negotiators' proposed second additional clause was particularly inadmissible, as it meant that anything in his Agreement with KOC might be overruled by a Political Agreement arrived at without his knowledge. Such clauses as those proposed by the negotiators in a concession Agreement which would of course become public property when concluded, were beneath his dignity to consider and he would never agree to them. As regards the matter of his appointment of a Director, he would not accept the Company's proposal and insisted on his right to appoint a Director to the KOC Board from the commencement of the concession. As regards the financial terms, he could not accept the Company's suggestion, and his final terms for all payments, including Royalty, remained as he had already told the negotiators on 29th March. There was no point in discussing these matters further, said the Shaikh, for he had given a full hearing to the negotiators' arguments and, while of course appreciating their sincerity, they too must appreciate his own point of view and inform their principals accordingly. There was another aspect of the matter which he asked the negotiators to take into serious consideration and communicate to their London principals. They must realise that, although as he had often told them he would prefer to grant his oil concession to their Company, he was being pressed by other companies to deal with them. For instance, only the previous day he had received a long telegram from London on behalf of a 100 per cent British Company asking him if he would negotiate an oil concession with them and offering better terms all round than KOC. Despite this, he still preferred KOC and would especially like to give this concession to such old friends of his as its joint negotiators. Therefore, if they would prepare a concession draft containing the various points and terms for which he had asked, he would gladly sign it; but it was useless to continue discussions of any reductions or changes in his demands, and he would confirm this position by letter to them that same day.

Before leaving the Shaikh, the negotiators had reminded him of a previous promise that, if any London Director or other representative there was agreed by the Company, he would not appoint anyone to the position for three months after the effective date of the concession. That point was therefore included in his letter which the negotiators received as follows later in the day:

My friends Major Holmes and Mr Chisholm
Representatives, Kuwait Oil Co. Ltd.

After greetings,

In reply to your letter of 10th April 1934 and about the discussions which took place between us this morning, I have pleasure to inform you with regard to the two articles mentioned in your letter under reference, that we cannot agree to them.

As regards the Director that we wish to appoint in London, should this appointment be postponed until three months after the effective date of the Agreement, as requested by you, we would not mind it.

As regards the other points that we have already explained to you, we feel their insertion necessary, therefore when you have done accordingly please submit your Concession Agreement for our observation and we shall sign it if we would find its terms agreeable and acceptable to us.

In conclusion may God preserve you,

Yours sincerely,
Ahmed Al Jabir as-Subah

The negotiators were impressed by the firmness and seriousness with which the Shaikh had conducted this interview on 11th April and rejected their proposals and arguments; and even more so by his confirming his attitude in such blunt terms by letter on the same day. When telegraphing the outcome of their interview to London on 12th April they attributed his abruptness partly to preoccupation with discussions in which he was then involved with the Iraqi and Saudi Arabian Governments, but mainly to his dislike of having any reference at all in his oil concession agreement to the British Government/KOC agreement and, still more, of any open admission that the latter was 'subordinate to and controlled by' the former. The Political Agent, shortly after their interview with the Shaikh, had told them that the latter had also informed him that he refused to have any clauses relative to the Government Agreement incorporated in his KOC Agreement; and they recommended in their telegram to London that this question was one which would obviously have to be settled with the Shaikh by the political authorities, being informed by return that the British Government would take up the matter with the Shaikh forthwith. As regards the Shaikh's adamant attitude over the Royalty figure, and his appointment of a Director, they believed that some diminution of the former might be negotiable, and that a non-Board representative might be eventually accepted instead of the latter, once the difficulty over the Political Agreement was solved.

Shaikh Ahmad's statement regarding another British competitor for concession disbelieved by KOC and H.M.G., although in fact true

As regards Shaikh Ahmad's statement that a 100 per cent British company had recently telegraphed him proposing to open negotiations with him for an oil concession offering better terms than KOC, the negotiators subsequently discovered that the message had come from Iraq, not London as stated by the Shaikh. Reporting this to London they stated that it might be from British Oil Development Co. Ltd. (see Note 111), which at that time was developing an oil concession in Western Iraq, obtained from the Iraq Government in 1932. But otherwise this intimation by the Shaikh that an unknown new competitor might be ready to outbid them, was not taken seriously either by the negotiators, or by KOC in London (who soon ascertained that BOD had not approached the Shaikh), or by H.M.G. to whom KOC immediately referred it, especially as the subject was not mentioned again by Shaikh Ahmad. H.M.G. and their representatives in the Gulf, believed that the Shaikh's statement had been an exaggeration, or more probably invention, designed to extract better terms from KOC. Their belief was reinforced when, on H.M.G.'s instructions, the Political Agent on 23rd April officially reminded the Shaikh of his treaty obligation not to negotiate any oil concession without the prior consent of H.M.G., who also expected him to consult them before even opening any such negotiations, without any reaction or further reference by him to the subject.

But, in fact, it was a genuine warning by Shaikh Ahmad of a matter which was to develop in a very awkward way six months later from mid-October onwards. Up till then, neither the negotiators, nor KOC, nor the political authorities in Kuwait or London knew of the existence of this potential competitor for the Kuwait oil concession. Only Shaikh Ahmad knew of it. As it was that knowledge which underlay both his firm attitude at the KOC negotiators' interview on 11th April and his uncompromising insistence thereafter on having his full demands met by KOC, an account is necessary at this point of how this unexpected intervention by an unknown competitor in the negotiations had come about.

Traders Ltd., the new and unknown competitor

The '100 per cent British Company' referred to by Shaikh Ahmad on 11th April was Traders Ltd. **(107)**. Registered in London two years previously, its owners were a group of very substantial British companies the nature of whose business, and the new company's own registration date, were of special significance in Traders Ltd.'s approach to the Shaikh for his oil concession.

The three and a half years' efforts by Standard Oil Co. of California, Gulf Oil Corporation, and the U.S. State Department from November 1928 onwards to secure from H.M.G. the 'open door' for American oil companies first in Bahrain and then in Kuwait, had aroused great interest in British political and commercial circles. Their successful culmination on 6th April 1932 when, as already described, the British Cabinet decided in favour of the 'open door' for Gulf Oil Corporation in Kuwait caused corresponding interest among a group of right-wing 'imperialist' Members of Parliament led by Lord Lloyd (see Note 109) and like-minded British businessmen who believed, and were not backward in expressing their views on the subject, that Britain should not share, even with their (already oil-rich) American friends, such oil resources as might be discovered in those Arabian States in the Gulf which Britain alone had fostered and protected for so many years. At the same time, the successful operations of the Anglo-Persian Oil Company in Persia and the Iraq Petroleum Company in Iraq were adding to the interest aroused in British trading and commercial circles by the international petroleum industry's increasing attention to the oil possibilities of Arabia.

As one consequence of the politico-commercial climate thus created, a group of British companies including oil refining, shipping and marketing interests decided to make a joint bid for the Kuwait oil concession, if it otherwise appeared likely to be obtained by American or partly American interests. For that purpose they formed Traders Ltd., which was registered in London (with an initial authorised capital of £1,000, of which £210 was issued) on 5th April 1932, the day before the British Cabinet decision was taken in favour of the 'open door' in Kuwait. As will be shown later **(108)**, it was to be Lord Lloyd **(109)** the new company's main political protagonist, who was eventually to disclose to H.M.G. in November 1934 the existence and purpose of

Traders Ltd. and its negotiations with Shaikh Ahmad. But until then neither H.M.G. nor their political representatives in the Arabian Gulf nor the Kuwait Oil Company (who were only made first aware on 12th December 1934 by H.M.G. that they had this new and formidable competitor) knew anything of Traders Ltd.'s activities. These, as from April 1934 and especially, as will be seen, from October onwards, were seriously to affect the final stages of KOC's negotiators' discussions with Shaikh Ahmad.

Traders Ltd. **(110)** was wholly owned and controlled by Hunting & Son Ltd. (tanker owners, oil merchants, and brokers) in association with Charles Tennant, Sons & Co. Ltd. (merchants) and Berry Wiggins & Co. Ltd. (oil refiners). Both Mr P. L. Hunting, head of Hunting & Sons Ltd., and also of Traders Ltd., and Lord Glenconner, head of Charles Tennant, Sons & Co. Ltd., were also directors of British Oil Development Ltd. **(111)**, which in 1934 was developing an oil concession in Western Iraq. Berry Wiggins & Co. Ltd.'s Middle East representative at that time was Lt-Col W. J. Bovill who previously, from 1919 onwards, had held various appointments in the military and civil administration in Iraq, and he became secretary of Traders Ltd. on its formation in 1932. Thus the new Company's directors were well acquainted with the oil-concession situation in the Gulf area, and it was on the strength of Bovill's previous experience there that in 1934 it appointed Mr J. Gabriel (see Note 70) of Basrah as its local representative to conduct negotiations in Kuwait with Shaikh Ahmad. Mr Gabriel, Armenian by birth, was a leading Iraqi lawyer in Basrah whose partner had been a colleague of Bovill's in 1921 in the civil administration of Iraq. By 1934 he had been for many years Shaikh Ahmad's legal adviser, making his appointment as Traders Ltd.'s representative an exceedingly shrewd one. For in that capacity he had been intimately concerned in the Shaikh's oil-concession negotiations with KOC and previously with APOC and EGS/Gulf, having been continually consulted by Shaikh Ahmad on their terms and documentation. As the Shaikh was eventually to reveal to H.M.G. in December 1934, it was Gabriel who had informed him in April 1934 that Traders Ltd., a 100 per cent British company, was prepared to offer him terms for his oil concession better than those so far offered by KOC and including such advantages as the directorship which KOC was not willing to concede.

Such was the real background of Shaikh Ahmad's warning to the KOC negotiators on 11th April 1934 that a British competitor was ready and willing to offer him better terms than theirs; a warning for which neither KOC nor H.M.G. could find any justification and therefore continued to ignore until the existence of Traders Ltd. was disclosed to them seven months later. It was the Shaikh's secret knowledge of this hidden weapon in his armoury which enabled him to resist from April onwards the KOC negotiators' efforts to persuade him to compromise on his final demands until eventually, with one minor exception, he obtained them in full in December.

Negotiators propose their return to London for discussion

When telegraphing to London on 12th April the outcome of their previous day's interview Chisholm and Holmes expressed the view that, unless Shaikh Ahmad's financial and other demands were conceded and the references to the Political Agreement omitted from their concession document, no further progress could be made by them in Kuwait pending settlement by H.M.G. of the Political Agreement question, which they anticipated would take a considerable time. In which case they suggested that they should arrange to suspend their negotiations for two or three months, while they returned to London to discuss the position of the various issues still outstanding between KOC and the Shaikh. These, in addition to the major matters of the Royalty and his appointment of a Director, included the several other amendments **(112)** which he had requested on 23rd March and to meet which the negotiators, on the basis of subsequent discussions, had incorporated alterations in a redraft of their concession document which had been airmailed on 9th April for consideration in London.

Apart from the prompt reply that H.M.G. would discuss and settle with Shaikh Ahmad what, if any, references to the Political Agreement would be contained in their concession document, or how the two should be linked, the negotiators received no reply from London until 5th May regarding their proposals for meeting the Shaikh's demands or returning to London for discussion. Their concession redraft incorporating the Shaikh's proposed amendments had arrived in London on 14th April and between 18th and 26th it, and the possibility of dissuading the Shaikh from his demand for a Director and of reducing his Rs. 3.4 demand for Royalty (including tax exemption) to Rs. 3 or below, were the subject of a voluminous exchange of letters and telegrams between KOC in London and the negotiators. Such matters as the Shaikh's wish to be consulted over the appointment of the Company's Chief local Representative in Kuwait provided for in the concession document (an appointment reserved for H.M.G. approval in their Political Agreement with KOC), a visit by him to London as KOC's guest, his request for a labour recruiting officer, his demand that imported food supplies should not be among the items covered by the company's tax-composition payment and his proposed free petrol allowance were among those whose pros and cons were closely examined without any instructions to the negotiators about what line they should adopt to meet the Shaikh's ultimatum, as they increasingly regarded it, of 11th April. Without such instructions, as they reiterated to London, they could not again approach Shaikh Ahmad, and on 26th April, when telegraphing their news of the position reached by the Political Agent in his negotiations over the Political Agreement with the Shaikh (who remained adamant in his refusal to have it mentioned in the concession document) they emphasised their increasingly difficult position. In view of the Political Agreement impasse, they anticipated that it would be several weeks at least before they could resume discussions with Shaikh Ahmad satisfactorily, and as meanwhile their inactivity was embarrassing both to him and to themselves and to public opinion, they telegraphed urgently for either instructions or permission to fly home for discussions. When this elicited only a suggestion from London that a meeting with KOC representatives might be arranged for them in Cairo, pending the outcome of H.M.G.'s argument with the Shaikh or perhaps a suspension of negotiations during the approaching hot weather in order to review their proposed textual alterations and prepare another redraft, but with no guidance on the major points which they knew Shaikh Ahmad was impatiently awaiting, the negotiators' reply on 29th April showed that their own view of the position was a very different one.

Until the political atmosphere had been cleared, they telegraphed, there was no point in meeting in Cairo which in any case would not take appreciably less time than a visit to London with its direct contacts with the KOC Board and H.M.G.; also a London visit would impress the Shaikh as indicating a serious effort by the negotiators to adjust his differences with their Company. In any event the redrafting of textual alterations should be deferred until political agreement had been reached and outstanding major differences settled. In the negotiators' opinion the Shaikh would persist in his demand for the exclusion of the political clauses and also for all his monetary conditions, though Chisholm was more hopeful than Holmes that his royalty demand could be slightly decreased if all his other stipulations were met. The hot season must not be used as an excuse to delay the negotiations. To do so would create distrust in the Shaikh's mind, as the summer was by no means intolerable in Kuwait. The political clauses, in which Shaikh Ahmad's pride was now involved, were the crux of the position and the deciding factor as to whether the Shaikh's signature to the concession would be obtained without endless delay or not. If they could be omitted from the concession document and be separately agreed between Shaikh Ahmad and H.M.G., the present tension would be relieved and an atmosphere of give and take engendered when, subject to the Shaikh's monetary conditions being 100 per cent or 99 per cent accepted, there were good prospects of all other issues, including his request for a directorship, being satisfactorily settled.

Shaikh Ahmad requests urgent reply to his ultimatum of 11th April

This telegram from the negotiators was followed three days later, while it was still under consideration in London, by another showing that Shaikh Ahmad also was now chafing at their lack of response to his ultimatum of 11th April. For on 1st May the negotiators received from him a curt communication, regarding which they telegraphed to London next day:

> We have received letter from Sheikh of Kuwait as follows: 'I have informed you of the terms and additions which I require in the Concession and the amount of payments and royalties which are final and I cannot reduce or alter any of them. Up to now I have not received any reply and as the matter requires urgency I forward this letter that you may consult your Company urgently telegraphically and inform me of your Company's opinion as I see no good in delaying.'
>
> We feel this letter is due to our continued inactivity since interview 11th April. Request you telegraph us as soon as possible either your final reply to Sheikh's demands or to arrange with Sheikh for us to confer with you over outstanding differences before final decision taken.

The peremptory tone of this letter from the Shaikh must have owed much to his secret knowledge, as already described, that in Traders Ltd., he had another competitor for his concession ready to offer the terms which he was demanding from a reluctant KOC. If KOC had shared that secret they would doubtless have closed with him at once, thereby saving themselves eight further months of fruitless negotiating. Even as it was they realised the need for prompt action, though they were still so confident in their negotiating strength as, they imagined, the Shaikh's only bidder that they were not yet prepared to discontinue the efforts to compromise on his royalty and directorship demands; and on 4th May at last telegraphed specifying new terms on which the negotiators were authorised to resume discussions with Shaikh Ahmad where they had been left since 11th April.

KOC's new instructions to negotiators fail to alter Shaikh's major demands

Following a conference in the India Office on 3rd May (they informed the negotiators) they were of the opinion that political matters would be adjusted to the satisfaction of both the Shaikh and H.M.G. without any extensive delay. After exceedingly careful consideration of the terms demanded by the Shaikh for his concession they could not agree to his appointment of a director of KOC, but as an alternative they would agree to his appointing an official representative in London whose salary and expenses not exceeding Rs. 1,250 per month KOC would pay. In their considered opinion, royalty (including taxation exemption) of Rs. 2.14 was about the highest which would enable Kuwait Oil to compete advantageously in available markets with oil from other countries. Provided that the Shaikh would agree to that royalty figure they would be willing to agree his demands for payment of Rs. 475,000 on signature and Rs. 95,000 and Rs. 250,000 annually thereafter before and after declaration of commercial production respectively. While it would be difficult (for official protocol reasons) to invite Shaikh Ahmad to visit London as a KOC guest, they would be willing when payment on signature of the concession was made to pay him also Rs. 25,000 for the expenses of any trip he might desire to make. They wished to avoid an obligation to supply the Shaikh with free petrol, but were willing to increase any annual sums payable to him under the concession by Rs. 5,000 annually instead. They were unwilling to agree his request for a Kuwait labour recruiting officer at KOC expense, as his local representative already provided for in the concession proposals would be available to advise the company on such matters, and they required the Company's imported food-supplies retained among the other imports covered by their tax-exemption payment. If the Shaikh agreed to all thefore going, they believed that there would be no difficulty in agreeing all the other alterations required by him in the text of the concession. They were airmailing to the negotiators a redraft of the text embodying all their suggested amendments and meanwhile the negotiators were authorised to present these revised terms to Shaikh Ahmad.

On receiving this telegram Chisholm and Holmes requested an interview with Shaikh Ahmad on 8th May, when a long discussion of London's new suggestions took place. In their less important aspects, they were accepted or only slightly amended by the Shaikh. He agreed to the proposed contribution of Rs. 25,000 for travelling expenses, and also to accept Rs. 5,000 yearly instead of free petrol supplies, but only until declaration of commercial production whereafter 10,000 gallons yearly should be supplied. And he accepted KOC's views as regards a Kuwaiti labour broker and for imported food supplies being covered by the tax-exemption element in his royalty. But on the major questions of the royalty and directorship he refused to make any concession or even to hear any arguments, saying that his demand had been final and that KOC must either take it or leave it. As the negotiators realised he was upset by the complex discussions proceeding between himself and the Political Agent concerning the political aspects of their agreement, they considered it inadvisable to try to argue the point and adjourned the interview after arranging to meet again shortly. During the following week they had several meetings with Shaikh Ahmad, at which they were able to make progress in securing his agreement to most of the minor amendments contained in London's airmailed redraft of the concession text, which they had by then received; and also in the major matter of his directorship. This was because London's covering letter of 4th May, enclosing the redraft, added some important details, not included in their telegram of the same date, regarding the Official Representative in London proposed by KOC for the Shaikh in lieu of a Director. These were that such a representative would be entitled to be present at all meetings of the Directors convened to consider any question arising between the Shaikh and the Company regarding the concession agreement, as well as to receive reasonable information concerning all matters affecting the Shaikh's interest under the agreement. On that basis Chisholm and Holmes were successful in persuading Shaikh Ahmad at their meeting on 16th May that at any rate in the early years of the concession a director was unnecessary for him, and he agreed to accept such a representative (though stipulating a salary of Rs. 2,000 monthly instead of London's proposed Rs. 1,250) for twelve years or up to 'declaration' whichever was the shorter period, after which however he still required the right to appoint a director.

Despite the now more favourable atmosphere which the negotiators mentioned when reporting to London the latest details by letter and telegram on 16th and 18th May, Shaikh Ahmad proved immovable on the royalty question, refusing even to discuss any reduction from his demand for Rs. 3.4 (including taxation exemption) to or near London's proposed Rs. 2.14. On this crucial point, and its handling, Chisholm and Holmes in their letter of 16th May stated as their joint opinion:

> we feel that the decision must shortly be taken either to accept the Sheikh's demands, or to offer him a lesser figure with the abandonment of the negotiations as the only alternative if he refuses. It is useless to attempt further reasoned arguments with the Sheikh as we have already put them to him many times.

Two days later, however, when telegraphing their views on the latest developments to London on 18th May, the negotiators were to show themselves for the first time to be in disagreement (see below) as to KOC's best future strategy.

Meantime their joint letter of 16th had reported good progress, in the Shaikh's discussions with the Political Agent, towards solving the political problem by embodying the clauses relating the H.M.G./KOC Agreement to the KOC concession in an exchange of letters between the Shaikh and H.M.G. which KOC would also endorse. This solution, eventually to be accepted by all concerned, had originated with Shaikh Ahmad himself after he had rejected three other forms of solution put forward for his consideration by H.M.G. The negotiators also enclosed in their letter of 16th May a new redraft, based on London's latest draft but embodying all the Shaikh's minor and major amendments as provisionally agreed in his discussions with Chisholm and Holmes during the previous week.

Shaikh Ahmad demands royalty on tonnage 'won and saved' not 'exported or sold'

Apart from the major differences still remaining over the royalty and directorship none of these amendments were destined to cause substantial difficulty during the rest of the negotiations with one important exception. This was an entirely new demand by Shaikh Ahmad on a matter never previously discussed since the negotiations began, that his royalty should be payable *not* on oil tonnage 'exported from or sold in' Kuwait, the formula used in the concession document originally presented to him on 17th February and all subsequent discussions, *but* on oil tonnage 'won and saved' in Kuwait, the formula used by both APOC and EGS/Gulf in their previous Kuwait proposals, and also in the current Persian, Iraq, Bahrain, and Saudi Arabian concessions. This new and unexpected request was attributed by the negotiators to the influence of the Shaikh's lawyer Mr Gabriel, who they knew had been again called to Kuwait to give his opinion on their latest proposals and who was also known to base most of his advice on precedents culled from the Iraq and Persian oil concessions. They were probably right in their assumption, but their annoyance at Gabriel's influence on their affairs would have been much greater had they known the important role which he was already playing behind the scenes, as Traders Ltd.'s representative (see page 55), in Shaikh Ahmad's calculations.

In their joint letter of 16th May to London, Chisholm and Holmes had written that they were shortly telegraphing their reactions to the latest situation as requested in London's letter of 4th May, which had stated that failing the Shaikh's complete agreement with London's own latest proposals, they would be glad to know if the negotiators still felt that a conference in London or Cairo was desirable.

Their telegram, despatched accordingly on 18th May, stated first the main results of their negotiations since 8th May and that full details of these, with a concession document redrafted to show all points as far as they had been able to agree them on a provisional basis with the Shaikh, had been posted to London the previous day. They went on to say that as they considered that they could get no further by discussions or arguments with Shaikh Ahmad, they would state their views separately as to the best course now to pursue, this having been considered desirable by Holmes. In Chisholm's view, unless London could agree all the Shaikh's financial and other demands as incorporated in the latest redraft, *either* the negotiators should be withdrawn to London for consultation as they had recommended in their telegram of 29th April, *or* London should instruct the negotiators to make a definitely final counter-offer stating that if it was not accepted they must break off negotiations. In Holmes's view KOC should *either* accept the Shaikh's demands as embodied in the latest redraft in their entirety, which course was the one he recommended, *or*, if KOC could not do so, they should neither accept nor reject the Shaikh's demands but instruct the negotiators to arrange with Shaikh Ahmad to suspend negotiations for two or three months (during which the Shaikh would undertake not to enter into negotiations with others) explaining to him that, with several political questions outstanding and hampering free discussion of the concession agreement, it was preferable for these questions to be disposed of before continuing negotiations, as by then any differences that might exist could be considered and adjusted without any third party influence. It would not, stated Holmes, be good policy for both negotiators to leave Kuwait before making some such arrangement, as to leave the field open would be to court trouble; Chisholm and he had reached a point where further discussion with Shaikh Ahmad was inopportune and he considered rejection of any of his demands would mean a long delay as to procure his consent to any considerable mitigation of them would prove most difficult.

Holmes apprehensive of the risk of there being another bidder for the concession

The reason for Holmes's disinclination, compared with Chisholm, to force the issue with the Shaikh, and for his insistence on the danger of either continuing to bargain with him or leaving the field open for another bidder, is plainly indicated by a telegram he sent two days later to Gulf Oil Corporation's representative in London, Major Davis. In this he expressed his conviction that, if the Shaikh's demands were not accepted unaltered

> the risk of bringing in outside applicants is very great. There is no doubt about it in my mind that the applicants who telegraphed the Sheikh that they would give better terms than us are a serious menace, and such an offer has definitely made a deep impression on the Sheikh. He holds the fixed idea that better terms are possible from the crowd of the above telegram. My opinion is that alterations desired in the Agreement would be more rapidly secured after signature operating under provisions of its Article 17 [this provided for alterations in the Agreement by mutual consent of its two parties] than to obtain the smallest revision before signature. Also that if you reject the Sheikh's latest draft its terms will be immediately increased against us. Continuous political activity creates a most unfortunate atmosphere for our negotiations especially as the Sheikh considers he has his back to the wall, straining to prevent his position being diminished, resulting in his stubbornness being unlimited.

There can be no doubt that Holmes's appreciation of the situation at that time was more perceptive than Chisholm's. It is not possible, as later events including his own correspondence will show, that Holmes could have been then aware of the actual existence of the rival bidder already so well placed in the Shaikh's favour through its representative, and his own legal adviser, Mr Gabriel. But some sixth sense, resulting from his eleven years' close acquaintance with Kuwait affairs and Shaikh Ahmad's personality, must have prompted him that there was danger in the air. There were also other more personal motives both for his views and for his growing disinclination, then first evidenced, to work smoothly in double harness with Chisholm as an equal co-negotiator.

Then aged sixty with a record of remarkable achievements behind him, while for Chisholm at thirty-two these negotiations were his first big assignment, to secure the Kuwait Concession had long been the final goal of Holmes's career, both for prestige and for more mundane reasons. Not a wealthy man, since EGS's first agreement in 1927 with Gulf Oil Corporation to try to obtain for them a Kuwait Concession he had counted on his share of their rewards for its successful negotiation and also on the Shaikh's directorship in its concessionaire company, for which he knew he was to be nominated by Shaikh Ahmad, to be the financial bulwarks of his old age. But first the intervention of APOC and Chisholm's arrival in 1932, followed by the alliance of Gulf Oil Corporation with APOC in 1933 and their consequent settlement with EGS, involving an unexpectedly small payment in respect of the cancellation of their obligations under the 1927 agreement, had disappointed some of his hopes. And now KOC were resisting the Shaikh's demand for a Director and offering instead only moderately paid representation in London. Furthermore, if his instinct was correct and another bidder for the Shaikh's concession was actually in prospect, the Kuwait Concession, once apparently so certainly in his sole grasp and recently even more certainly so as a co-negotiator for KOC with Chisholm, might perhaps be lost by KOC and, with it, his prospect of the Shaikh's directorship or London representation.

There is evidence in contemporary archives that considerations of this sort were then exercising Holmes's mind and that Chisholm was aware of them, a situation which began to generate a tension between the two men which was to be intensified, as described later, towards the end of their joint negotiation. Chisholm was beginning to believe, as also were KOC in London quite independently, that Holmes was less whole-hearted in trying to reduce the Shaikh's royalty and directorship demands than he himself was on behalf of KOC, half of which was owned by his own company while Holmes himself had no such direct stake in the success of the negotiations apart from any favours to come from his old friend Shaikh Ahmad. Holmes in his turn had become suspicious (as he was to inform the Political Agent seven months later **(113)**) that APOC were inclined to reduce his reputation with his KOC employers, because of a telegram sent by KOC in London on 10th May to the negotiators (who in fact had decided to ignore it). This informed them that KOC had arranged for APOC's legal adviser in Basrah (who by coincidence was a partner in the same law firm as Mr Gabriel; see page 55) to be available for the Arabic translation of revisions in their concession draft which hitherto, as arranged by Holmes, they had been sending to Cairo to the firm which for many years he had employed on similar work for EGS. This reasonable suggestion, meant only to save valuable time in the then uncertain state of the Kuwait–Cairo air service, Holmes suspected of implying antagonism to himself on the part of APOC.

KOC regards Shaikh's 'won and saved' royalty formula as unacceptable and queries proposed salary for London representative

On 24th May KOC in London telegraphed their preliminary views on the negotiators' letter and telegram of 16th and 18th; adding that as these also had to be considered by New York, final views and instructions could not be sent until early June. In their view, KOC could not possibly accept a 'won and saved' formula for royalty, especially as the Shaikh's letter of 1st May, and their own response to it, had been based on the 'export and sales' formula as contained in their concession document, which had been in his possession since 17th February and never since questioned. Shaikh Ahmad's stipulations in respect of the Company's Chief Local Representative and his own representatives in both Kuwait and in London were open to serious objections, while the proposed salary for his London representative was too high; 'have you any idea for whom the Sheikh is creating this lucrative position? Would it not be well to provide specifically that he should be a Kuwait subject?' (These questions, and their implications, were particularly resented by Holmes.) They favoured Chisholm and Holmes remaining in Kuwait until the negotiations were successfully concluded. They expected the political issues would shortly be settled between the Shaikh and H.M.G., and they did not believe that he would decline to compromise the few remaining differences in view of the generally very favourable terms already offered to him which they were confident that no responsible outside party would better.

The negotiators were disappointed at London's disagreement with their views and the probability that they would shortly be instructed to reopen the bargaining process which Shaikh Ahmad's letter to them of 1st May had so categorically closed. Holmes was especially apprehensive that this would seriously damage their future prospects of reaching agreement with him. They expressed their feelings in an urgent telegram to London on 27th May, stressing that if after a further ten days' wait they had to confront the Shaikh with a refusal of his terms it would merely harden his resistance, and again requesting permission to return immediately to London for consultations before any final decision was communicated to them for the Shaikh. The psychological effects of such a move would be to assure him that a maximum effort was being made to appreciate his point of view, and it was essential to remember that owing to the mixed political and commercial aspects of the negotiations Shaikh Ahmad had become distrustful of London's attitude towards him. KOC in London merely telegraphed on 28th May that the negotiators should not leave Kuwait and to expect further instructions in about a week's time, meanwhile sending their comments and answers in detail regarding the various points, including the Shaikh's ideas for his London representative, raised in London's telegram of 24th May.

Shaikh Ahmad informs negotiators that failing complete acceptance of his demands he will break off negotiations until September

While Chisholm and Holmes were still trying to decide how best to meet this awkward request, involving as it did matters on which they each held very differing opinions, their dilemma was solved on 30th May by an urgent message from Shaikh Ahmad. They had not discussed business with him since 16th May, though they had sent him a message on 24th that London's answer to his demands could not be expected before 5th June at earliest. But he now requested them to visit him on 2nd June, a visit which, as their following telegram of 3rd June to London shows, was radically to alter in their favour their argument with London as to future procedure:

> As requested by Sheikh of Kuwait we interviewed him yesterday. He informed us that if your reply to his demands does not constitute full acceptance as in his last draft and financial terms, he does not propose to discuss concession further with us until September. He definitely instructed us in that case to return to London forthwith and stated that in our absence he will undertake not to entertain or discuss any offer from other parties. He says there is no need to return before September or later if circumstances justify. We asked his reasons for this move and he said it was:—
>
> (1) to avoid prolongation of negotiations into hot weather which is uncomfortable for him and for us, further he has other pressing business to attend to, which our negotiations are holding up.
>
> (2) because he is convinced points now in question can only be settled by us personally in London, system exchanging letters and telegrams having proved unsatisfactory to both parties.
>
> In our opinion Sheikh Ahmad's attitude confirms what we have telegraphed on 27th May. Sheikh is most anxious to come to an agreement with KOC but feels you do not appreciate his position which can only be rectified by verbal discussion with us. Owing to our long-standing relations with Sheikh personally he believes we will advocate his reasonable demands. He has therefore broken off negotiations with us and refuses to further talks on details concession until we have represented his views and his reasons to you and brought back your final reply with full powers to conclude. If Sheikh's instructions as above are not complied with it will render our position impossible. Therefore it is our opinion that failing full acceptance last draft you do not present any new suggestions until you have discussed position and Sheikh's ideas with us in London.
>
> Early instructions requested.

In confirming this telegram in their weekly airmailed report to London on 6th June, Chisholm and Holmes added that they had been informed by the Political Agent that the matter of the political clauses had been satisfactorily settled between the Shaikh and H.M.G., on the basis of an exchange of letters; and that the related problem of the Shaikh's interest in the appointment of the Company's Chief Local Representative was also nearing satisfactory solution.

KOC misgivings about Holmes

The negotiators' telegram caused misgivings among their KOC principals in London, not so much because of the Shaikh's latest ultimatum, but because of its terms. These, they considered, were so remarkably identical with the attitude and advice which they had been receiving but refusing to adopt from the negotiators, and especially with Holmes's advice, that they were inclined to suspect that the latter had taken advantage of his great influence with Shaikh Ahmad and instigated his action in order to force their hand. These suspicions, in line with their previous feeling that Holmes might be behind the Shaikh's insistence on a London director or highly paid representative, were intensified by information from two other separate sources. H.M.G. had informed them that the Political Resident in the Gulf who, when KOC had originally been formed, had advised them not to make Holmes one of their negotiators because of his known commitments and intimacy with the Shaikh, had recently reported his belief that the latter's obstinacy was being encouraged by Holmes, whom he recommended should be withdrawn, leaving Chisholm to complete the negotiations. APOC had also received reports from Qatar, where their representatives were then discussing an oil-exploration licence and where Chisholm was well known because he had visited there twice in 1933 to initiate those discussions, that it was common gossip both there and in Bahrain that the Kuwait negotiators were in difficulties, 'because Holmes, who was the Sheikh of Kuwait's friend, encouraged him to demand high terms, which made it very difficult for The Tall One' (the latter was Chisholm's Arab nickname, Holmes's being The Father of Oil, a title well-earned by his successful discovery of oil in Bahrain). Against such considerations, however, London had to balance Holmes's great influence with the Shaikh and the reactions of both if their relationship was questioned.

London's initial reaction was to telegraph the negotiators on 8th June not to arrange their return to London pending full instructions which would shortly be sent. On the same day the Gulf directors of KOC agreed that Holmes should be recalled for consultations leaving Chisholm in Kuwait, but telegraphed instructions to Kuwait including that decision were delayed until 12th June by the necessity of consulting both New York and H.M.G. on all aspects of the next steps to be taken.

By that time, however, Shaikh Ahmad had intervened again on 10th June, when he summoned the negotiators to meet him and asked them why they were still in Kuwait and what answer London had sent regarding their last meeting with him eight days previously. As they telegraphed the same day to London, when they replied to him that no instructions had yet arrived doubtless because every avenue was being explored with a view to meeting his demands, the Shaikh had stated most emphatically that it was no case for exploring avenues; failing complete acceptance now of all his demands he would not discuss anything at all with them until after the summer. The negotiators therefore requested London to be informed by return if complete acceptance was being considered and, if not, requested permission to return to London on 14th June. As no air passages were available they would travel by the fastest alternative, which was by train from Basrah to North Iraq, thence by car to Turkey, thence by train via Istanbul to London, an eight-day journey. In a separate telegram Holmes said that the negotiations were deadlocked, as the Shaikh demanded unqualified acceptance or rejection of all his demands and until he received one or the other it was hopeless to remain in Kuwait; the Shaikh insisted KOC should have first-hand information regarding his attitude from Chisholm and himself, and Holmes intended to leave Kuwait on 14th June for London and stay there as long as circumstances necessitated; he considered Chisholm should return also.

London acknowledged the negotiators' telegram on 12th June in acid terms which showed their dislike of the situation confronting them. Unqualified acceptance of Shaikh Ahmad's demands was simply out of the question, they stated, and as the points of difference were so few and so minor from his own standpoint they were wholly unable to understand his point of view as expressed in the negotiators' recent telegram. They could only conclude for example that there must be some reason not made clear to them for his insistence upon a representative and director in London with such powers as demanded. If the Shaikh's insistence on the negotiators' return to London implied expectation that they would induce KOC directors to unqualified acceptance, any such result was not to be expected. However in all the circumstances they thought it desirable for Holmes to return to London for the psychological effect, if any, on the Shaikh and to clarify KOC's views for the negotiators. Chisholm should remain in Kuwait and before Holmes left the negotiators must jointly remove from Shaikh Ahmad's mind any misconceptions regarding the probable results of his discussion with KOC directors. To leave him under the impression that Holmes's trip might result in unqualified acceptance would only make matters more difficult when Holmes returned and advised him otherwise. They desired the negotiations to be continued and concluded without further interruption than Holmes's trip to London and quick return to Kuwait would necessitate.

Shortly after London had sent this telegram they had second thoughts on the advisability of leaving Chisholm in Kuwait without Holmes. To do so would not only ignore the Shaikh's request for both negotiators to leave Kuwait for the summer, but also might, in the Gulf directors' view, have the bad effect of giving him an impression that KOC differentiated (as they in fact had now come to do) between the credibility of their two negotiators; also Chisholm's views, which evidently differed in some respects from Holmes's, were desirable when considering the latter, and as both negotiators would have to be given very full new instructions on several complex issues, both should be given equally first-hand briefing.

Negotiators return to London for instructions

A further telegram was therefore despatched to Kuwait instructing Chisholm also to return to London and consequently, after the two negotiators had briefly interviewed the Shaikh on 14th June, Holmes, whose passage was already booked, left for London that afternoon, Chisholm following him three days later. This farewell interview with

Shaikh Ahmad was formal but friendly, with hope expressed by both sides that when they met again after the summer their remaining differences would be resolved and the way open for a successful conclusion of their long negotiations. When they were driving together to the Palace Holmes had told Chisholm that the interview had better be kept short, as the previous evening he had 'had a long talk with Shaikh Ahmad, nothing serious but just background chat, so we need not bother him today with any business'. They agreed, however, that they should get the Shaikh to confirm what he had previously said to them on 2nd June about his suspension of their negotiations, and have it recorded in writing for future reference; which they did, and on 16th June Chisholm addressed a formal letter on the subject to the Political Agent in terms which, as they were to form part of the controversy which arose in the following December between Shaikh Ahmad and H.M.G. must be quoted in full:

> I have the honour to inform you that my colleague Major Frank Holmes left here for U.K. on 14th instant, and I myself am leaving for U.K. tomorrow 17th instant. These movements are the result of interviews with H.E. the Sheikh of Kuwait who has expressed the desire that, as he and our Company are not yet in complete agreement on the final terms of a Concession, further discussions should be postponed until after we have consulted our principals in London. His Excellency was good enough to state on 14th instant that for three months as from that date he would not receive or consider offers for an oil concession from any party than ourselves.

Holmes arrived in London on 22nd June and Chisholm on 25th. It is interesting to note that the American Legation in Baghdad, in one of their routine reports **(114)** to the U.S. State Department on oil-concession affairs in Kuwait and elsewhere in the Gulf (where the U.S.A. in those days had no official representation) reported on 25th June that Holmes, Chisholm and Dickson, when passing through Baghdad *en route* from Kuwait to London, had all given different reasons for the Shaikh's suspension of negotiations. Holmes, the legation reported, attributed it to British Government interference; Chisholm to APOC and Gulf not authorizing certain points required by Shaikh Ahmad; Dickson to the Shaikh's requiring more remunerative terms than those already offered.

Chisholm and Holmes had a first meeting with the KOC board in London on 2nd July, and for the next three weeks were involved in continuous discussions with them or separately with the APOC and Gulf directors. In view of the Traders Ltd. situation, as already described, it is of interest that at the first meeting both Chisholm and Holmes expressed their conviction that Shaikh Ahmad was not negotiating with any other company although he certainly had been approached to do so. Also that when on 6th July the Political Agent (Lt-Col Dickson), who with his family had arrived in England on leave, was entertained by the KOC directors, he expressed the same conviction, adding that Shaikh Ahmad (with whom his long and close acquaintance rivalled or even excelled Holmes's, with the great additional advantage of his being an exceptionally fine Arabic linguist) was 'a man of great integrity, whose word was his bond, and intolerant of sharp practice'.

By the end of July, the KOC directors' discussions with Chisholm and Holmes, and also with H.M.G., had sufficiently covered all aspects of the situation to enable them to put in hand for the negotiators, before their expected return to Kuwait in September, yet another draft of the concession proposals and comprehensive briefs and guidance as to how to achieve its acceptance by Shaikh Ahmad. There was no doubt that by then all political questions would have been satisfactorily settled between H.M.G. and the Shaikh so far as the H.M.G./KOC political agreement and its relationship to his concession negotiations with KOC were concerned. As regards their concession document, KOC had decided that although the Shaikh's directorship could not be conceded and extreme efforts must be made to persuade him that to reduce his royalty demand and accept the 'export and sales' formula were in his own long-term interest, both those major points could ultimately be conceded if, but only if, they became the only two obstacles to a complete agreement. As all the minor points required by the Shaikh could be either accepted or only very slightly modified, both the KOC board and the two negotiators felt that nothing more remained for discussion towards setting the scene for success on their return to Kuwait. On 18th July Chisholm and Holmes had jointly telegraphed **(115)** Shaikh Ahmad from London noting that one month had elapsed since they had left Kuwait, looking forward to renewing his friendship on their return, and saying that a letter would follow with their news. Their jointly signed letter, despatched on 20th July, in addition to the usual salutations and personal greetings, included the following:

> We arrived here after a comfortable journey and without delay began discussions with your friends of the KOC in London. The English and American Directors of the Company participated in the discussions which have been very helpful. We are preparing things so that we can be ready to return to Kuwait when we receive your command to do so. When that time comes, we sincerely hope and believe that an agreement will be reached between you and our company. If you have any other commands for us, besides those which you gave us when we left Kuwait, we shall be very glad to receive them.

Towards the end of July Holmes and Chisholm went on a month's leave, with instructions to be available for further discussions with the KOC board from 1st September onwards. During their absence Shaikh Ahmad's letter in reply to theirs was received by KOC; dated 1st August 1934, it read as follows:

> Dear Friends,
>
> I have received with many thanks your kind letter dated 20th ultimo bring me your good news of your safe arrival in London in the best of health, and I am very pleased to know that the agreement will be done accordingly. I have nothing to add at the present except what I told you about. I am looking forward to your proceeding to Kuwait in the appointed time.
>
> Yours sincerely,
> Ahmad Al Jabir Al Sabah

On the strength of this correspondence with the Shaikh, and with everything so far as could be foreseen arranged to secure his early agreement on their concession proposals, the KOC board in early September embarked on the final stage of the two negotiators' London briefing and documentation for their return to Kuwait in optimistic mood. But at exactly the same time in Kuwait Shaikh Ahmad made an important move in his secret dealings with Traders Ltd. which, had they known it, would have confounded their optimism as thoroughly as it was to complicate and confuse the negotiations when it eventually came to light in mid-October.

Shaikh Ahmad's letter of 2nd September to Traders Ltd.'s representative Gabriel

On 2nd September 1934 Shaikh Ahmad addressed the following letter to Mr Gabriel in his capacity as Traders Ltd.'s representative in Basrah:

> I received two copies of the concession, retained one and returned you the other, that you may despatch same to the Company and inform them to communicate with His Britannic Majesty King George's Government in London and produce to them the copy of the concession; that if His Majesty's Government consent to it, they may telegraph their assent to the Political Agent in Kuwait.
>
> We have accepted the conditions of the concession, and I will affix my signature thereunto on hearing the results from the Political Agent in Kuwait to the effect that His Majesty's Government have sanctioned the Agreement of the concession and assented thereto, provided that you undertake for us that the company agrees to the points undermarked in red pencil in the Arabic version of the exact text of the concession.
>
> Yours sincerely,
> Ahmad Alsabah.

The consequences, and Shaikh Ahmad's explanation, of this letter to Traders Ltd., contrary both to his treaty obligation (recently re-confirmed on 23rd April **(116)** to the Political Agent) to consult H.M.G. before even opening the negotiations which must have preceded it, as well as to his promise to the KOC negotiators on 14th June not to receive or consider oil-concession offers from any other party before 14th September, will be detailed in due course. At this point the Shaikh's action needs only be recorded as the most unexpected twist and best kept secret in the long course of the eleven-years-old Kuwait negotiations; and also the starting-point for their rapid acceleration to a quick climax diametrically opposed to Traders Ltd.'s aspirations.

Chisholm and Holmes leave London on 26th September arriving in Kuwait on 13th October

The three months period of Chisholm's and Holmes's absence from Kuwait during which, as agreed at their last meeting with Shaikh Ahmad on 14th June, he was not to consider oil-concession offers from others, would end on 14th September, and their return to Kuwait had been programmed for about that date. But in the first week of September it became evident during KOC's London discussions with H.M.G. that as the latter's final instructions to the Political Resident in the Gulf regarding his settlement of political details with the Shaikh could not reach him before the end of

September, the negotiators should not arrive back in Kuwait before then. Consequently Major Holmes instructed Mohammed Yatim to return from Bahrain to Kuwait on 10th September (Mr Chisholm's staff were already there getting his house ready) and inform the Shaikh that the negotiators would soon be on their way; and on 18th September, following a final KOC/H.M.G. meeting the previous day, the negotiators telegraphed from London to Shaikh Ahmad:

> With your Excellency's kind permission we propose to leave London on 26th September for Cairo and after a few days there to leave for Kuwait arriving on October 4th. We much look forward to seeing Your Excellency again and enjoying your kindness and friendship. Your sincere friends Chisholm Holmes.

to which he replied on the following day:

> Thank you for your telegram. Welcome. Ahmad.

Chisholm and Holmes had to stay in Cairo **(117)** *en route* to Kuwait because of Arabic translations to be done there of the amended concession document drawn up in London, and of alternative drafts for some of its clauses (such as the proposed 'export and sales' royalty formula, and that concerning a London representative for the Shaikh, for use in their negotiations. Leaving London on 26th September they reached Cairo on 2nd October, after a delay at Alexandria owing to the late arrival of their boat from Genoa. The translations were completed in time for them to leave Cairo for Gaza in Palestine (whence they were to fly to Basrah) on 7th October. While in Cairo London telegraphed them not to arrive in Kuwait before 12th October, because the Political Resident's discussions with Shaikh Ahmad (concerning his last point of difference with H.M.G., namely his interest in the appointment of the Company's Chief Local Representative) would not be finished until then; and they had telegraphed the Shaikh in Kuwait accordingly. After a delay in Gaza before they could get air passages and accommodation for their baggage (including a large radio installation for presentation to Shaikh Ahmad) they eventually reached Basrah on 10th October, and proceeded thence to Kuwait on 13th.

Holmes learns of Traders Ltd. competition and convinces Shaikh Ahmad that it represents an APOC plot to block KOC negotiations

During the negotiators' three days' stay in Basrah there had been a most unexpected development.

Holmes had heard from a friend of his when launching at the British Consulate of rumours that during his absence a competitor company had begun negotiating in Kuwait with Shaikh Ahmad, offering better terms than KOC and almost reaching agreement. Without mentioning the matter to Chisholm, on his arrival in Kuwait he questioned Mohammed Yatim and his Kuwait friends and advisers (including Mullah Saleh, Shaikh Ahmad's private secretary) on the subject, and was astonished to learn that the rumours which he had heard in Basrah were substantially true. They told him that during the negotiators' absence from Kuwait Shaikh Ahmad had had several visits from Mr Gabriel who was acting on behalf of a company named The Trade Arts Co. Ltd. (Traders Ltd.; see page 54 *et seq.* and Note 139), which was offering higher payments and better terms than KOC for an oil concession. According to Mullah Saleh, the Shaikh had very recently written to Gabriel in Basrah that if the KOC negotiators on their return accepted all his demands he would have no further dealings with Gabriel's company, but if they did not, he would ask for Traders Ltd.'s final terms and if they were satisfactory negotiate a concession with them. The Shaikh had informed Gabriel that he would require to know who were the backers of Traders Ltd., about whom he so far only knew that they were 100 per cent British. Mullah Saleh and his friends according to a long and detailed letter written by Holmes on 15th–16th October **(118)** to Gulf in London reporting the matter, also added two further remarkable (and, as was eventually to emerge, wholly untrue) pieces of information about Traders Ltd. These were that that company was backed and supported by APOC and that Gabriel, its representative and negotiator, was acting under the instructions of Mirza Muhammad **(119)** (APOC's legal adviser in Basrah and a partner in Gabriel's law firm there) who had even, said Mullah Saleh, tried to obtain the services of Holmes's assistant Mohammed Yatim for the new company.

A moment's reflection should have made Holmes realise the impossibility of any such connection as that suggested between APOC and Traders Ltd. Quite apart from its inherent improbability, he was well aware that both APOC and Gulf, under their agreements of 14th December 1933 **(120)** providing for the formation of KOC, were both legally bound not to make any attempt except through or for KOC to negotiate a Kuwait Concession. But it is a measure of his alarm at the existence of a new competitor, his resentment at KOC's refusal since April to accept his advice to close the negotiations and obtain the concession by agreeing all the Shaikh's demands, and his belief that this was due to his being suspect by APOC that he forthwith, as his letter of 15th–16th October to Gulf makes clear, adopted his Kuwait friends' belief that Traders Ltd.'s intervention represented an APOC scheme to double-cross their American partners in KOC and obtain the Kuwait Concession for a wholly British company of their own contriving. 'Now I am certain where the trouble lies,' he wrote in his letter, meaning by this the trouble which he had had for months past in getting KOC to accept his advice that all the Shaikh's demands should be agreed,

> If we are to prevent being double-crossed my opinion is that the terms as per draft 4 should be accepted by the American group... The APOC certainly are praying for a breakdown in our negotiations and living in hopes of Gabriel Co's success. Your Associates are certainly beyond the pale... This letter must be treated as 100% confidential—if the British Government became aware that I had given the game away I would be in a very awkward position... I am confronted with a most difficult task... I have added to my trouble perfidy and deceit and defection from such a quarter. I have not given the slightest hint to Chisholm that I know of the perfidy of his crowd.

In a postscript of this letter of 15th–16th October, Holmes wrote that Shaikh Ahmad had just then visited him at his house and, when asked about Gabriel's company, confirmed their approaches to him, adding that he had avoided finding out who were behind Traders Ltd. because he wished to keep an open mind until KOC accepted or rejected his demands. In a further postscript Holmes wrote that a member of the Shaikh's Council had just confirmed the position reached by Traders Ltd. and had promised to keep him fully informed of further developments. A letter shown to Holmes which had been received that day from Gabriel by the Shaikh, requested him to begin negotiations with Traders Ltd., as KOC's negotiators had contradicted their promises to return in September and Traders Ltd.'s terms had been agreed with H.M.G. who had sanctioned their negotiations (statements, as will be shown later, which were completely untrue). The Shaikh, wrote Holmes, was replying in a letter which Holmes himself had drafted, that H.M.G. had not communicated anything to him about Gabriel's company either before or after Chisholm's and Holmes's return to Kuwait on 13th October, and that he could not consider negotiating with Gabriel's company until their negotiations had ended unsatisfactorily. 'This letter,' wrote Holmes, 'is 100 per cent confidential: if the British Government become aware that I had given the game away, I would be in a very awkward position.'

Thus, on 16th October 1934, Holmes came to adopt the extraordinary (and wholly baseless) theory which was to govern his conduct during the rest of the negotiations, namely that APOC, while ostensibly negotiating with Shaikh Ahmad through KOC in good faith with their Gulf partners, were simultaneously and secretly double-crossing the latter by having him offered better terms by Traders Ltd. He also convinced Shaikh Ahmad of the existence of such a scheme and of the necessity of their working together, secretly from Chisholm and the British political authorities (both of whom he considered must be privy to such an APOC scheme) to prevent it and thereby thwart what he claimed to be a covert and last-minute APOC attempt to get the Kuwait Concession into their own 100 per cent British hands instead of those of the British/American KOC. Why Holmes should have taken this extraordinary course of action, instead of at once informing his co-negotiator and KOC of the existence of their new competitor (whose true nature would thereupon have been immediately elicited from Shaikh Ahmad by H.M.G., as it was eventually to be two months later) is difficult to explain. It probably resulted from his first reaction to two highly unpalatable facts; that for the first time he had been outmanoeuvred by Shaikh Ahmad, and that the intrusion of an obviously powerful new competitor meant at best a long extension of KOC's negotiations and at worst their failure, involving both a defeat in what had looked like being the crowning success of his career, and the loss of the Shaikh's KOC representation or directorship in London which meant so much for his personal fortunes. At such a juncture Holmes can well have reasoned that only prompt and drastic measures could clinch the concession for KOC quickly enough to eliminate Traders Ltd.; that only his own exceptional influence and intimacy with Shaikh Ahmad could provide those measures; and that if they had to entail some temporary deviousness and deception of his co-negotiator, his KOC employers, and the Shaikh, these could be well justified by snatching immediate victory for KOC from the jaws of delay or defeat, while they would also achieve for Shaikh Ahmad not only his full demands from KOC but also something with which Traders Ltd. could obviously never provide him, namely the American participation in his concession for which he had so long schemed. As the remaining course of the negotiations will show, Holmes's plan of action, however devious, completely succeeded in achieving these objectives.

Negotiators present new draft concession to Shaikh Ahmad

Having arrived in Kuwait on 13th October, the two negotiators paid their first call on Shaikh Ahmad on 15th. He returned their calls on the following day when they were also his guests for dinner, and next morning, 17th October, at their first business interview, presented the new draft concession (see Note 117). Shaikh Ahmad said he would discuss it with them after he had studied it, but this was unlikely to be before the following week.

These events, and their information from the Political Agent that the Shaikh and H.M.G. had reached agreement on the clause relating to the Company's Chief Local Representative in the concession, were detailed in the negotiators' weekly airmailed report to KOC in London dated 17th October **(121)** signed as always by both of them. This report, in accordance with Shaikh Ahmad's and Holmes's agreement the previous day (because of Chisholm's supposed involvement in the supposed APOC/Traders Ltd. plot) to work together in secret from Chisholm and KOC, made no reference to the existence of any competitor, which had been similarly ignored both at the two negotiators' interview with the Shaikh and at their dinner party with him the previous evening. And although during the following two weeks Holmes's private discussions with the Shaikh and his advisers were actively continued, no mention of them appeared in the negotiators' joint communications to KOC in London. Their brief airmailed report of 24th October **(122)** stated that the situation showed no development since 17th October, and that Shaikh Ahmad had not indicated what his reaction was to be on the draft which they had then presented to him; it also briefly noted that Colonel Dickson, on return from leave, had resumed his duties as Political Agent as from 18th October (an event which was later to assume some significance). Their even briefer, only eight-line letter of 31st October **(123)**, again regretted that the negotiators

> had as yet no development in our negotiations to report. The Sheikh indicated that he would see us today to discuss our latest draft, but he has now put us off until Saturday 3rd November. We hope to telegraph you some news of the situation after seeing him then. So far we have been able to gather no indication whatsoever of his intentions.

Although Holmes's signature was added as usual to Chisholm's under that final sentence, it was very far from the truth so far as he was concerned. For a few days previously, on 25th October **(124)**, at a private interview with Shaikh Ahmad, with only Mohammed Yatim present as interpreter, he had reached (see below) what seemed to be a satisfactory compromise on the Shaikh's two major demands as yet unsatisfied in KOC's draft concession, namely his right to appoint a Director and a 'won-and-saved' royalty basis.

Holmes's theory that Traders Ltd. represents an APOC plot regarded as incredible by Gulf

Meanwhile, Holmes's letter of 15th–16th October to Gulf in London, detailing his discovery of Traders Ltd.'s unexpected intrusion into the negotiations and his theory that it represented an attempt by APOC to double-cross their Gulf partners in KOC had been received there on 23rd October and immediately sent on to Gulf in New York where it arrived on 31st October. There his theory was at once dismissed as too fantastic to be credible. Gulf wrote to their head office in Pittsburgh on 1st November **(125)** 'We cannot believe APOC would take the steps suggested.' Holmes's attitude was regarded as yet another example of those exhibitions of his headstrong temperament with which they had become familiar in the course of their acquaintance with his negotiating technique since 1927. They did not doubt that some competitor might have made overtures to the Shaikh, as he himself had told the negotiators in April, but as any such competitor was bound to declare itself initially to H.M.G. and had not yet done so, no serious threat from such a quarter could, they thought, exist, while Holmes's own description of the Shaikh's continued preference for KOC pointed in the same direction. This opinion was further strengthened on 2nd November when

New York received from Gulf in London a telegram (see Note 124) reporting the receipt there on 25th October by Stevens of an important cable from Holmes in Kuwait. This said that at a private interview that morning with Shaikh Ahmad the latter had said he would not 'permit the English to stab the Americans in the back'; he had given up his demand for a Director, and requested Holmes to ask Gulf to give up in return their objection to his demand for a 'won-and-saved' royalty basis which, according to Holmes, the Shaikh had promised King Ibn Saud that he would have in his concession, on the same lines as in the concession granted by the latter in 1933 to the Standard Oil of California. Shaikh Ahmad had promised Holmes, at the same interview, not to make any deal with Gabriel's company, and Holmes suggested that Gulf had better not raise the matter of that company with APOC (to which suggestion Stevens had cabled agreement on 26th October).

This information from Holmes reassured Gulf that whatever threat might have existed from any so far undiscovered competitor had been successfully parried by Holmes and also that his anti-APOC suspicions had subsided. With the obstacle of the Shaikh's insistence on a director removed, and with only two substantial differences remaining, the royalty basis and amount (on both of which, as they had instructed the negotiators in London, they were willing to yield in the last resort) they considered that a favourable termination of the negotiations was at last imminent. This view seemed to be confirmed when the negotiators telegraphed KOC in London on 4th November **(126)** the results of their interview with Shaikh Ahmad when at his request they had visited him on the previous day 'for the first time since 17th October'; a phrase which, though correct as regards any joint visit by the negotiators, clearly ignored Holmes's discussions with the Shaikh during the previous three weeks in which, unknown to Chisholm, he had made such progress towards solving their remaining problems.

It is remarkable that in the next six weeks of the negotiations the existence and nature of Traders Ltd.'s competition, of which Gulf had been so categorically informed in Holmes's letter of 15th–16th October, was not mentioned by them to their APOC partners in KOC; which was only to become aware of its existence when H.M.G. disclosed it to them on 12th December. It is less remarkable that they did not even then, or indeed ever, inform APOC either of Holmes's extraordinary theory that Traders Ltd.'s intervention could have represented an APOC attempt to double-cross their KOC partners or of his consequent single-handed and successful dealings with Shaikh Ahmad. For not only had Gulf come to believe that Holmes had abandoned his theory, as a result of which in any case he had successfully advanced KOC's case with the Shaikh, but they must also have felt that to disclose his suspicions would have been to endanger the harmonious co-operation existing in the KOC board and apparently also between the two negotiators in Kuwait. But much more remarkable is the fact that never, even after the negotiations had ended and during their subsequent friendly personal acquaintance only ended by Holmes's death in 1947, did Holmes mention to Chisholm his APOC-plot theory of October 1934 or his consequent dealings with Shaikh Ahmad. There is some evidence that he came to regret and tried to redress the damage which he had thereby done to Chisholm's personal relations with the Shaikh which, though he continued outwardly cordial until his death in 1950, were never the same after October 1934. It was only to be in 1970, when the Political Agent's letter of 28th December 1934 to the Political Resident **(127)** came to their notice in the British Foreign and Commonwealth Office archives, that either APOC (now BP) or Chisholm were to learn of Holmes's curious theory and actions in October 1934.

Shaikh Ahmad disappointed with KOC proposals and gives negotiators his alternative terms

When telegraphing to London on 4th November the results of their previous day's interview with Shaikh Ahmad (Note 126), and in their fuller airmailed report on 7th November, the negotiators reported him as expressing considerable disappointment, after careful study of their draft, at its failure to meet so many of his demands as discussed before they had gone to London. He was especially surprised that he was still being offered less as regards the royalty basis than had already been offered both by APOC and EGS/Gulf in their previous negotiations with him. He intended therefore to send them a written reply, in which he would again make plain his requirements, and also detail certain new demands which he would have to make if he was to accept the 'export-and-sales' royalty basis proposed by KOC instead of the 'won-and-saved' basis which he wished for. When the negotiators tried to persuade him that their proposed basis was in the best interests of both parties, the Shaikh replied that he could not understand how they could reconcile this with their previous preference for a 'won-and-saved', and it was eventually agreed that Chisholm should visit him on the following morning to explain the KOC position more fully. When Chisholm did so, although Shaikh Ahmad did not recede from his position, the negotiators reported that he showed by his questions and remarks a good understanding of the technical points involved and also 'gave quite a fair reason for his preference for a won-and-saved basis'. Apparently 'won-and-saved' as translated into Arabic gave an impression more like 'won-and-stocked', so Shaikh Ahmad had in mind three destinations for oil produced from the wells, namely stocks, exports and sales. Probably with recollections of the large stock and storage tank-farms which he had seen at Abadan, he felt (wrote the negotiators in their report to London on 7th November) that, if he agreed an 'export-and-sales' basis, he might be deprived of royalty on a conceivably very large quantity of oil stocked in Kuwait. As regards Chisholm's argument that refining operations in Kuwait could be penalised or cramped on a 'won-and-saved' basis by its non-allowance for refining wastage and recycling to reservoir, the Shaikh had said that to meet this point his letter would propose that a deduction from royalty paying tonnage might be allowable for oil returned to reservoir up to a certain percentage of production. He asked Chisholm what he thought of such a proposal, to which the latter replied that any such percentage would be most difficult to compute as the quantities involved would depend on market conditions from time to time.

The real reason for Shaikh Ahmad's preference for the 'won-and-saved' basis, as Holmes, but not Chisholm, had known since 25th October **(128)**, was that he had promised King Ibn Saud to obtain it, but this was not mentioned by the Shaikh to Chisholm or in the negotiators reports to London. Pending receipt of Shaikh Ahmad's promised letter defining his position, they reserved further comment on his attitude apart from saying that the other major points of difference, namely his demands for Rs. 3.4 as royalty (including tax exemption) instead of Rs. 3 offered by KOC and for a director in London instead of the representative offered by KOC, had not been raised by the Shaikh, the former probably because no payment figures had been included in their draft concession presented to him on 17th October. As regards his demand for a director, Holmes alone knew that Shaikh Ahmad had already agreed to waive it on 25th October.

The Shaikh's promised letter **(129)** received by the negotiators on 7th November, made his position clear on all these major points of difference between himself and the KOC negotiators.

Shaikh Ahmad wrote that, after fully considering the negotiators' arguments for the altered terms of their last concession draft as compared with the draft containing all his requirements which they had discussed with him in June and taken to London for consultation with KOC, he had come to the conclusion that further discussions with them would not be to the interest of either party. Therefore, as he wished to bring matters to an end, he was now making two alternative suggestions, either of which would be acceptable to him if agreed by KOC, regarding the major matters of difference between them; there were also a few minor points, but these were mostly agreeable to him. *Either* the royalty basis should be 'won-and-saved', with a royalty of Rs. 3.4 (including tax exemption), with payments as already agreed of Rs. 475,000 on signature, Rs. 95,000 minimum p.a. before and Rs. 250,000 after declaration of commercial production; *or* the royalty basis should be 'exports-and-sales' in which case royalty (including tax exemption) should be Rs. 3.6, with payments of Rs. 600,000 on signature and Rs. 100,000 p.a. minimum before, and Rs. 300,000 p.a. after declaration of commercial production. As regards the question of the Shaikh's appointment of a director in London this would be dealt with if and when KOC had accepted either of these suggestions.

When telegraphing the terms of this letter to London on 8th November the negotiators commented that they were now evidently faced with the necessity to settle the remaining financial aspects of the concession (i.e. both the amount of royalty and its basis), whereafter all other outstanding issues should present no difficulty. If the Shaikh's 'won-and-saved' royalty basis was accepted, they considered that he might possibly be persuaded to accept a royalty (including tax exemption) of Rs. 3.2 (it is significant in this connection that Rs. 3.2 was the royalty being offered to the Shaikh in Traders Ltd.'s concession document **(130)**, as Holmes was then aware). But they warned that if KOC's final decision was that figure or less they must be prepared for the risk of rupturing the negotiations. In conclusion, they requested London to telegraph them authority to negotiate on a 'won-and-saved' basis and also their definitely final maximum royalty figure.

London replied on 12th November that full instructions would be telegraphed shortly, and in their weekly report airmailed on 14th this was acknowledged by the negotiators, who stated that they had informed Shaikh Ahmad accordingly, and hoped that London's telegram would enable them at last to reach complete agreement, as he obviously wanted to conclude the negotiations before Ramadan began in the first week of December. Despite the guarded reference to the director question and certain other points in his letter of 6th November, the negotiators felt that these would present little difficulty once the financial issues were settled. A proviso in his letter that, if royalty was to be paid on an export-and-sales basis, the Company's returns of oil to the reservoir must be restricted to a maximum 7½ per cent of production, was one, wrote the negotiators, which they had previously understood that he might attach to his demand for a 'won-and-saved' basis. This, they considered, was because the Shaikh had become extremely suspicious, through bad advice, of the export-and-sale basis as being one which might enable KOC to exploit his oil with maximum profit to themselves, and minimum to him. For instance, they believed he had been told that on an export-and-sales basis, the Company by refining in Kuwait would be able to export and sell only the higher-priced fractions of his crude oil, returning all the balance to the reservoir, thus gaining large profits and paying a minimum royalty.

London accepts Shaikh Ahmad's royalty and royalty-basis terms

London's telegram of instructions was received on 18th November, when the negotiators wrote to inform Shaikh Ahmad, adding that they believed there should now be no further delay in bringing the negotiations to an early conclusion as requested in his letter on 6th November, and asking him to appoint a time for them to visit him. Their optimism was because London had at last accepted both the 'won-and-saved' royalty basis requested by the Shaikh (and urged on Gulf by Holmes in his telegram on 25th October) and also that the royalty including tax exemption should be Rs. 3.4. As regards the 'won-and-saved' royalty basis London's telegram assumed that this would allow for both refining losses and recycling to reservoir in accordance with the relevant draft clause **(131)** brought from London by the negotiators. Although that text had not yet been discussed by the negotiators with the Shaikh who, as pointed out in their letter of 14th November to London, was suspicious of such matters, they considered that such a minor issue could hardly delay an early end to their negotiations now that all of Shaikh Ahmad's major demands were in sight of settlement.

Shaikh Ahmad received the negotiators on 22nd November, when they gave him their news from London; which he received with satisfaction, though with less enthusiasm than might have been expected at such a complete realisation of his negotiating aims. This was only partly because he had already been told by Holmes privately about London's decisions; the main reason was that he was beginning to feel uneasy about his dealings with Traders Ltd. in secret from H.M.G., as is indicated by a letter dated 17th November **(132)** (see page 71) from the Political Agent to the Political Resident in Bushire. The interview soon terminated, after the negotiators had given him a copy of their draft clause **(133)** relating to a won-and-saved royalty basis which he said he would examine and then inform the negotiators within a few days by letter of his decision and also of minor alterations which he wished to have included in the concession text. It was not however until 9th December, nearly a fortnight later that this letter from Shaikh Ahmad was sent to the negotiators, specifying his final demands. And meanwhile, from 19th November onwards, important developments were taking place in London following the first disclosure there to H.M.G. on that date of Traders Ltd.'s existence and activities since April.

These must first be described before returning to events in Kuwait.

H.M.G. informed by Lord Lloyd of Traders Ltd.'s activities

On 19th November Lord Lloyd, Traders Ltd.'s main political protagonist (see page 54) in its attempt to obtain the Kuwait Concession for a wholly British company, who a month previously had publicly criticised both H.M.G. and APOC 'for not keeping Arabian Gulf oil under 100% British control', had informed G. W. Rendel, one of his friends in the Foreign Office **(134)** that Traders Ltd. (whose formation and intentions in Kuwait he described) had successfully, according to his friend Mr Hunting its Chairman, concluded an agreement with the Shaikh of Kuwait granting them his oil concession, subject only to the approval of H.M.G.

When the Foreign Office communicated this surprising information to the India Office (the British Government department then principally concerned with Arabian Gulf affairs) they immediately telegraphed to the Political Resident at Bushire and the Political Agent in Kuwait describing Lord Lloyd's information and enquiring if they knew anything to confirm it. They were instructed, however, not to broach the matter with Shaikh Ahmad because, as the telegram pointed out 'a serious situation would arise if we find that the Shaikh has without our knowledge been in discussion with a would be concessionaire' contrary to his treaty obligation to H.M.G. The Political Resident replied on 20th November that he knew nothing of the matter, but on 22nd November telegraphed that he had just received from the Political Agent in Kuwait a long letter dated 17th November containing relevant information of importance **(136)**. The gist of this he telegraphed to the India Office, to the effect that the Shaikh had recently told the Political Agent that in August an Iraqi representative of an oil company had approached him with a draft oil concession more favourable than KOC's, when he had replied that the company concerned must first approach H.M.G. and secure their approval to negotiate with him. Although the Shaikh would not divulge the name of the company or its agent, the Political Agent suspected that the company was probably Frank Strick & Co., of Basrah (APOC's agents there) with the Basrah lawyer Gabriel as its representative; if the company was Frank Strick & Co. presumably it was working on behalf of a larger group oil company or associate.

This information from the Political Agent was very shortly to be disproved as regards the identity of the company concerned (though not as regards its representative Mr Gabriel) when Traders Ltd. disclosed its identity and Kuwait activities to H.M.G. on 28th November. But the full text (see Note 136) of the Political Agent's letter of 17th November is also significant for the negotiations in a different way. It contained precisely the same suspicions that Gabriel's company represented an attempt by APOC to double-cross its Gulf partners in KOC, couched in exactly the same terms, as Holmes had expressed in his letter to Gulf in London on 15th–16th October **(137)** two days before the Political Agent had arrived back in Kuwait from leave in England (see Note 122). As the Political Agent's letter mentioned his having heard 'one or two remarks' from Holmes on the subject, there can be no doubt that his suspicions of a possible 'double-cross' attempt by APOC of its Gulf partners in KOC had been instigated by Holmes as part of the latter's strategy from 16th October onwards to thwart Traders Ltd.'s intrusion on KOC's negotiations. Having convinced Shaikh Ahmad on 16th October that there was an APOC 'plot' it was necessary for Holmes to persuade the Political Agent similarly in view of his close relationship and influence with the Shaikh. Holmes's success in doing so is evident in the Political Agent's letter of 17th November; and even more so in his subsequent letter of 28th December **(138)** to the Political Resident.

Traders Ltd. (Hunting and Bovill) visit India Office and describe their Kuwait negotiations

When the India Office in London received the Political Resident's information on 22nd November that Shaikh Ahmad had admitted having been approached in August for an oil concession by some unidentified company, they requested the Foreign Office to ask Lord Lloyd to put his friends of Traders Ltd. in touch with them as soon as possible if, as he had said, they were interested in obtaining an oil concession in Kuwait. Lord Lloyd agreed to do so **(135)**, and on 28th November Mr Hunting and Lt-Col Bovill (Chairman and Secretary respectively of Traders Ltd., see page 55) called at the India Office and discussed their position with Mr Laithwaite, the official in charge of Kuwait affairs, giving full details of their company and its dealings up to that date with Shaikh Ahmad.

Mr Hunting opened the proceedings by presenting a formal letter **(139)** of that day's date from Traders Ltd., signed by himself as Chairman, informing the India Office that, subject to H.M.G.'s approval, Shaikh Ahmad had granted an oil concession covering the whole of Kuwait to themselves, a British company 'entirely owned and controlled by Messrs. Hunting & Son Ltd. and their associated companies.' The letter requested that H.M.G. should formally and officially approve the grant of the concession to Traders Ltd. and that Shaikh Ahmad should be informed accordingly. Attached to the letter, to prove Traders Ltd.'s bona fides and claim to have been granted the concession by the Shaikh, were three appendices described as:

A. A detailed list of Companies and persons associated with Traders Ltd.
B. A copy of the concession terms agreed between Shaikh Ahmad and Traders Ltd.
C. An English translation of an autograph letter dated 2nd September, granting the concession subject to H.M.G. approval, from the Shaikh to Mr J. Gabriel, Advocate of Basrah, local representative of Traders Ltd. who had arranged details on the spot, the whole letter being in the Shaikh's own handwriting and signed by him.

Three facts emerging from these appendices (reproduced in Note 139) are worth noting at this point. Appendix A shows that Holmes's and the Political Agent's suspicions that APOC (or Frank Strick & Co., its Basrah Agents) were behind or connected with Traders Ltd. were groundless. Appendix B shows that Traders Ltd.'s concession terms included Rs. 500,000 on signature and Rs. 500,000 minimum annually thereafter (against KOC's Rs. 475,000 on signature and Rs. 95,000/Rs. 250,000 annually before/after commercial production), and the right to appoint a director. Appendix C was Shaikh Ahmad's letter dated 2nd September, already quoted on page 64 above, but repeated here for easier reference:

2nd September 1934.

I received two copies of the Concession, retained one and returned you the other that you may despatch same to the Company and inform them to communicate with His Britannic Majesty King George's Government in London and produce to them the copy of the Concession; that if his Majesty's Government assent to it they may please telegraph their assent to the Political Agent in Kuwait.

We have accepted the conditions of the Concession and I will affix my signature thereunto on hearing the result from the Political Agent in Kuwait to the effect that His Majesty's Government have sanctioned the Agreement of the Concession and assented thereto, provided that you undertake for us that the Company agrees to the points undermarked in red pencil in the Arabic version of the exact text of the Concession.

Yours sincerely,
Ahmad Alsabah.

After Laithwaite had received and read this letter, he requested further details from Hunting and Bovill, who told him that their dealings with Shaikh Ahmad through Gabriel had been going on since the latter had first approached him on their behalf in April (when, it will be recalled, Shaikh Ahmad had told the KOC negotiators and the Political Agent that a 100 per cent British Company had contacted him). Laithwaite asked them why they had not informed H.M.G. of their dealings with the Shaikh, which they must have known was the required procedure, either then or in September as specifically requested in the Shaikh's letter, and even now had only done so after Lord Lloyd's informal intervention. To have left it so late was so irregular and also, with the Shaikh's protracted negotiations with KOC on the point of satisfactory conclusion without any indication from Traders Ltd. or the Shaikh that he was also negotiating with them, so inconvenient that he was bound to ask Hunting and Bovill for an explanation. They explained that they had deliberately delayed informing H.M.G. to the last moment, being apprehensive that an earlier disclosure of their position would have enabled KOC, through its influence with H.M.G. through APOC and also because of its Anglo-American make-up which H.M.G. policy evidently favoured, to bring irresistible pressures against them on Shaikh Ahmad. Now, however, as they knew that final terms were on the point of settlement between KOC and Shaikh Ahmad, as their own terms were in important respects more favourable to him than KOC's, as they represented reputable British interests, and as Shaikh Ahmad was clearly, as evidenced by his letter, ready to negotiate or even conclude an agreement with them, they felt entitled to request H.M.G.'s approval, as they now formally did, of Traders Ltd.'s negotiations with the Shaikh, and for the latter to be informed accordingly. They also made clear that if their application was refused, their political friends were likely to ask questions in Parliament as to why British interests should be denied the H.M.G. approval and support which Anglo-American interests had so long been receiving.

The interview ended after Laithwaite had replied that Traders Ltd.'s letter and views would now have to be considered in all its aspects by H.M.G.; that this would take a little while because their company and its seven-months-old dealings with Kuwait, which clearly should have been referred to H.M.G. by both Traders Ltd. and the Shaikh at a much earlier stage, were hitherto unknown to H.M.G., and also because the Shaikh would have to be consulted and his views ascertained; and that H.M.G. would reply to Traders Ltd.'s letter as soon as possible thereafter.

The position disclosed by this interview and Traders Ltd.'s letter of 28th November was considered during the next few days by the India Office and Foreign Office, who also consulted Lord Lloyd on the company's background and political backing and investigated the financial and commercial status of the company and its associated concerns and supporters. They concluded that, despite the irregularity of Traders Ltd. not having disclosed its contacts and negotiations with Shaikh Ahmad at a much earlier stage, as a British company of good standing and undoubted commercial probity it must be regarded as fully entitled to the support and approval of H.M.G. in conducting negotiations with Shaikh Ahmad *if*, as was essential to discover as soon as possible, he confirmed that company's view that he wished to conduct, or even conclude negotiations with them. Other points for investigation with the Shaikh were why, contrary to his agreement with H.M.G., he had never informed them of his negotiations with Traders Ltd.; and whether his letter of 2nd September to Gabriel was confirmed by him to be genuine and meant, as Traders Ltd. asserted, as granting his concession (subject to H.M.G. approval) to them, apparently contrary to his undertaking to Chisholm and Holmes not to consider concession offers from others for three months from 14th June. Traders Ltd.'s proposed concession terms, as attached to their letter of 28th November (see Note 139) to the India Office, would also have to be thoroughly examined and if necessary amended, in the same way as KOC's, APOC's and EGS/Gulf's had been in the past by H.M.G.'s experts, who had already noted that they were contrary to the Shaikh's and H.M.G.'s interests in several respects.

H.M.G. requests Political Resident to ask Shaikh Ahmad to confirm Traders Ltd.'s account of their negotiations with him

On 4th December the Secretary of State for India telegraphed to the Political Resident at Bushire, detailing Traders Ltd.'s account of their dealings to date with Shaikh Ahmad and instructing him to visit Kuwait as soon as possible and ascertain from the Shaikh whether he confirmed or otherwise Traders Ltd.'s account. He was also to request the Shaikh to explain the circumstances in which he had apparently, without informing H.M.G., been carrying on secret negotiations with Traders Ltd., including writing to their representative Gabriel on 2nd September a letter (which the Political Resident was instructed to obtain and inspect) granting them an oil concession, contrary both to his obligations to H.M.G., of which he had been reminded as recently as 23rd April, and to his promise to Chisholm and Holmes of KOC on 14th June.

This India Office telegram was repeated for information to the Political Agent in Kuwait who, as the Political Resident was prevented by illness from going to Kuwait, was instructed on 12th December himself to make the necessary enquiries from Shaikh Ahmad and telegraph the result immediately to both London and Bushire. The Political Agent accordingly addressed three letters **(140)** to the Shaikh on 14th December conveying his India Office instructions and requesting immediate replies; which were received from Shaikh Ahmad on 15th and 16th and their gist telegraphed **(141)** by the Political Agent to London and Bushire. Their contents throw much light on the complications resulting from Traders Ltd.'s intrusion on the negotiations as from April 1934, but before describing them it is necessary to refer back to Chisholm's and Holmes's dealings with the Shaikh from 22nd November onwards. These had developed so rapidly, especially since 9th December, that by 14th December complete accord had been reached, and on 15th December Shaikh Ahmad had agreed to sign KOC's concession as soon as the necessary number of documents could be prepared and translations obtained from Cairo. On 15th December he had also telegraphed to Gabriel in Basrah that as he had granted his concession to KOC he was consequently no longer questioning in Traders Ltd.'s proposals, thus lessening the urgency of his questioning that same day by the Political Agent and effectually solving the India Office's problem of how best to deal with Traders Ltd.'s belated application for official approval of its Kuwait negotiations.

Shaikh Ahmad's letter to the negotiators with draft concession including all his final requirements

Following their interview of 22nd November with Shaikh Ahmad the negotiators had not expected to have to wait more than a week for his promised letter. When more than a fortnight had passed they informed London on 7th December in reply to the latter's telegrams pressing them for news, that they would make enquiries forthwith and telegraph within a day or two how matters stood; though Ramadan would begin on 8th December, the Shaikh had told them that it would not interfere with their business, and they hoped 'in spite of the unaccountable delay since 22nd November' that his letter would contain nothing to prevent a speedy conclusion.

The 'unaccountable' delay had two good reasons, of both of which Holmes was aware. Firstly, the drafting of Shaikh Ahmad's letter, for which Holmes's assistance had been required, had been unexpectedly complex and difficult. It had to include all the minor points still required by the Shaikh, based on his original list of 23rd March (see Note 105), with others added since by reference to Traders Ltd.'s offer of terms in August. For this Gabriel's legal drafting ability was no longer available owing to his position with Traders Ltd. Also the two major points to be covered in the letter, namely, the Shaikh's decisions regarding the precise nature of his won-and-saved royalty formula and the status and responsibilities of the London representative which he was to accept instead of a Director, were neither of them easy to formulate. Secondly, when the letter's drafting was nearly completed Shaikh Ahmad had received on 6th December a letter from Gabriel in Basrah, informing him of Traders Ltd.'s interview with the India Office in London on 28th November and that he might expect to be queried on the subject shortly by H.M.G.; reminding him of the terms of his letter of 2nd September; and urging him not to be dissuaded from resuming his negotiations with Gabriel as soon as possible. This communication, which the Shaikh was to delay answering until 12th December by when he could reply that further negotiations with Traders Ltd. were rendered impossible by his imminent agreement with KOC, was very disturbing to both him and Holmes. It was evidently desirable to reach complete agreement with KOC before H.M.G.'s questioning of Shaikh Ahmad, as this might result in Traders Ltd. being authorised to continue their negotiations with all that could entail in further delays and difficulties for KOC. This required the Shaikh's letter to KOC not being framed in terms likely to be either refused by KOC or only accepted after further negotiation. But at length on 9th December Shaikh Ahmad's letter **(142)** was finalised and despatched to the negotiators, together with a copy of their concession draft as presented to him on 22nd November amended so as to include all his final requirements.

On 10th December the negotiators telegraphed a précis of the letter to London (followed up by full copies and analyses on 12th December by air letter, destined never to be answered as by the time they arrived the negotiations had ended), and their views and recommendations on the following day. Shaikh Ahmad's letter had agreed to renounce his claim to appoint a director, and to accept instead a London representative as already offered by KOC on 17th October, except that the salary figure was to be Rs. 2,250 monthly instead of Rs. 2,000; as regards the 'won-and-saved' royalty basis, he would not agree to KOC's request that allowance should be made for refining losses and tonnage recycled to reservoir; and he required the right of selection, in consultation with H.M.G., of only the company's first Chief Local Representative in Kuwait. Apart from these three matters, the rest of his letter covered matters of minor importance, but the negotiators stressed to London the significance of its assurance that if the amended draft of the concession was completely accepted they might inform KOC that 'we shall sign it and grant the concession' and especially of a verbal intimation from Shaikh Ahmad (which Holmes told Chisholm he had received) that he wished for a 'speedy conclusion'.

Although the negotiators asked the Shaikh on 11th December for an interview before sending their views and recommendations to London, he replied that it would be impossible to grant one for several days. When telegraphing this to London that afternoon they expressed as their joint view that apart from the conditions of the 'won-and-saved' royalty basis and the Shaikh's interest in the selection of KOC's first Chief Local Representative, all the Shaikh's other alterations were of little importance and settlement of those two questions, now that the Shaikh had agreed to have a London representative instead of a director, would most probably result in their being easily negotiable and complete agreement being reached. As regards the Chief Local Representative, they recommended that the Shaikh's request was acceptable and that KOC should so advise H.M.G., whose previous agreement with him on the subject it only very slightly altered, while in any case the matter was well covered by the H.M.G./KOC political agreement which had been accepted as binding by Shaikh Ahmad. As regards the Shaikh's request regarding the conditions of the 'won-and-saved' royalty basis the two negotiators gave differing recommendations. Chisholm believed it would still be possible to argue the Shaikh into allowing both refinery losses and recycled production to be royalty-free if the rest of his alterations were agreed. Holmes thought KOC would be well advised to accept the Shaikh's demands on that as on the other points, and thus to obtain the concession at once, as he was 'not optimistic as regards obtaining any mitigation of the terms demanded by the Shaikh'. In expressing this view and sense of urgency Holmes must have had prominently in mind the possible consequences of H.M.G.'s imminent questioning of Shaikh Ahmad regarding his Traders Ltd.'s dealings.

The negotiators' telegram was received by KOC in London on 12th December. On the same day Gulf's senior representative and KOC director there, Mr Stevens, received an urgent telegram **(143)** from Holmes. It described his private and confidential interview (one result of which was Shaikh Ahmad's reply to Gabriel that no further negotiations with Traders Ltd. were possible; see page 74) that morning with the Political Agent who, as already related, had just received H.M.G.'s instructions to question Shaikh Ahmad on his dealings with Traders Ltd. The Political Agent, reported Holmes, after telling him of his instructions from London, had said that the interests behind Traders Ltd. were so powerful, and prepared to raise the matter in Parliament if their application for approval of their negotiations were refused, that in view of Shaikh Ahmad's letter to Gabriel of 2nd September H.M.G. would evidently have to approve their formal application. The Political Agent had also advised H.M.G. that KOC should be informed of Traders Ltd.'s activities.

In retrospect it might appear extraordinary that the Political Agent did not inform Chisholm as well as Holmes of the Traders Ltd. situation and its ominous bearing on the prospects of KOC, of which they were the authorised joint negotiators with whom he was in almost daily contact and whose company he was under instructions from H.M.G. to assist. But the reason is clear.

He had been so firmly convinced by Holmes, as shown by his letter of 17th November (see Note 136) to the Political Resident, and as his similar letter of 28th December (see Note 127) was to confirm even more clearly, that Traders Ltd. represented a covert attempt by APOC (and Chisholm on its behalf) to thwart KOC's negotiations, that he must have felt obliged in KOC's interests to exclude Chisholm from his confidence about it. Holmes's successful influence over him in this respect even at this particular moment, when H.M.G.'s information and instructions regarding Traders Ltd. should have removed any suspicion that they could conceivably be connected with APOC with which H.M.G. were so intimately concerned, is remarkably attested in a telegram received by the India Office from the Political Resident on 14th December **(144)**. This reported his having just received a telegram from the Political Agent saying that he was suspicious that Traders Ltd. represented an effort by APOC to block the success of KOC.

KOC learn from the India Office on 12th December of Traders Ltd.'s competition and instruct negotiators to accept unaltered Shaikh Ahmad's terms

Stevens, on the strength of Holmes's telegram, enquired from the India Office on 12th December if they had anything to confirm some recent rumours from Kuwait that another company had applied for H.M.G. permission to negotiate with the Shaikh in competition with KOC. When the India Office replied in the affirmative, Stevens forthwith informed Fraser (see Note 92), APOC's Deputy Chairman and Chairman of KOC, who on telephoning the India Office was greatly surprised (unlike Stevens, whom Holmes had warned on 16th October of Traders Ltd.'s competition, Fraser had no previous intimation of its existence) to be told of Traders Ltd.'s application of 28th November and that the matter was at that moment under urgent reference to Shaikh Ahmad in Kuwait.

The KOC Board in London appreciated at once the urgency of the new situation caused by the appearance of a powerful new competitor, and its vital bearing on their attitude towards Shaikh Ahmad's letter of 9th December, and the negotiators' recommendation of 11th December which they were then considering. They decided to close immediately with the Shaikh by accepting in their entirety all his terms and conditions on acceptance of which, as his letter had informed the negotiators, he would sign and grant KOC the concession. By doing so they would end the negotiations and *ipso facto* remove the otherwise dangerous threat to their interests inherent in the impending possibility of Traders Ltd. entering the lists against them. Having informed the India Office accordingly, an urgent telegram was despatched to the negotiators on 13th December **(145)** instructing them to inform Shaikh Ahmad immediately by letter that all his proposals were accepted and that the negotiators were authorised to sign the concession accordingly. No attempt was to be made to discuss any alterations in the draft text received from Shaikh Ahmad, as KOC's purpose was to accept forthwith the terms upon which the Shaikh had stated in his letter of 9th December that he would sign and grant KOC the concession. The negotiators were requested to express to Shaikh Ahmad in suitable terms the KOC board's satisfaction at reaching agreement, and to inform London when they expected that the concession would be formally signed.

Shaikh Ahmad agrees to sign KOC concession when H.M.G. approval received, and informs Gabriel that no further discussions with Traders Ltd. possible

On receipt of this telegram on 14th December, the negotiators wrote **(146)** as instructed to Shaikh Ahmad, who invited them to visit him next day when they expressed again their London Principals' compliments and congratulations and their own gratification that their lengthy discussions were so satisfactorily ended. The Shaikh expressed his own satisfaction and desire to sign the formal documents as soon as the negotiators could have them prepared. Reporting this to London on 16th December, the negotiators also telegraphed that they had arranged for the necessary documents to be prepared in Cairo and hoped to receive them by airmail on 19th when duplicate copies were also to be sent from Cairo to London. As Shaikh Ahmad would undoubtedly require an assurance from H.M.G. that his signature of KOC's concession had their approval, the negotiators requested London to ensure that this was communicated to him through the Political Resident or Political Agent without delay. If that was done, they anticipated that the signing formalities would take place a day or two after the documents arrived in Kuwait, when they would also deliver to the Shaikh, as agreed on 8th May, letters **(147)** confirming the annual payments to be made to him in respect of petrol supplies and a payment in respect of travel expenses. On 16th December Holmes also sent a private telegram **(148)** to Gulf in London informing them that on his advice Shaikh Ahmad had telegraphed on 15th to Gabriel in Basrah that he had granted his concession to KOC and would therefore have no further discussions with Traders Ltd.

The highly satisfactory news from their negotiators in Kuwait was reported on 17th December at a meeting in the India Office in London by the KOC directors, who requested that H.M.G.'s agreement to his granting their concession should be communicated as soon as possible to the Shaikh so that its formal signing could take place. They were assured that this would be done although, as such a matter required Cabinet approval, a few days' delay would be involved. The India Office representatives also informed the KOC directors of the telegrams which they had received on 16th December from the Political Agent in Kuwait reporting how H.M.G.'s questions on 14th December regarding his dealings with Traders Ltd. (see page 73) had been answered by Shaikh Ahmad; whose interesting explanation of his actions since April must now be related.

Shaikh Ahmad's letters to Political Agent explaining his relations with Traders Ltd. since April 1934

Shaikh Ahmad replied to the Political Agent's three letters (see Note 140) dated 14th December in three letters (see Note 141), one dated 15th and two 16th December. They answered all the points raised by H.M.G. regarding the Shaikh's relations since April with Traders Ltd., covering them so comprehensively that their one omission, namely any reference to his and Holmes's joint discussions about Traders Ltd. and suspicions of APOC regarding them since mid-October, must be supposed to have been due to advice from Holmes. This would have been given on 15th December (see Note 148) when, the Shaikh having granted the concession to KOC and informed Gabriel accordingly, it must have been obvious to Holmes that in future the least said the better regarding those aspects of the matter, which in any cases were unknown to, and therefore had not been queried by H.M.G.

Shaikh Ahmad's letters of 15th–16th December to the Political Agent recalled how, at an early stage in his discussions with the KOC negotiators, his lawyer Gabriel had informed him that there was a British firm that wished to obtain the Kuwait Concession on terms satisfactory to himself (Shaikh Ahmad was here referring to the offer which he had received from a '100% British company' and of which he had on 14th April informed the negotiators and also the Political Agent, as he had reminded the latter in November as recounted in the Political Agent's letter of 17th November (see Note 136) to the Political Resident). Shaikh Ahmad had told Gabriel at that time that he was negotiating a KOC concession with Chisholm and Holmes, and subsequently that the negotiators were leaving for consultation in London, having agreed with him to postpone negotiations 'until the end of August', up to when he would not open negotiations with anyone else. The Shaikh had told Gabriel that, when that period expired, if the company which Gabriel represented was a purely British company, he would be prepared to negotiate with them, but only on condition that their London representatives should apply to H.M.G. for the latter to inform the Shaikh of their approval, without which he could not negotiate or grant a concession. After the KOC negotiators had left for London in June and their agreed time limit had expired 'on 30th August', wrote the Shaikh, he had examined and found suitable two copies of Traders Ltd.'s proposed terms for a concession which had been brought to Kuwait by Gabriel (on 20th August according to his information to the Political Agent). He had thereupon written to Gabriel his letter of 2nd September (of which he enclosed a copy for the Political Agent; showing it to be exactly similar to that which had been shown to the India Office by Traders Ltd. on 28th November). 'Was there,' wrote Shaikh Ahmad, 'anything in the wording of that letter from me to justify the Company in assuming (as they have done) that I had actually granted them the concession?'

The Shaikh's letters denied that he had not kept H.M.G. properly informed of his dealings with Traders Ltd.; and explained why he considered that the three-month period, during which he had promised the KOC negotiators not to consider concession offers from others, had expired on 30th August and not 14th September as recalled by the negotiators and recorded in Chisholm's letter of 16th June (see page 63) to the Political Agent which the latter had shown to the Shaikh.

He had, wrote Shaikh Ahmad, informed the Political Agent as well as the KOC negotiators in April that a British company had approached him; and had also told Gabriel (as above) that his company must inform H.M.G. of their wish to negotiate with him and arrange for H.M.G. approval to be communicated to him, as without it he could not negotiate with them. Again in his letter of 2nd September he had stipulated that Gabriel's company must inform H.M.G. in London of their proposed concession terms, as he could go no further until he had been informed by H.M.G. of their approval. If the Political Agent (who was absent from Kuwait on leave from 18th June to 18th October) had been in Kuwait when that letter was written the Shaikh would have informed him. But the then acting Political Agent did not know enough Arabic to talk with him without interpreters, and as the negotiations were of a very secret nature, 'I kept the knowledge of them to myself', wrote Shaikh Ahmad, 'for fear of being betrayed.' As soon as possible after the Political Agent's return from leave, the Shaikh had told him of the whole matter for H.M.G.'s information (here Shaikh Ahmad was referring to his conversation with the Political Agent as recorded in the latter's letter of 17th November (see Note 136) to the Political Resident).

Shaikh Ahmad wrote that he had carefully told the KOC negotiators, as regards the period of their absence from Kuwait during which he had promised them not to consider concession offers from others, that it would terminate at the end of August. He agreed that he had seen Chisholm's letter of 16th June to the Political Agent recording the negotiators' understanding that the termination date would be 14th September. But he insisted that his own view was the correct one, adding that he had warned the negotiators that if they had not reached agreement with him by the termination date (i.e. by 30th August 1934) he would be compelled to open negotiations with the British company whose approaches he had disclosed to them, and to the Political Agent, over two months previously on 11th April.

Shaikh Ahmad's letter dated 15th December (see Note 141) to the Political Agent, which enclosed a copy as requested of his letter to Gabriel of 2nd September, also stated that he had that day telegraphed to Gabriel that he had come to terms with KOC. He also enclosed, to show his latest position as regards Traders Ltd., a copy of his recent letter to Gabriel of 12th December (see Note 141). This acknowledged a letter from Gabriel of 6th December which had contained information received from Traders Ltd. in London, presumably regarding their interview of 28th November at the India Office; and went on to inform Gabriel that the Shaikh was on the point of final agreement with KOC, Chisholm and Holmes having agreed practically all his demands;

> I cannot now refuse them at this stage. I have been so long alleging pretexts to them for my delaying their question that I have become truly ashamed of the abundance of my promises. All this time I have been waiting for your company to make a move but up to date, unfortunately, I have not received from them anything of a satisfactory nature that I can depend on. It is not I, therefore, who have failed your company by delay.

This was the last communication to be sent by Shaikh Ahmad to Traders Ltd. although, as will be seen, their disappointment was to impel them to pursue him with further correspondence over the next eight months. Two deductions from it may reasonably be made. If Traders Ltd. had not delayed their approach to H.M.G. until the Kuwait Concession had been as good as granted to KOC but, after receiving Shaikh Ahmad's letter of 2nd September to their representative Gabriel, had applied at any time before the KOC's negotiators return to Kuwait on 13th October for H.M.G.'s approval to negotiate for his concession it would undoubtedly have been granted, as the Shaikh was then evidently willing to receive them. Equally undoubtedly, Shaikh Ahmad would sooner or later have rejected their offer in favour of KOC, after having obtained from the latter the increased financial terms necessary to outmatch those on offer from Traders Ltd., as must have been his shrewdly calculated objective after receiving and examining the latter's terms in August. Not only would his eleven-year-old acquaintance with KOC's two owning companies and their negotiators (especially his old friend Holmes) as referred to in his letter of 12th December to Gabriel, have made him do so. There was also the decisive and overriding consideration that only from the KOC could he obtain the American participation in his concession which he had so long and resolutely maintained as a prime negotiating requirement; a consideration which Holmes had turned to such good account when making an imaginary anti-American plot by APOC the main feature of what may fairly be described as his 'shock tactics' from 16th October onwards to thwart Traders Ltd.'s approach to the Shaikh.

Their main points having been telegraphed to London on 16th December, the full texts of Shaikh Ahmad's three letters (see Note 141) of 15th–16th December explaining his dealings with Traders Ltd., were despatched by the Political Agent on 20th December to Bushire and thence forwarded to London. There they were not received by the India Office until 7th January 1935, by which time their interest had decreased following Shaikh Ahmad's grant and formal signing on 23rd December of the KOC concession. They were subsequently considered by H.M.G. to provide an insufficient explanation for his not having given them earlier and more direct information of his relations with Traders Ltd., and he was officially informed to that effect by the Political Resident in April 1935. But the Shaikh's views as to the date by which he considered his promise to the KOC negotiators not to negotiate with others to have terminated were accepted as valid, in spite of the terms of Chisholm's letter of 16th June to the Political Agent recording the negotiators' understanding of the matter. Both H.M.G. and KOC considered that there could have been misunderstanding between the Shaikh and the negotiators at their interview of 14th June, especially as the negotiators' telegram of 3rd June **(149)**, first informing KOC in London of the Shaikh's proposed suspension of negotiations, had stated that 'he does not propose to discuss concession further with us *until September*'.

British Government departments concerned recommend H.M.G. approval of Shaikh Ahmad's concession to KOC and rejection of Traders Ltd.'s application

The contents of Shaikh Ahmad's letters, as telegraphed on 16th December (see Note 141) by the Political Agent to the India Office (and communicated by the latter to the KOC directors on 17th December) were considered by the India Office (particularly the Shaikh's letter of 12th December to Gabriel and subsequent telegram of 15th December, informing him that he had granted his concession to KOC, and his denial that his letter of 2nd September meant the actual 'grant' of a concession to Traders

Ltd.) as decisive evidence that Traders Ltd.'s application of 28th November for approval of such a 'grant' could not be accepted. This conclusion and also that Shaikh Ahmad and KOC should be informed that H.M.G. approved his decision to grant the concession, on the terms agreed, to KOC were discussed in all their aspects on 19th December by a meeting **(150)** of representatives of the British Government departments concerned (India Office, Foreign Office, Admiralty and Petroleum Department) who recommended to H.M.G. (for the necessary Cabinet decision) that Shaikh Ahmad, KOC and Traders Ltd. should all be informed accordingly.

Meanwhile in Kuwait Chisholm and Holmes, since their telegram to London on 16th December, had been in communication with Cairo arranging the translation and final details of the concession documents being prepared there. On 17th December London had informed them that H.M.G.'s approval of Shaikh Ahmad's granting their concession was expected to reach him a few days later; on 20th December **(151)** they reported to London that the completed concession documents had arrived from Cairo on 19th December, and were ready for signature as soon as H.M.G.'s approval was received. On 20th December they also wrote to Shaikh Ahmad **(152)** enclosing copies of the concession document, informing him that they had been checked and found exactly similar to the text enclosed with his letter to them of 9th December, and requesting to know when he wished the formalities of signature to take place. To which the Shaikh replied on the same day that he would have the texts examined and inform them further as soon as possible.

H.M.G. approve Shaikh Ahmad's grant of concession to KOC, and inform KOC and Traders Ltd. accordingly

On 21st December H.M.G. decided to approve Shaikh Ahmad's grant of his concession to KOC and that the Political Agent should inform him accordingly. On the same day the India Office wrote to KOC **(153)** confirming the decision 'and to say that it is assumed that you will now take the necessary steps to complete the signature of your Agreement with his Excellency'.

They also wrote (see Note 153) to Mr Hunting, Chairman of Traders Ltd., setting out at some length why they could not act as requested in his letter and subsequent discussion of 28th November concerning the grant of a Kuwait oil concession to his company. As his letter had been the first intimation received by them of Traders Ltd.'s interest in the matter, they wrote, they had queried Shaikh Ahmad as to his relations with them and had been informed by him that his letter of 2nd September to their representative Gabriel was not in any way intended by him to represent the grant of a concession. The Shaikh had also stated that, having heard nothing further in respect of that letter from Traders Ltd. after 2nd September, he had informed Gabriel on 12th December that he was about to come to terms with the KOC, and on 15th December had telegraphed him that he had done so and that further conversations with Traders Ltd., must cease. The India Office letter then explained in detail why for some considerable time, as was well known to the Shaikh and to applicants for his oil concession, although the responsibility for deciding as to the grant of a concession was his, this was conditional on H.M.G. being consulted and kept fully informed by both the Shaikh and any such applicant both before any concession negotiations began, and also during them, in order to safeguard his interests and H.M.G.'s. The Kuwait Oil Company, which as Traders Ltd. were aware had been in negotiation with the Shaikh for a long time, had fully conformed with this procedure. But as regards Traders Ltd.'s negotiations with the Shaikh, H.M.G. had had no approach or information from that company until their letter of 28th November, although that letter indicated that their discussions with the Shaikh had reached an advanced stage by 2nd September. Despite the request in Shaikh Ahmad's letter of that date that Traders Ltd. should communicate with H.M.G., a delay of almost three months thus took place before Traders Ltd. made any communication to H.M.G., by which time the negotiations between Shaikh Ahmad and the KOC had been completed except for one or two points. In such circumstances, and as the Shaikh in the exercise of his choice had elected on 15th December to grant his concession to KOC and had then informed Traders Ltd.'s representative Mr Gabriel accordingly, H.M.G. were not prepared to interfere with his decision and had informed him of their approval.

This letter was to be acknowledged on 31st December **(154)** by Mr Hunting on behalf of Traders Ltd., stating that the company could not accept the India Office's decision as final and were referring the matter to their legal advisers. On 24th July 1935 (see Note 154) a further letter on behalf of Traders Ltd. was addressed by Mr Hunting to the India Office enclosing for information a copy of his letter of the same date to Shaikh Ahmad in Kuwait. This, after referring to the oil concession 'granted to Messrs. Traders Ltd. by your letter of 2nd September 1934 and since alleged to have been given to the Kuwait Oil Company by your Excellency', stated that Traders Ltd. would not accept, as they had informed H.M.G. on 31st December 1934, the Shaikh's decision to grant the concession to KOC in view of their correspondence and 'direct concluded negotiations' with him through their accredited agent; and that Traders Ltd. had placed the matter in the hands of their legal advisers with a view to taking proceedings 'to contest the validity of this concession'. Shaikh Ahmad received this letter in September 1935 and, after discussion with the Political Agent, sent no reply. Nothing further was heard on the subject.

Shaikh Ahmad and the negotiators sign the KOC concession agreement, 23rd December 1934

On 22nd December the Political Agent received H.M.G.'s instructions to inform Shaikh Ahmad of their approval of his grant of the concession to KOC. He informed the Shaikh accordingly, and subsequently visited Chisholm and Holmes to confirm that he had done so. That afternoon they received a message from Shaikh Ahmad requesting them to meet him next morning for the formal signing of the concession document at the Political Agency where, at 11 a.m. on Sunday, 23rd December 1934, the Kuwait Oil Company's concession agreement **(155)** was signed by Shaikh Ahmad, and by Chisholm and Holmes on behalf of the Company, the Political Agent (Lt-Col H. R. P. Dickson) witnessing their signatures.

In their final report to KOC giving details of the concession-signing ceremony, the negotiators wrote (on 10th January 1935 after their return to London) that it was followed by half an hour's general conversation when congratulations were exchanged, 'the Shaikh being evidently delighted at so satisfactory a termination of the negotiations'. They had also given him two letters (see Note 147) dealing respectively with the company's gift of Rs. 25,000 for his projected visit to London (which took place in the summer of 1935) **(156)** and annual payments in lieu of free petrol supplies. During the conversation Shaikh Ahmad urged them to return to London as soon as possible and to accelerate work on the concession, 'stressing the unpleasantness of hot-weather working and the desirability of making a start this spring', advice which KOC followed by starting its work in Kuwait on 16th March 1935 (see below).

Chisholm and Holmes, their joint negotiations for KOC at last completed, left Basrah together by air for London on 27th December 1934. Three weeks later Holmes received a letter dated 5th January 1935 **(157)** from Shaikh Ahmad appointing him to be the Shaikh's representative in London as provided for in Article 6 of the concession agreement, which position he was to hold until his death in 1947 (see Note 15). The two other appointments provided for in Article 6 were made by Shaikh Ahmad in February 1935 (when Abdullah, son of Mullah Saleh, was appointed as his Official Representative in Kuwait; Mohammed Yatim had first been offered the position by the Shaikh, but declined); and in February 1936 (when Lt-Col Dickson—see Note 39—became the Company's Chief Local Representative in Kuwait, an appointment which he held until his death in 1959).

With these appointments noted, and also that on 16th March 1935 senior Gulf and APOC geologists arrived in Kuwait on behalf of KOC (being met on arrival by Holmes and Abdullah Mullah Saleh **(158)** as the Shaikh's official representatives) to initiate the Company's concessionary operations, this record of the 1934 Agreement's negotiations is complete. But the following brief description of the Agreement's subsequent operation (it was to remain unaltered until 1951), and note of some significant conclusions emerging from this first full account of its long and complex negotiations, can usefully supplement it.

Kuwait Oil Company's operations (1935–72)

The KOC's initial geological surveys in 1935 and evidence derived from Holmes's water-wells drilled in 1927 (see page 13) decided the company to drill first at Bahra in north Kuwait with Burgan in south Kuwait as a second choice. Administrative and technical staff, equipment and material began arriving in Kuwait in December 1935 and the first well started drilling at Bahra in May 1936, but eventually reached 7,950 feet without producing oil. Meanwhile preparations to drill at Burgan had been proceeding and the first well there, begun on 26th October 1937, struck high-pressure oil at 3,672 feet in large quantity on the night of 23rd–24th February 1938, which can be regarded as the birthday of Kuwait's enormous oil industry **(159)**. Eight more producing wells were drilled at Burgan and an exploratory well at what was later the Magwa field, between then and April 1942; but in July 1942 the KOC's operation had to be suspended, and all completed wells plugged with cement, on instructions of the Allied Governments because of the then very critical war situation in the Gulf area.

By early in 1945 the war situation had so improved that the company's operations could be resumed. The plugged wells were reopened, development of the oilfields went rapidly ahead, and in June 1946 the first oil exports were made, 797,350 tons being exported by the end of that year. With twelve drilling-rigs in operation, one hundred additional wells were completed during the next five years, all of them producers, so that oil production rose as follows:

1947	2,185,300 tons
1948	6,291,600 tons
1949	12,183,700 tons
1950	17,018,700 tons

and as in 1949 seismic exploration at Magwa north of Burgan had revealed favourable prospects, from 1951 onwards development drilling was divided between the main Burgan field and its subsidiary field in that area, forty-five wells being drilled in 1951 and 1952 along the Ahmadi ridge between the two fields.

In the summer of 1951 the cessation until 1954 of oil exports from Iran (due to the dispute between the Iranian Government and APOC) led to a great increase in demand for oil from Kuwait; whose production rose from 17,018,700 tons in 1950 to nearly 28,000,000 tons in 1951 and thereafter as follows:

1952	37,000,000 tons
1953	42,600,000 tons
1954	47,000,000 tons
1955	53,900,000 tons

The number of producing wells in the Burgan, Magwa, and Ahmadi fields in 1955, when production also began in north Kuwait near the site of the company's first and unsuccessful well at Bahra in 1936, was 185, subsequent production being:

1956	54,100,000 tons
1957	54,400,000 tons
1958	69,100,000 tons
1959	68,400,000 tons
1960	80,600,000 tons

since when annual production has continued to increase (first reaching 100,000,000 tons in 1964) to 135,200,000 tons from 700 producing wells in 1970, and an estimated 150,000,000 tons in 1972.

This remarkable expansion in the Kuwait Oil Company's activities, bringing with it a corresponding increase in Kuwait's oil revenues and prosperity, was conducted in accordance with its 1934 concession agreement with Shaikh Ahmad (who died in January 1950 when the late Shaikh Abdullah as Salim as Sabah succeeded him as Amir) until 1951, when the first of its several subsequent revisions took place.

Misconceptions about the negotiations corrected

Some conclusions emerging from this first full record of the negotiations preceding the 1934 Kuwait Oil Concession Agreement corrected various misconceptions current about them during recent years.

The negotiations and their outcome were not, for instance, a case of Kuwait's then British protecting power first preventing Shaikh Ahmad from awarding his oil concession to either the Eastern General Syndicate (Holmes) or Gulf Oil Corporation, then encouraging its favoured APOC to intervene and subsequently pressuring the Shaikh into awarding it cheaply to an Anglo-American combine. Nor was Shaikh Ahmad's policy, either during the competitive stage of the negotiations or when the combined competitors approached him, controlled or directed wholly or mainly either by H.M.G. or by his friend Holmes; despite the undoubted credit deserved by the latter for his unique faith in the oil potentialities of Kuwait and the rest of the Arab coast of the Gulf. The full facts and documents tell a different story.

From 1911 until 1923 only the APOC was interested in obtaining an oil concession in Kuwait, and between 1921 and 1923 would have undoubtedly been granted it by Shaikh Ahmad had not the British Government, on his behalf, made conditions for it too stiff for the company to accept. That British policy towards Kuwait was then, as so often later in major world events, directed by Sir Winston Churchill, is an interesting footnote on history.

During the next ten competitive years which followed Holmes's appearance in Kuwait in 1923 to compete with APOC, first on behalf of EGS and from 1927 on behalf of Gulf Oil Corporation, the British Government, although understandably reluctant in those days to facilitate American commercial penetration of the Gulf area, went even to that length in its endeavours to hold the ring fairly between the two competitors and to get the best terms possible for Kuwait. And even in 1934, on the last-minute appearance of Traders Ltd. as a new competitor, the British Government was ready to sanction its intervention as being in Shaikh Ahmad's interest in the auction of his concessionary rights if the Shaikh himself had not rejected it.

An important and hitherto little appreciated point emerging from the record of the negotiations is its convincing evidence that Shaikh Ahmad from start to finish did not, as is often supposed, either play second fiddle to Holmes as their conductor, or only reluctantly conclude that his two British and American competitors should become joint grantees of his concession. On the contrary, the Shaikh seems to have manoeuvred towards that conclusion, as being in his State's best interest, from the time when Holmes first informed him in 1928 that he had Gulf Oil Corporation's powerful American backing against APOC's equally powerful British competition. The key factor in Shaikh Ahmad's policy from 1928 onwards, apparently motivated by disappointment in 1922 at British treatment of Kuwait interests at that year's Ojair Conference (see Note 22), emerges from this record as having been to add a valuable new American interest to the invaluable long-standing British support of his State. Because of his achievement both of that political aim, difficult and novel as it was at that period of Middle East history, and also of excellent commercial terms for his State's oil concession, Shaikh Ahmad emerges as the prime architect of the 1934 Agreement and the vast benefits it has brought to Kuwait. That his objectives were achieved despite the existence of a definite division in Kuwait 'public opinion' (commonly considered by historians as non-existent or ineffective in those days but clearly shown in this record to have been active and effective in Kuwait from 1922 onwards) regarding the merits of the two competitors for his concession, is further evidence of Shaikh Ahmad's adroit handling of both its political and commercial aspects. And the fact that his dealings with Traders Ltd. from April 1934 onwards came as a severe surprise, when first revealed in the following October, even to his close friend and confidant Holmes, disposes of any theory that Holmes may have deserved more credit than Shaikh Ahmad for the 1934 agreement.

Conclusion

In concluding this record of the 1934 agreement negotiations the author, now their only survivor, may be permitted to remind its readers that it is a matter over which it is even more fallible than usual to indulge in hindsight, judging events of those days by today's standards. Nowhere during the last forty years have living conditions, money values, communications, political relations, oil exploration techniques, almost everything in fact except the climate, changed more radically than in the Gulf area. As one example, the amounts offered for and the monetary terms finally agreed in the 1934 Kuwait concession agreement, trivial in comparison with the wealth it was destined to produce, were at that time generally regarded as on the lavish side of normal for so speculative an undertaking (even after the Bahrain oil-strike of 1932) in such a small territory.

This record's author regards himself as fortunate in having taken part in these negotiations, which laid the foundations of the modern State of Kuwait and through which he has enjoyed the friendship of its Rulers and leading citizens for the past forty years.

NOTES

NOTES

The originals of all documents quoted are now either in public archives (principally the India Office Library of the British Foreign and Commonwealth Office in London, and the American National Archives in Washington) or in private archives (principally those of The British Petroleum Co. Ltd. in London and of the Gulf Oil Corporation in Pittsburgh, U.S.A.). The documents are printed in ordinary type, while the author's editorial comments are printed in slightly larger italicized type.

1. *The following is the English translation of the Agreement dated 23rd January 1899 between the Shaikh of Kuwait (Shaikh Mubarak) and the British Government, which was eventually to terminate in 1961 when Kuwait became an independent State; similar Agreements had previously been made by the British Government with the Shaikh of Bahrain in 1880, the Sultan of Muscat and Oman in 1891, and the Trucial Shaikhs in 1892.*

(Translation)
Praise be to God alone (lit. in the name of God Almighty)
('Bissim Illah Ta'alah Shanuho').

The object of writing this lawful and honourable bond is, that it is hereby covenanted and agreed between Lieutenant-Colonel Malcolm John Meade, I.S.C., Her Britannic Majesty's Political Resident, on behalf of the British Government, on the one part, and Shaikh Mubarak-bin-Shaikh Subah, Shaikh of Kuwait, on the other part, that the said Shaikh Mubarak-bin-Shaikh Subah, of his own free will and desire, does hereby pledge and bind himself, his heirs and successors, not to receive the agent or representative of any Power or Government at Kuwait, or at any other place within the limits of his territory, without the previous sanction of the British Government; and he further binds himself, his heirs and successors, not to cede, sell, lease, mortgage, or give for occupation or for any other purpose, any portion of his territory to the Government or subjects of any other power without previous consent of Her Majesty's Government for these purposes. This engagement also to extend to any portion of the territory of the said Shaikh Mubarak which may now be in the possession of the subjects of any other Government.

In token of the conclusion of this lawful and honourable bond, Lieutenant-Colonel Malcolm John Meade, I.S.C., Her Britannic Majesty's Resident in the Persian Gulf, and Shaikh Mubarak-bin-Shaikh Subah, the former on behalf of the British Government, and the latter on behalf of himself, his heirs and successors, do each, in the presence of witnesses, affix their signatures, on this the 10th day of Ramazan, 1316, corresponding with the 23rd day of January 1899.

(sd/-) M. J. Meade, Lieut-Col.
Political Resident in the Persian Gulf.
(L.S.) (sd/-) Mubarak-Al-Subah.

Witnesses:
sd/- E. Wickham Hore, Captain, I.M.S.
sd/- J. Calcott Gaskin.

(L.S.) Muhammad Rahim-bin-Abdul Nebi Saffer.
(Sd.) Curzon of Kedleston.
Viceroy and Governor-General of India.

Ratified by His Excellency the Viceroy and Governor-General of India at Fort William on the 16th day of February 1899.

(sd.) W. J. Cuningham,
Secretary to the Government of India
in the Foreign Department.

(SEAL)

The British Political Resident (who from June 1897 to March 1900 was Lt-Col M. J. Meade) was at that time the British Government's senior political representative in the Gulf, stationed at Bushire in Persia. He was appointed by and responsible to the British

Government of India (until India became independent in 1947, whereafter he was appointed by the British Government and responsible to the British Foreign Office in London), which was responsible to the British Government through its India Office in London. The Political Resident had subordinate representatives (with the title of British Political Agent) elsewhere in the Gulf. A British Political Agent was not appointed in Kuwait until 1904; the list below gives the names and dates of British Political Agents in Kuwait from that date until 1961 when, on Kuwait becoming an independent State, the position and title was abolished. In 1972 the position and title of British Political Resident was also abolished, his duties having by then been taken over by British Ambassadors in the different States adjoining the Gulf.

British Political Agents, Kuwait

1904–1909	*Major Knox*
1909–1915	*Captain Shakespear*
1915–1916	*Lt-Col Gray*
1916–1918	*Lt-Col Hamilton*
1918	*Captain Loch*
1918–1920	*Captain McCollum*
1920–1929	*Major More*
1929–1936	*Lt-Col Dickson*
1936–1939	*Captain de Gaury*
1939–1941	*Major Galloway*
1941	*Lt-Col Dickson*
1941–1943	*Major (later Sir T.) Hickinbotham*
1943–1944	*Mr Pelly*
1944–1945	*Mr Jackson*
1945–1948	*Mr Tandy*
1948–1951	*Mr Jakins*
1951–1955	*Mr Pelly*
1955–1957	*Mr (later Sir G.) Bell*
1957–1959	*Mr Halford*
1959–1961	*Mr (later Sir J.) Richmond*

---

2. *Anglo-Persian Oil Co. Ltd. was founded in 1909, to exploit an oil concession in Persia (now Iran) granted by the Persian Government in 1901, after oil had been struck in 1908 at Masjid-i-Suleiman in south-west Persia. By 1911 the oilfield was connected by pipeline with the Company's refinery which in 1913 came into operation at Abadan, thereafter its operating headquarters in Persia, on the Shatt-el-Arab near Mohammerah (now Khorramshahr). The Company, in which the British Government became the majority shareholder in 1914 by investing £2,001,000 (see Note 9), changed its name to Anglo-Iranian Oil Company Ltd. in 1935 and to The British Petroleum Company Ltd. (its present name) in 1954. In 1914 it became a large shareholder in the Turkish Petroleum Company (formed to exploit the oil reserves of Turkish Mesopotamia, now Iraq) which in 1925 was reconstituted and in 1929 renamed the Iraq Petroleum Company Ltd. In 1934 APOC became an equal partner with Gulf Oil Corporation of Pennsylvania in Kuwait Oil Company Ltd.*

In 1911 APOC's Managing Director was Mr (later Sir Charles, and Lord) Greenway, subsequently its Chairman from 1914 to 1926, when he was succeeded as Chairman by Sir John (later Lord) Cadman (see Note 77).

---

3. *In 1911 the British Political Resident (see Note 1) was Lt-Col P. Z. (later Sir Percy) Cox (1864–1957). Formerly Political Agent (see Note 1) at Muscat 1899–1904, he was Political Resident in the Gulf 1904–20; Chief Political Officer in Mesopotamia (later Iraq) 1914–17 and Civil Commissioner 1917–18; British Minister in Teheran 1918–20; and British High Commissioner in Iraq 1920–23, during which period he presided in 1922 at the Ojair Conference (with King Ibn Saud) whose decision regarding the southern boundary of Kuwait displeased Shaikh Ahmad (see Note 22).*

On 3rd November 1911 Mr Charles Greenway (see Note 2), Managing Director of APOC wrote from London to Lt-Col Cox as follows:

Lt.-Col. P. Z. Cox, C.S.I., C.I.E.
Bushire
My dear Cox,

It has come to my knowledge that Mr. Reynolds—our ex-employee—is endeavouring to induce the Shell Co., which is now only a share-holding Company in the Royal Dutch Co., to secure a Concession for Oil in Koweit.

I do not know what reason Reynolds has for assuming that there are any oil deposits of value in Koweit but if in the course of his duties for this Company he discovered any oil indications there he should have brought them to our notice, and have recommended our applying for a Concession, because he was well aware that it would be very prejudicial to our interests to have a powerful foreign rival close on our heels in the Persian Gulf.

However that may be, it would, of course, not at all suit our book to allow the Shell Co. to get an Oil Concession at this place, nor would it, I think, any more suit the policy of the British Government, because the Shell and Royal Dutch Companies are closely associated with the Deutsche Bank, and it is quite probable that they, knowing that they would receive the cold shoulder from our Government, would, if they desired to obtain a Concession at this place, endeavour to obtain it from Turkey through the Deutsche Bank or some other German channel.

I should therefore be much obliged if you could lend us your help in this matter, and let me know whether you think a valid Concession for working Oil in Koweit is obtainable from the Sultan without reference to the Turkish Government. If so, I should like you to put forward an application on behalf of the Anglo-Persian Oil Co. for a Prospecting License, the draft terms of which I will send to you later on if you think there is any chance of a Concession being obtainable.

The question of whether or not there are oil deposits of any value in Koweit is, of course, entirely problematic, and consequently we should only in the first place be prepared to take out a Prospecting License, which in view of the difficulties of prospecting in this place should be for a fairly long period, say 2/3 years.

Yours, sincerely,
(Sgd) C. Greenway

---

4. *In 1912 Mr (later Sir) Winston Churchill, as head of the Admiralty (see Note 9), was converting the ships of the British Navy from coal-burning to oil-burning. In that connection he appointed a Commission, headed by Admiral Sir Edmond Slade, to visit the Persian oilfields of the Anglo-Persian Oil Company (see Note 2) in 1912/13 and report on oil production prospects there and elsewhere in the Gulf. The other members of the Commission were:*

Professor J. (later Sir John, and Lord) Cadman; then a well-known mining-engineer, and Adviser to the Home and Colonial Offices of the British Government. Later (see Note 77) he was Chairman of APOC from 1926 to 1941.
Mr Blundstone; geologist and oil expret.
Mr E. H. Pascoe; geologist, Assistant-Superintendent of the Geological Survey of the Government of India (see Note 5 for his report dated 7th May 1913 on the oil prospects of Kuwait).
Mr James; geologist of the APOC.
Mr Clarke; Secretary of the Commission.

On 26th November 1913 the Political Agent at Kuwait (Captain Shakespear) reported on the Commission's visit there from 11th to 15th November as follows:

MEMORANDUM

on visit of Sir Edmond Slade and Oil Experts' Commission to Kuwait

Arrangements having been made beforehand, Admiral Sir E. J. W. Slade and party consisting of:

Professor J. Cadman, Mining Engineer and Adviser to Home and Colonial Offices.
Mr Blundstone, Geologist and Oil Expert.
Mr E. H. Pascoe, Geologist, Government of India Geological Survey.
Mr James, Geologist to the Anglo-Persian Oil Company.
Mr Clarke, Secretary to the Commission.

arrived at Kuwait in H.M.S. 'Sphinx' in the morning of the 11th November 1913. Admiral Slade landed and called on Sheikh Sir Mubarak-as-Sabah. After completing arrangements the Political Agent joined the party on board H.M.S. 'Sphinx' in the afternoon and the vessel proceeded some 30 miles down the coast to opposite Sadaiba village.

On the morning of the 12th November the whole party with their baggage landed and marched to the Political Agent's camp near Burgan Hills. A preliminary inspection was made of the site of the bituminous seepage in these hills the same afternoon.

The following day, 13th November, was devoted to a thorough examination of the hills, their rocks and a geological survey made of the site and its immediate surroundings by the party.

On the 14th November the Commission having decided that nothing further was to be learned at the site marched back to the coast and re-embarked in H.M.S. 'Sphinx', which left immediately for Kuwait.

On the 15th November the Sheikh of Kuwait called on Admiral Slade and Professor Cadman at the Political Agency and provided a guide and boats to facilitate reaching the site of a seepage reported to exist near M'daira (Lat 29°. 37'. 30" N and Long 18°. 2' 0" E). Mr Pascoe and Mr James accompanied by the Political Agent were, however, unable to reach the spot before sunset and had consequently to return unsuccessful. The Commission then left in H.M.S. 'Sphinx' for Mohamerah.

Later arrangements were made to obtain, if possible, samples from this seepage reported near M'daira. On receipt of the samples, which proved that a seepage actually does exist at the spot described, they were forwarded by post to Admiral Sir E. J. W. Slade with a letter from the Political Agent giving such further details as he had been able to obtain, for the information of the Commission.

The Commission were given every facility by Sheikh Sir Mubarak-as-Sabah, who placed horses at their disposal and extra baggage camels for their transport to and from the Political Agent's camp at Bargan Hills and co-operated in the most friendly way.

W. Shakespear
Captain
Political Agent, Kuwait.

Kuwait
26th November 1913

---

5. *Geological report on Kuwait's oil prospects: no. 2392–758, dated Calcutta, the 7th May 1913*

From—H. H. Hayden, Esq., C.I.E., I.C.S., Director, Geological Survey of India,
To—The Secretary to the Government of India, Department of Commerce and Industry, Simla.

In continuation of my letter No. 2047, dated the 3rd April 1913, I have the honour to forward herewith a copy of a report by Mr. E. H. Pascoe, Assistant Superintendent in this Department, upon the prospects of obtaining oil near Kuwait, Persian Gulf.

Prospects of obtaining oil near Kuwait, Persian Gulf.

Introduction

Some 24 miles south of Kuwait, a small area of Tertiary rocks projects above the surface of the Desert, and is known as Burqan. I examined it between the dates 2nd to 5th March, 1913. Burqan is close to the rock known as Wara, and consists of some small knolls and rising ground. The knolls are ridge-like and are grouped along or parallel to the perimeter of a broad ellipse, which represents in all probability a low flat dome with its longer axis running approximately from north to south. The monotony of the flat plain occupying the centre is relieved by a few additional knolls. Conglomerate or conglomeratic

Rocks

sandstone, exactly similar to those at Wara, caps all these knolls and ridges, and in all cases has a scarcely appreciable dip. There is much sagging at its edges and a great deal of scree, but here and there it is possible to observe what lies beneath it. Below it, in the easternmost ridges, I found bands of ochreous sandstone, sandy clays of variegated colours—red, pink, yellow, etc., a white kaolin-bearing sandstone, and many large fragments of selenite or crystalline gypsum. Flakes of gypsum occur in other parts of the hills, and the efflorescent salt in the central plain. All this points to the exposure, or immediate proximity beneath the surface of Dr. Pilgrim's 'Fars' series. I am inclined to think the Fars beds are actually exposed, the caps of conglomerate and conglomeratic sandstone representing the basal bed of the Bakhtiari series. The Fars exposure would be about a mile long by 3/4 mile broad.

Structure

The geotectonic features appear to run, in a general way somewhere between north, south and north-west–south-east. This is borne out (i) by the orientation of the Barqan dome and the relative position of Wara Hill (a little west of north from Burqan), (ii) by the general strike and trend of outcrops in Persia and further south in Arabia, (iii) by the trend of the Persian Gulf and Euphrates and Tigris Valleys, (iv) by the direction of the low flat swell-ridges observed and mapped by Captain Shakespear further inland. The eastern slope of these flat swells is considerably more gradual than the western, a circumstance conformable with the chart of the Gulf which shows a very gentle eastward shelving along the Arabian side but a comparatively steep drop into deep water along the Persian coast. From this we may deduce that easterly-dipping beds cover a greater proportion of the total area than do westerly-dipping beds. The dome at Burqan is a very gentle one. A little bituminous earth is found in the central plain.

Oil

The prospects of obtaining an oil supply depend principally upon (i) the age and the capacity for storage of the rocks, (ii) structure, (iii) position with regard to the oil 'belt'. The age of the beds is favourable, if it is correct to assume they belong to the Fars, since this is the oil-bearing series in Persia. There is no reason to believe the nature of the beds beneath Burqan should be different from that of the Fars in Persia, and they may be looked upon therefore as sufficiently porous to retain oil in workable quantities. The structure, if correctly diagnosed, is also favourable, a gentle dome in fact being an excellent type of folding suitable for an oil-field. As to its geographical position, this locality is not on the line of strike of the rich oil deposits now being worked above Ahwaz, and is, in fact, over 170 miles to the south-west of this line. This does not necessarily mean that oil in commercial quantity does not occur below Burqan, but it adds a decidedly speculative element to any operations. A 'floating tract of naphtha', coinciding no doubt with a submarine seepage was described by Captain Constable as occurring between the islands of Qaru and Kubbar, some 56 miles south-east of Burqan. On the whole, my opinion is that the chances are not unfavourable, and that an oil company would not require much inducement to test it, if protection and permanency of concession were guaranteed. The central plain of Burqan would be the best site for tests, and these should be as deep as possible, as the uppermost layers of the Fars are probably all that are exposed. Under present conditions there would be great difficulty in obtaining water for the boiler; possibly an internal combustion engine would have to be employed.

---

6. *Translation of a letter dated the 27th October 1913, from Lt-Col Sir Percy Cox, Political Resident in the Persian Gulf, to Shaikh Sir Mabarak-as-Subah, Ruler of Kuwait:*

After compliments, and with reference to the conversation which took place between us yesterday, if Your Excellency sees no objection therein I wish with Your Excellency's consent to inform the British Government that Your Excellency is agreeable to the visit of Admiral Slade in order that he may inspect the places (showing traces of) bitumen at Burgan and elsewhere and that if there seemed in his view a hope of obtaining oil therefrom Your Excellency agrees not to give a concession in this regard to anyone other than a person nominated and recommended by the British Government.

This is what was necessary to state to Your Excellency and may you be preserved.
Dated the 26th Za-al-Kada 1331 (27th October 1913).

Translation of a letter dated the 26th Za-al-Kada 1331 (27th October 1913) from Shaikh Sir Mabarak-as-Subah, Ruler of Kuwait to the Political Resident in the Persian Gulf:

After compliments

With the hand of friendship we received your esteemed letter dated the 26th Za-al-Kada 1331 and in it you stated that with reference to the conversation which passed between us yesterday if we saw no objection therein it would be profitable to your Honour to inform the British Government that we were agreeable to the arrival of His Excellency the Admiral—we are agreeable to everything which you regard advantageous and if the Admiral honours our (side) country we will associate with him to be in his service, to show the place of bitumen in Burgan and elsewhere and if in their view there seems hope of obtaining oil therefrom we shall never give a concession in this matter to anyone except a person appointed from the British Government.

This is what was necessary and I pray for the continuance of your high regard and may you be preserved.
Dated 26th Za-al-Kada 1331.

---

7. *The British Government department in London principally responsible for dealing with Kuwait affairs (in close collaboration with the India Office; see Note 1) was the Foreign Office prior to 1921, when that responsibility was transferred to the Colonial Office. APOC's letter to the Foreign Office dated 12th May 1921 enquiring about Kuwait concluded as follows: 'My Board are aware that the responsibility for these territories has recently been transferred to the Secretary of State for the Colonies [then Mr Winston Churchill; see Note 9], but in view of the fact that all previous correspondence has been with the Foreign Office they have ventured to address you in the first instance.'*

Thus from 1921 to 1933, covering all but the last eighteen months of the Kuwait concession negotiations, the Colonial Office was principally responsible for H.M.G.'s interests in them, although various other government departments had to be consulted (including the Foreign Office, India Office, Admiralty, and Petroleum Department of the Board of Trade) as members of a Cabinet Committee first set up in 1915 to deal with Middle East affairs. That so many different government departments were interested, and entitled to be consulted, inevitably entailed delays which at several stages of the negotiations were the cause of complaints by Shaikh Ahmad, APOC, Gulf, and EGS.

In 1933 (see Note 73) the Colonial Office's Kuwait responsibilities were taken over by the India Office, which was eventually to transfer them to the Foreign Office in 1947 when India became independent of Britain.

---

8. *Born in 1885, Shaikh Ahmad was thirty-six years of age when he succeeded his uncle Shaikh Salim following the latter's death in February 1921. Lt-Col Dickson (see Note 39), Political Agent in Kuwait from 1929 to 1936 and a close friend of Shaikh Ahmad until the latter's death in 1950, writes as follows regarding his succession in his book* Kuwait and her Neighbours *(Allen & Unwin, 1956):*

It had always been the custom for Sheikhs of Kuwait to rule personally and autocratically, and to avoid all delegation of authority. It was also their pride that they were accessible to all their subjects, even to the most humble, it being their usage to give audience for two or three hours each morning in the market-place, and to decide great and small cases by direct judgment.

On the death of Sheikh Salim, however, the townspeople, tired of the unnecessary war [with Ibn Saud] into which they had been led against their will, determined that in future they would have some say in the affairs of the State. They informed members of the Sabah family that they would accept as their ruler only one who would assent to a council of advisers. The choice fell on Ahmad Al Jabir, the popular eldest son of the late Sheikh Jabir and already a favourite of Ibn Saud. On his return from Nejd [where Sheikh Ahmad had been sent by Sheikh Salim early in February to seek a truce with Ibn Saud; when the news of Sheikh Salim's death reached them Ibn Saud said at once that there was no longer any quarrel to settle] on 29th March he acceded to the throne. An agreement was drawn up between him and the people of Kuwait to the effect that all criminal cases would be decided in accordance with the shariah, the religious law of Islam. In case of appeal, the written statements of both parties and the Qadhi's judgement would be submitted to the Ulema, whose decision would be final. If both parties to a dispute agreed beforehand for a third party to arbitrate between them, his decision would hold. The Ruler would seek advice in all matters, external as well as internal, affecting the town. If anyone had any suggestion to make for the benefit of the town or people, he would lay it before the ruler, who would consult his people and adopt it if they so desired. Under the presidency of one of the leading merchants, Hamed ibn Abdullah al Sagar, a council of twelve members, six from the eastern half of the town and six from the western half, was duly elected. It rarely met, however, and in practice Sheikh Ahmad followed the older system still beloved of his people, and ruled in much the same way. Sheikh Ahmad was a strong good-looking and pleasant man of the same type as his father. Thick-set rather than tall, his bluff, jovial manner and peculiarly charming smile made him an eminently suitable person in so far as the ceremonial part of his duties was concerned.

---

9. *Mr (later Sir) Winston Churchill (1874–1965) was head of the British Colonial Office from April 1917 to October 1922. As Head of the Admiralty from 1911 to 1915 he had been responsible for appointing the Slade Commission (see Note 4) which visited Kuwait in 1912–13, and for H.M.G.'s investing £2,001,000 in APOC (see Note 2) in 1914 in order to safeguard oil-fuel supplies for the British Navy.*

---

10. *APOC would have preferred to have opened negotiations in Kuwait on the basis of one comprehensive document. It only reluctantly deferred to the Colonial Office's insistence on its old-established form of 'model mining lease' combining separate exploration and prospecting licences with a mining lease. It deferred equally reluctantly to the Colonial Office's requirements that (1) although the exploration and prospecting licences could cover the entire territory of Kuwait, the mining lease must only cover selected areas.*

(2) that Customs exemption for imported plant and material must be for the first ten years only and subject to reconsideration thereafter, despite the Company's argument that exemption during the whole concession period was usual and already enjoyed by the Company in its Persian and Iraq concessions.

This H.M.G. requirement as regards customs exemption (the other two were soon to be abandoned) was to prove a very inhibiting factor (especially after the Colonial Office had waived its similar requirement in 1925 in APOC's Muscat concession and also the EGS's Bahrein Concession) for many years in APOC's attitude towards Kuwait owing to its possible repercussions on the terms of its major Concession in Persia.

In view of the effect on the negotiations of H.M.G.'s insistence from 1928 onwards on a 'British Nationality Clause' in any Kuwait oil concession agreement, it is interesting to note that a Colonial Office letter to APOC of 29th March 1922, embodying its principal requirements for prior agreement by the Company before 'Mr Secretary Churchill . . . will cause the local authorities to be instructed to open negotiations with the Sheikh of Koweit', included the following paragraph:

I am further to explain that the grant and operation of this concession must be conditional upon the strictly British character of the lessees, and that your Company will be required to give an undertaking to this effect before His Majesty's Government can proceed in the matter.

This requirement was, of course, one which the APOC had no difficulty in agreeing.

---

11. *In 1922 Sir Arnold Wilson (1884–1941; he was Knighted in 1920 when acting British Civil Commissioner in Iraq) was Resident Director for APOC in Persia at Mohammerah/Abadan. This followed his appointment by Sir C. Greenway, APOC's chairman (see Note 2) in 1921, on retirement from British Government service, to be Managing Director in Persia, Mesopotamia, and the Gulf of Strick Scott and Co., Managing Agents of APOC there until 1923. From 1923 he was General Manager of APOC (jointly with T. L. Jacks; see page 11) at Mohammerah/Abadan until his transfer in 1926 to the Company's head office in London. He retired from APOC in 1932, becoming a Member of Parliament from 1933 to 1940, when he was killed in action with the Royal Air Force which he had joined in 1939.*

Wilson, as Consular Assistant at Ahwaz in Persia, had been present at APOC's first oil discovery in 1908 at Masjid-i-Suleiman (see Note 2); he was subsequently deputy Political Resident in the Gulf from 1912 to 1920, and deputy Chief Political Officer and acting British Civil Commissioner in Iraq from 1915 to 1920.

---

12. *The Political Agent at Kuwait in 1923 was Major More (see Note 1).*

---

13. *The terms which, unknown to and unauthorised by APOC, were described in January 1923 to Shaikh Ahmad by the Political Agent (on the instructions of the Political Resident, Lt-Col Trevor) were based on a précis of the Colonial Office's early suggestions to the Company, although these had since been modified as a result of London discussions since early 1922.*

They included a payment of Rs. 10,000 (approx. £750) for a two-year exploration licence (Rs. 1,000 had been the first Colonial Office proposal) covering all Kuwait territory; followed by a two-year prospecting licence, for a similar payment, covering selected areas; followed by a mining lease, covering a maximum of 100,000 acres in not more than three blocks, for a period to be negotiated during which a royalty would be payable of Rs. 3.8 per ton of crude oil produced with a minimum of Rs. 30,000 p.a. This royalty would be subject to review after ten years. Imports necessary for operating the concession were to be 'free of all taxes and custom-house duties' for the first ten years only.

---

14. *The following draft agreement was given on 8th March 1923 by APOC to the Political Agent in Kuwait for further negotiations with Shaikh Ahmad. Although its Article 8 omitted the Colonial Office's stipulated time-limit of ten years for tax and customs exemption for the Company's imports, this was to be agreed in a separate exchange of letters. By this procedure it was hoped to prevent its becoming a precedent in the Company's concession negotiations elsewhere. This, and other aspects of the draft including its new 'combined' form, were objected to by the Political Resident (both by Lt-Col Trevor and by Lt-Col Knox who succeeded him in April 1923) as being contrary to their current instructions from the Colonial Office; but their objections were over-ruled by H.M.G. in London following Major Holmes's appearance in the negotiations in May 1923.*

AGREEMENT between Shaikh Ahmad al Jabir al Sabah, C.I.E., Ruler of Koweit of the one part, hereinafter called 'the Sheikh' and the D'Arcy Exploration Company, Ltd. of the one part, hereinafter called 'the Company'.

Article 1:

With approval of H.M's Government, the Sheikh grants to the Company by these presents a special exclusive privilege under the conditions set forth below to search for natural gas, petroleum, asphalte and ozokerite throughout the whole extent of the territories under his control for a period of 60 years from the date of this agreement.

Article 2:

The Sheikh grants in the first place to the Company an Exploration License for a period of two years covering the whole extent of his territories. The License is subject to extension for a further period of two years on the Company showing to the satisfaction of the Sheikh acting on the advice of the Political Resident in the Persian Gulf that such extension is justified.

Article 3:

The Sheikh with the approval of H.M's Government hereby grants after the expiry of the above-mentioned period of two years or any extension of renewal thereof a prospecting license for a further period of two years over areas to be selected by the Company with the approval of the Sheikh and with the cognizance of the Political Resident in the Persian Gulf.

Article 4:

At the expiration of the prospecting License the Sheikh with the approval of H.M's Government hereby grants to the Company a Mining Lease over an aggregate area not exceeding 100,000 acres divided into not more than three blocks.

Article 5:

In virtue of the above-mentioned Mining Lease the Company shall have the exclusive right to prospect and drill for, extract, treat, refine, manufacture, transport and deal with petroleum products, naphtha, natural gases, tar, asphalt, ozokerite and other bituminous materials and for water and to make all investigations necessary for the purpose of their business in any part of the said territories and shall have the exclusive right to construct and operate pipe-lines, railways, refineries, storage-tanks, wharves and jetties and to construct and erect tramways, roads, buildings, machinery and telegraph apparatus of all kinds in any part of the said territory so far as may be necessary for the purpose of their business.

The Company shall have the right to collect and use free anywhere in the Sheikh's territory stone, gypsum, salt, sulphur, clay, wood and water whether from rivers or springs for the purpose of their work.

The selection of routes and sites for such work and the course of the pipe-lines shall rest with the Company.

Article 6:

All lands whatever required by the Company for the purpose of their preliminary exploration for oil including all or any natural springs of oil in the said territories together with an adequate supply of water so far as available shall be granted by the Sheikh to the Company free of charge and without lot or hindrance from any of their relatives or dependants or any of their tribesmen or of third parties. And it is hereby agreed that in consideration of the sums to be paid to the Sheikh laid down in Article 12, any claim for compensation, reward or subsidy put forward by a third party in respect of such land or of the services to be rendered to the Company by the Sheikh shall be met by the said Sheikh or his heirs and successors and that under no circumstances shall the Company be made to pay compensation for the same.

Article 7:

The Sheikh grants gratuitously to the Company all uncultivated lands belonging to him which may be needed by the Company for their operations.

Lands so granted shall be the property of the Company for the period of their agreement and the Company shall have full proprietary rights therein.

The Sheikh also recognises that the Company has the right to acquire all and any other lands and buildings necessary for the said purpose with the consent of proprietors on such conditions as may be arranged between the Company and the said proprietors without their being allowed to make demands of a nature to surcharge the prices ordinarily current for lands and buildings in their respective localities.

At the expiry of the agreement the ground and all immovable property left by the Company shall be handed back to the Sheikh.

Article 8:

All lands granted by this agreement to the Company or that may be acquired by the Company in the manner provided for in Articles 3 & 4 of this agreement as also all products exported shall be free of all imposts and taxes during the tenure of the present concession.

All material and apparatus of every kind necessary for the exploration, working and development of the deposit and for the construction and development of the pipe-lines including also all the works specified in the first paragraph of Article two (2) hereof and including the reasonable personal requirements of employees and of office requirements shall enter the said territories free of all taxes and custom house duties.

Article 9:

The Sheikh undertakes to give all the protection in his power to the Company and to their Staff and labourers in any part of his territory from thefts, highway robbery, assault, &c.

Similarly the Sheikh undertakes to protect all the property of the Company and of its employees which may be exposed to wilful damage or to possible loss by theft.

Article 10:

For the above purpose it is agreed that the Company in communication with the Sheikh shall appoint permanent trustworthy guards under a reliable man of good family whose pay and that of the guards shall be provided by the Company and who shall remain there summer and winter.

If in spite of the efforts of the said guards thefts should occur, the Sheikh undertakes to recover the property stolen and to compensate the Company for any damage sustained in his territory. In the event of non-recovery of the stolen property the value thereof shall be deducted by the Company from the same sums due to the Sheikh.

The number of guards at any place where the Company may operate shall be fixed by the Company.

It shall be the duty of the above-mentioned head of the guard to keep order in the areas in his charge to such extent as the Company may order him; he shall have no right to interfere in any disputes among the Company's servants unless asked by the Company to do so.

Article 11:

In the case of misconduct the said guards are subject to dismissal or fine by the Company. In the event of any heavier punishment being needed the offender shall be handed over to the representative of the Sheikh and the Sheikh undertakes that punishment shall be inflicted and to use his utmost endeavours to uphold the authority of the Company's employees thereby avoiding unnecessary troubles to both parties of this contract.

Article 12:

Should the Company succeed in finding oil in commercially exploitable quantities they agree to pay half-yearly to the Sheikh a Royalty of Rs. 3–8–0 per ton of net crude oil got and saved (i.e. after deducting water and foreign substances and oil required for the customary operations of the Company's installations in the Sheikh's territories).

This rate of royalty to be subject to revision by mutual agreement at the end of *twelve years from date of this agreement*; in default of agreement either party shall have the right to demand that the question at issue shall be submitted to arbitration as provided in Article 17.

Article 13:

The Company hereby undertake that the amount received by the Sheikh in respect of Royalties shall not be less than Rs. 30,000/- in any completed calendar year in which the Company continues work, beginning with the year after the date on which the Company shall have declared that oil has been found in commercially exploitable quantities.

Article 14:

Within one month from the signature of the Agreement the Company shall pay to the Sheikh the sum of Rs. 10,000/- in consideration of the assistance and protection to be afforded to their employees. With the exception of this single sum they shall not be liable to pay any further sums to the Sheikh except as provided above after oil has been found, *but on the grant of a prospecting license* under Article 3 the Company shall pay to the Sheikh a further single sum of Rs. 10,000.

Article 15:

In the event of the Company failing to discover oil in sufficient quantity they shall have the right to stop work at their option and the Sheikh shall have no claim to any payment or compensation.

Similarly in event of the Company failing to conform to the terms of this Agreement the Sheikh acting on the advice of the Political Resident has the right to terminate the agreement provided that in default of agreement on this question either party have the right to demand that the question at issue shall be submitted to arbitration as provided in para 17 below.

Article 16:

The Sheikh solemnly declares that he will carry out all the conditions of this contract willingly and faithfully and that he and his heirs and successors shall be its guarantors and shall accept full responsibility for any claim or by any person questioning the validity of any stipulation on this contract.

This contract cannot be cancelled except as provided as above and shall remain in force for the period of the Company's concession from the Sheikh or any extension or removal thereof.

Article 17:

In the event of there arising between the parties to the present Concession any dispute or difference in respect of its interpretation or the rights or responsibilities of one or the other of the parties therefrom resulting such dispute or difference shall be submitted to two Arbitrators one of whom shall be named by each of the parties and to an Umpire who shall be appointed by the Arbitrators before they proceed to arbitrate. The decision of the Arbitrators or in the event of the latter disagreeing that of the Umpire shall be final.

---

15. *Major Frank (Francis) Holmes (1874–1947), who was to play a major part in the Kuwait oil-concession negotiations from his first intervention in them in May 1923, was then forty-nine years old. Born on his father's farm in New Zealand in 1874, as a youth of twenty he went to Johannesburg in South Africa to work for a gold-mining company of which his uncle was General Manager, and thereafter until the 1914–18 War gained world-wide experience as a mining engineer. In 1896 he transferred to Western Australia, working there as manager and metallurgist in various gold-mining companies until 1901. Soon after the death in 1900 of his first wife (he had married in 1897; his only child, a son, born in 1898 died in 1939 after serving in the Indian Army), Holmes visited London and was engaged by a major British mining group to be Asst. General Manager of the Penang Corporation in Malaya. In 1905 he returned to Australia and New Zealand, intending to become a mining consultant, but shortly afterwards accepted appointments by his former employers first in Mexico and thereafter, until the outbreak of the First World War in 1914, in gold and tin mining companies in various countries including Russia, Nigeria, and Uruguay, on return from which to London in 1914 he married for the second time.*

In August 1914 Holmes joined the British forces as a senior supply officer with the rank of Major. After serving in Egypt and the Dardanelles in 1915 and later in France, he was detached in 1918 for special duty in Abyssinia and Aden to arrange meat and other supplies for the British troops in Mesopotamia (Iraq). As their supply base was at Basrah, he there made his first contact with the Arabian Gulf where, as an experienced mining-engineer, he was interested in all he heard about the APOC's operations on its Persian side and oil seepages and water problems on its Arabian coast.

After demobilisation in 1919 in London, Holmes discussed with his pre-war friends and employers (P. C. Tarbutt and Co. and Charterland and General Exploration Co. Ltd.) business opportunities which he had noted in the Gulf. In August 1920 they formed Eastern and General Syndicate Ltd. (for details see Note 16; Mr, later Sir Edmund Davis, Chairman of Charterland was first Chairman of EGS) with the object of securing concessions, options and general business in the Middle East, for which Holmes and another former Charterland employee were engaged by the new Syndicate. In 1920 they went to Aden, where they started EGS's first venture, a pharmacy (see Note 16); it was eventually sold in 1925, by which time Holmes's colleague had left EGS. In October 1922 Holmes left Aden for Bahrain, thereafter his Middle East base until 1933, with possible oil concessions or options in mind particularly in Saudi Arabia from Ibn Saud (then not yet King, but known as Sultan of Nejd and its Dependencies), who had answered his letter of enquiry from Aden by requesting him to visit him for discussions.

Holmes was accompanied to Bahrain by Dr Mann, one of his London acquaintances on business there for Ibn Saud, who introduced him to its Arab merchant community including the Yatim family, of whom Mohamed Yatim was thereafter to be his interpreter-assistant until 1935. After a week in Bahrain they went to Hofuf, capital of the Saudi Arabian mainland province of El Hasa. There, with a favourable introduction from Abdul Aziz Al Qosaibi, Ibn Saud's trade representative in Bahrain, Holmes sufficiently interested Ibn Saud in an oil option for a draft agreement to be prepared for consideration by the Sultan's advisers. Ibn Saud was on his way to the El Hasa coast at Ojair for a conference there in November (see Note 22) with the British High Commissioner in Iraq, Sir Percy Cox (see Note 3), and Holmes accompanied the Sultan's party to the conference. He there continued discussion of his proposed draft agreement with the Sultan's advisers so successfully that on the last day of the conference, 2nd December 1922, Ibn Saud agreed provisionally to grant him an exclusive option for exploration of oil and mining rights in the province of El Hasa, subject to favourable results of further enquiries necessary to be made. This proviso resulted from Sir Percy Cox's action, when Ibn Saud asked him if H.M.G. would have any objection to his agreeing Holmes's request for the option, after replying that H.M.G. would have no objection, in extracting a letter from the Sultan that he would take no decision until he had considered a similar request which he had received on behalf of APOC from Sir Arnold Wilson (see Note 11). Although this prevented the immediate success of Holmes's negotiations, his option-contract on behalf of EGS covering the El Hasa province was subsequently agreed and signed five months later on 6th May 1923 when at Ibn Saud's request he revisited Hofuf to finalise the matter. There is little doubt that Ibn Saud never had any intention (see Note 21) of agreeing APOC's request for the same option. That Company's close association with H.M.G. was so disagreeable to him, not because of any enmity towards H.M.G. but because he disliked such a political element in what he wanted to be a purely commercial transaction, that before making his agreement with Holmes he required the latter to put in writing that 'the Syndicate shall not sell to the Anglo-Persian Oil Company Ltd. either as to the whole or part thereof, any oil or mineral concession or concessions that may be granted by Your Highness to the Eastern and General Syndicate Ltd.'

In October/November 1922 Holmes had discussed the prospects of his getting an oil concession in Kuwait with the Yatim family in Bahrain, and also with his friends among Ibn Saud's entourage at Hofuf and Ojair, including Amin Rihani (who was as well known to Shaikh Ahmad in Kuwait as he was to Ibn Saud) and a prominent Kuwaiti, Seyed Hashem, then employed in Ibn Saud's Amiri diwan. It was on their advice that, after returning to Bahrain from signing his El Hasa option-agreement with Ibn Saud, he despatched to Shaikh Ahmad on 9th May 1923 the telegram which initiated his Kuwait negotiations during the next eleven years as described in this Record.

Soon after his return to England on conclusion of the Kuwait concession negotiations in December 1934, Holmes received a letter dated 5th January 1935 from Shaikh Ahmad appointing him to be the Shaikh's London Representative as provided for in the Concession Agreement. His duties in that capacity were thereafter to be his main business occupation, and both by regular correspondence and frequent visits to Kuwait he maintained his close friendship with the Shaikh and connections with Kuwait affairs for the rest of his life. His other main occupation was his farming estate near Chelmsford in Essex, where a life-long interest in agriculture and animal husbandry, inherited from his child-

hood in New Zealand, had full scope in his later years. Holmes (with his wife, who survived him) was never happier than when showing his gardens and prize cattle to his friends, especially those from Kuwait; who included Shaikh Ahmad during a visit to England in 1935 and the writer of this Record on many occasions. Always a hospitable and generous host, with a wealth of remarkable anecdotes and stories drawn from his worldwide career and related with gusto over large cups of coffee (his favourite beverage 'but only the best coffee is good enough'), Holmes's robust physique and lively mind succumbed at last in February 1947 to an internal malady which had increasingly affected him from 1922 onwards. At his funeral an enormous wreath of flowers from Shaikh Ahmad was conspicuous.

Holmes's character, personality, and unique belief from 1922 to 1932, contrary to the full weight of expert world opinion, in the oil potentialities of Arabia, were briefly described shortly after his death in an obituary notice (contributed by the writer of this Record) in The Times *of London of 5th February 1947 as follows:*

Major Frank Holmes

Mr A. H. T. Chisholm writes:

Major Frank Holmes, who died last week, was uniquely responsible for discovering the vast petroleum resources of Arabia, whose development is of such current interest to-day. Though avoiding publicity in this country he was an outstanding British personality in the Middle East, especially among the Arab sheikhs of the western shore of the Persian Gulf, who appreciated both the formidable personality of this rugged New Zealander and the great riches which the initiative of 'The Father of Oil' (Abu-el-Naft), as they called him, brought to their coffers. A mining engineer by profession, Holmes staked his own opinion against that of experts in petroleum geology (who had pronounced Arabia 'oil-dry') when he obtained an oil concession for Bahrein Island in 1923. For over five years thereafter he sought in vain to find a British or American oil concern to back his fancy and exploit the concession. Eventually he succeeded, and the discovery of a major oilfield in Bahrein in 1932 was rapidly followed by still greater discoveries in Saudi Arabia, Kuwait (where Holmes himself was concerned in obtaining a concession from his old friend Sheikh Sir Ahmed as-Subah in 1934), and elsewhere. Of powerful physique, blunt speech, and great strength of character, Holmes had also those qualities of generosity, friendliness, and frankness which Arabia most admires.

---

16. *Eastern & General Syndicate Ltd. was incorporated in London on 6th August 1920 with an authorised capital of £50,300; of which £15,400 had been issued by 9th June 1922 when its first Directors Report and Accounts for the period ending 31st August 1921 was sent to shareholders.*

The Company's principal shareholders were the Chartered and General Exploration and Finance Co. Ltd., Allen and Hanburys Ltd. (Pharmaceutical Chemists) and Mr B. M. Messa (an Aden merchant). Its first directors were Mr (later Sir) Edmund Davis (Chairman of the Chartered Company and director of many other mining and railway companies); Mr F. W. Gamble (a director of Allen and Hanburys), Messrs. E. W. Janson and P. C. Tarbutt (both of P. C. Tarbutt & Co., Consulting Engineers, and directors of various mining companies) and Mr J. E. H. Lomas (mining engineer and director of mining and rubber companies).

According to the first Directors Report, the Company had been formed 'to deal with concessions in Arabia, with which object Major Frank Holmes and Lt-Cdr C. E. V. Crawford had proceeded to Arabia to look into several propositions that had been suggested. The Company has opened in Aden, under efficient European management, a chemist's and druggist's business [see Note 15] from which, after the initial period, a satisfactory revenue is expected'.

---

17. *Mohammed Yatim, a member of an important merchant family of Bahrain who was Holmes's interpreter-assistant from 1922 onwards, and Mullah Saleh, Shaikh Ahmad's secretary (and Secretary of his Council of State) throughout his long rule (he had previously been secretary to the Shaikh's two predecessors), were already well acquainted in 1923.*

When Holmes despatched his telegram from Bahrain of 9th May 1923 to Shaikh Ahmad in Kuwait, Yatim simultaneously sent the following telegram to Mullah Saleh:

I have today posted to you an important letter. I hope to arrive in Kuwait one week later to discuss important matters dealing with oil concessions. Strongly urge that you advise His Excellency the Sheikh to see the liberal terms offered by the Company which has been successful with

His Highness Ibn Saud before giving any oil concessions to other companies. I bring an offer from this Company based on the same liberal terms for His Excellency's consideration. I consider this question most important and vital for your country. May I request you to secure for me suitable furnished quarters during my stay.

On Holmes's arrival a week later in Kuwait he was introduced to Saleh by Yatim, and thereafter until the end of the negotiations in 1934 they were his constant companions and most active allies in Kuwait. They were also the most influential so far as Shaikh Ahmad was concerned, Saleh through his official position (as this Record of the negotiations shows on many occasions) and Yatim because the Shaikh appreciated his social and convivial habits. In March 1925 the fact that Yatim 'has engaged a band and some dancers and is giving private invitations to the Shaikh and to different leading people nearly every night at his house' was gravely recorded in APOC correspondence between Kuwait, Abadan and London as a significant factor in the propaganda conflict then in progress between the pro-EGS and pro-APOC parties in Kuwait.

Mullah Saleh's influence at that same period was regarded as so important that the APOC, whose policy, contrary to that of Holmes, was never to offer largesse or future benefits to potentially influential supporters, made a unique exception in his case in 1925, and authorised overtures for his support. But these came to nothing and his loyalty thereafter to his friends Yatim and Holmes remained constant. Saleh's valuable assistance was rewarded by various gifts from EGS during their negotiations, with the promise of a future pension if they were successful. Although this came to nothing when EGS's connection with the negotiations ceased on the formation of KOC in 1934, Mullah Saleh's influence and assistance remained at Holmes's disposal, and was to be especially useful to him during the last few months of 1934. Early in 1935, Shaikh Ahmad appointed Mullah Saleh's son Abdullah to be his Official Representative in Kuwait for matters connected with the KOC as provided for in Article 6 of the 1934 concession agreement.

Mohammed Yatim had no further connection with Holmes, or with the Kuwait Oil Company, after the signature of its Concession Agreement in December 1934. At one time in 1935, because of what he considered to have been inadequate fulfilment of promises made to him during the negotiations by Holmes, he considered taking legal proceedings against the latter but eventually took no action.

---

18. *Minute of 'a meeting between Sir A. T. Wilson and Shaikh Ahmad al-Jaaber as-Sabah, Ruler of Kuwait, on the 2nd June 1923', written by Mirza Mohammed (see Note 119) who accompanied Sir A. T. Wilson at the meeting:*

Sir A. T. Wilson asked the Shaikh if he had had time to peruse the draft of proposed agreement between the Shaikh and the d'Arcy Exploration Co. and to offer such remarks as he may have.

The Shaikh replied in the affirmative, adding that he will have to make many observations and suggestions for modifications. He pointed out that, seeing that the agreement affected the vital interests of his country and of its people, it would be unfair for him to expedite the conclusion thereof without due consideration and consultation. He would, on the other hand, be doing an injustice to the Company if he rejected any of their proposals offhand. Furthermore, he would be guilty of telling the Company an untruth if he promised to undertake anything which was beyond his power to perform. 'This affair is of a commercial nature,' he went on to say, 'and it needs considerable deliberation in order that full justice may be done to both parties.'

Sir A. T. Wilson stated that, in view of his friendship for the Shaikh and his loyalty to His Majesty's Government, he would in no wise attempt to force the Shaikh's hands by bringing the least pressure to bear upon him, and that the latter could consider the agreement with due care. Colonel Knox had already wired for a ship for his voyage and intended leaving Kuwait on Monday night after dining at the Shaikh's palace; and Sir A. T. Wilson would prolong his stay at Kuwait till Thursday and leave the place by the Slow Mail if the Shaikh was inclined to think that it would be possible for him to discuss the terms by that time. If the Shaikh did not think so, however, Sir A. T. Wilson would leave Kuwait in company with the Political Resident.

The Shaikh answered that, while he wished to settle the matter with all possible speed, he would beg of Sir Arnold not to urge upon him to hasten a decision. 'I will not make any contract with any Company except the one presided over by you, and of this you may rest assured.' He added that there were many modifications for him to suggest; that he would read one article of the agreement after another, suggest such alterations as may be required by the interests of his country and submit such new articles as he may deem fit.

Sir A. T. Wilson inquired if the Shaikh could think of discussing any of the articles in this meeting, so that those provisions on which both parties agreed may be known and those over which they disagreed may be reserved for future discussion.

The Shaikh rejoined by saying that it would surely take some time to do so and that he would request an ample respite to deliberate on the different points. 'But I tell you', he remarked, 'that I shall not enter into an agreement with any other company without your information.'

Sir A. T. Wilson asked the Shaikh if this assurance could be given to him in writing. The Shaikh replied to this:—'I give you my word of honour that I will not conclude my contract with any Company without informing you, at first, of the terms suggested by them. This word I give you in view of the friendship and regard I entertain for you personally, as I have regarded you as my father or even more.'

While thanking the Shaikh for the above assurance, Sir A. T. Wilson reiterated his remark with reference to his departure, and stated that his purpose in inviting the Shaikh to discuss such of the terms as he could was to avoid a further visit by him to Kuwait, as he, Sir Wilson, had to proceed to Baghdad and perhaps to Teheran and that he did not presume that it would be convenient for him to visit Kuwait again.

The Shaikh observed that he was aware of the fact that Sir Wilson was very busy with important matters, but that the Shaikh himself would be visiting Basrah in a few days, say a fortnight, and would take an opportunity of discussing the matter with Sir Wilson there, failing which he would no doubt refer to the representative of the Government at Kuwait. 'But you should not have the least doubt,' he added, 'that I will not enter into any agreement whatever with any Company except the one of which you are the head, without informing you beforehand.'

Sir Wilson explained to the Shaikh the addition made by him to article 10 of the proposed agreement. The Shaikh appeared to be pleased with the same and observed that it was useful for him. He repeated to Sir A. T. Wilson his intention of visiting Basrah and meeting Sir Wilson at Mohammerah, but he emphasised again that it was not advisable to discuss the conditions of the agreement without due meditation on his part.

---

19. *Sir Arnold Wilson's letter to the Managing Director of APOC, London:*

Mohammerah,
Persian Gulf

Confidential 17th August, 1923.

Dear Sir,

With reference to our letter of July the 16th 1284: Koweit. On the writer's return to Bushire, Colonel Knox kindly showed him further papers on this subject.

On the 29th June the Political Agent wrote to the Resident that the following was the gist of some of the more important Clauses (26 in all) of the Concession which the Shaikh desires to give the Eastern General Syndicate:

(1) Concession for 70 years.

(2) Company to pay the Shaikh £3,000/- per annum for protection.

(3) Company to pay £2,000/- cash down when work is started.

(4) If work is abandoned after 35 years, all machinery belongs to Shaikh; if before 35 years, to Company.

(5) Disputes to be settled by arbitration in London.

(6) 1% export duty to be charged on all oil leaving the country.

(7) Import duty to be charged on provisions, &c., but not on machinery.

(8) The Company to pay Shaikh 20% of their net profits. Shaikh to have option on shares in Company up to 20% of total amount issued.

(9) Salaries of all officials provided by Shaikh shall be paid by Company, i.e. all guards, &c., notwithstanding the £3,000/- paid for protection. Shaikh not responsible for raids, forays or attacks, but will merely assist Company to the best of his ability.

The Resident remarks that these terms were, ostensibly, so much more favourable than those that have hitherto been discussed between the A.P.O.C. and the Colonial Office, that it was not surprising that the Shaikh should now be seeking for excuses to justify his preference for the Eastern General Syndicate. A leading merchant of Koweit, Hamad es Sagar, visited Ahwaz on the 10th of August to see H.E. Shaikh of Mohammerah. Khan Bahadur Mirza Muhammad, who was present, informs the writer that Hamad informed the Shaikh of Mohammerah that the Shaikh of Koweit had considered the conditions put forward by the A.P.O.C. and found them far less satisfactory than those proposed by the E.G.S. and he was therefore anxious to close with the latter. He added that the Shaikh of Koweit was in this matter almost entirely in the hands of the Customs Mudir and of his Secretary, who are both in league with Holmes.

The Political Resident has written to the Shaikh of Koweit informing him definitely and finally that the Concession with the E.G.S. will not be sanctioned by His Majesty's Government, who do not regard this firm with approval. He is of opinion that if we sit tight and do nothing for some time, and then approach the Shaikh afresh with somewhat more favourable terms, we shall be able to overcome the Shaikh's opposition; but, as he remarks, the market has been spoilt by Major Holmes's proposals.

We venture to take this opportunity of pointing out that there seems little reason to think that the personality and past career of your General Manager or the Government share-holding in the A.P.O.C. have really been factors of importance in deciding the attitude of the Sheikh of Koweit. It will be clear to you from earlier correspondence that the Shaikh of Koweit up to the beginning of this year was willing and anxious to close with the company, subject only to a reconsideration of the amount of royalty per ton (vide our letter of January, 164). He had seen Sir Arnold Wilson and had expressed his anxiety to come to an agreement as soon as possible in 1922: and Major More the Political Agent, emphasised to Sir Arnold Wilson both verbally and in writing that it would not be difficult to get the agreement through if we pressed the Shaikh and met him to some extent. Your instructions on the other hand (the latest of which were dated the 14th of March, No. 317), were that it was not desired that we should take active steps in the matter of the agreement, and we had no option but to comply.

Yours faithfully,
For ANGLO-PERSIAN OIL CO. LTD.
(Signed) A. T. Wilson
General Manager

---

20. *The Neutral Zone was an area of approximately 2,500 square miles adjoining the southern boundary of Kuwait, its other boundaries being the Gulf coast to the east and Saudi Arabia to the west and south. This area was demarcated and agreed on 2nd December 1922 at the Ojair Conference (see Note 22) as one in which Kuwait and Saudi Arabia would have equal rights. Since 1970, following agreement between the two countries each to have sole jurisdiction over half the area (Kuwait the northern half, Saudi Arabia the southern), it has been known as the Divided (or Partitioned) Zone.*

21.

EASTERN AND GENERAL SYNDICATE LIMITED
Cablegram received from Major Frank Holmes dated
Bahrein 13th May 1923, received 14th May 1923.

Sultan Ibn Saud signed delivered to me the Hasa Concession fullstop The Neutral Zone Concession it is not yet signed Sultan Ibn Saud arranging with Koweit expect joint signatures on an early date fullstop Have given undertaking Sultan Ibn Saud not to sell the whole or part of any concession to The Anglo-Persian Oil Co. Ltd., fullstop Referring to your telegram of 8th in respect of The Territory of Nejd other than the Hasa Concession Area under The Hasa Concession total amount £7,000 £500 due after 60 days from the 6th of May fullstop If you could credit Sultan Ibn Saud this amount with the Eastern Bank Ltd. Baghdad without waiting until expiration of 60 days it would assist me materially to secure very strong hold over not only The Territory of Nejd other than the Hasa Concession Area but also over the Koweit territory(ies) and of Bahrein fullstop

---

22. *The Ojair Conference, 22nd November–2nd December 1922, was held to settle the boundaries, then largely undefined, between Iraq and Saudi Arabia, and between Kuwait and Saudi Arabia. Iraq, then under British Government mandate, was represented by Sir Percy Cox (see Note 3), Ibn Saud represented his own country, and Shaikh Ahmad was represented by the British Political Agent at Kuwait (Major More). Major Frank Holmes's presence during the Conference (see Note 15), enabled him to make considerable progress towards obtaining the El Hasa oil concession from Ibn Saud and also useful contacts for his subsequent concession negotiations in Kuwait. All these aspects of the Conference are fully described by Lt-Col Dickson (see Note 39), who was present throughout its proceedings as interpreter-assistant to Sir Percy Cox, in his book* Kuwait and her Neighbours *(Allen & Unwin, 1956) pages 267–80.*

Kuwait's new boundaries with Saudi Arabia were settled at the Conference by an Agreement dated 2nd December 1922, sealed by Ibn Saud and signed on his behalf by one of his officials, and signed on behalf of Shaikh Ahmad by Major More. The new boundaries, as well as setting up the Neutral Zone (now the Divided Zone; see Note 20) to the south of Kuwait, resulted in several thousand square miles of territory in that direction, hitherto claimed or partly claimed by Kuwait, being included in Saudi Arabia.

Lt-Col Dickson was also present shortly after the Conference when Sir Percy Cox informed Shaikh Ahmad in Kuwait that he had been obliged to give away to Ibn Saud nearly two-thirds of the total territory hitherto claimed by Kuwait (reducing it to its present approximately 6,000 square miles). When the Shaikh asked him why he had done this without even consulting him, Sir Percy replied that had he not conceded the territory, Ibn Saud would certainly have taken it, if not more, by force of arms; whereto Shaikh Ahmad rejoined that he believed Britain had fought the last (1914–18) war in defence of the rights of small nations. Although, thus faced with a fait accompli, *Shaikh Ahmad agreed to add his signature to the agreement, in Lt-Col Dickson's considered and very expert opinion 'the young Shaikh Ahmad Al Jabir, scarcely a year on the throne and very impressionable, received a blow to his faith in Great Britain from which he never really recovered'.*

---

23. *Letter from the Colonial Office to the Acting Resident in the Persian Gulf (the Duke of Devonshire was then Secretary of State for the Colonies):*

Downing Street,
6th September, 1923.

Sir,

I have the honour to acknowledge the receipt of your despatch of the 20th June, No. 299, transmitting the text of a draft Agreement submitted to the Sheikh of Kuwait by Sir Arnold Wilson on behalf of the D'Arcy Exploration Company and your despatches No. 378 and 405 of the 13th and 24th of July on the same subject. Before commenting upon the provisions of this draft Agreement, I desire to explain in some detail the attitude of His Majesty's Government towards this Concession and towards the question of the grant of Oil Concessions on the Arabian Littoral of the Persian Gulf in general. For this purpose it will be convenient to review briefly the circumstances which have led up to the present situation. In this despatch I do not propose to deal with the question of the Nejd Concession, since this will form the subject of a separate despatch.

2. As you are aware, His Majesty's Government as part of a considered policy have obtained from the Sheikhs of Kuweit and Bahrein and from other native rulers in the Persian Gulf, formal undertakings to the effect that they will not grant any concessions for the development of Oil in their territories to any person, or persons, not approved ('appointed') by His Majesty's Government. The object of obtaining these undertakings was twofold. In the first place it was intended thereby to protect the rulers themselves against the pernicious activities of unscrupulous Concessions hunters of the type which preyed so successfully upon the Turkish Empire before the war, and secondly it was desired by this means to restrict the grant of Concessions to reputable British firms, and thus to prevent the infiltration of foreign influence in the Persian Gulf, which was regarded as politically undesirable.

3. In order to avoid advantage being taken of the native rulers' inexperience in such matters, it was laid down that applicants for Concessions should not negotiate direct with the Sheikhs, but should, in the first instance seek His Majesty's Government's permission to apply for a Concession upon approved lines, and, having obtained such permission and agreed upon terms which His Majesty's Government would be prepared to recommend to the Sheikhs as acceptable, should then open negotiations either through the Political Resident or through the medium of the Political Agent concerned.

4. In the summer of 1921 the Anglo-Persian Oil Company expressed a desire to secure a series of exclusive prospecting licenses for Oil over the whole Arabian Littoral of the Persian Gulf. The Company's proposal was referred to this Department by the India Office and, as my predecessor was satisfied that the Anglo-Persian Oil Company were in a position to undertake and carry out satisfactorily the development of the oil resources of that area, the proposal was accepted in principle. After considerable correspondence and negotiation the mainheads of an Agreement covering exploratory and prospecting rights in the territories of Bahrein and Kuwait were drawn up in Agreement with the Company by the interested Departments of His Majesty's Government, and were communicated to the Political Resident for submission to the Sheikhs concerned. These draft heads of Agreement, which were adapted from the usual Colonial Model, were so framed as to provide the nucleus of an Agreement, which would, in my opinion, have been in every way favourable to the interests of the Sheikhs, in that they were designed not only to provide for the expeditious and systematic development of the oil resources of their territories, but also to produce for the Sheikhs very handsome royalties.

5. However, before any agreement could be concluded Major Holmes, avowedly, and probably actually, in ignorance of the necessity for obtaining the prior permission of His Majesty's Government, approached the Sheikhs of Kuwait and Bahrein direct and succeeded in inducing them to grant or undertake to grant to the Eastern and General Syndicate, which he represented, exclusive Concessions for the exploitation of oil and other minerals in their territories.

6. These concessions contain several unsatisfactory features (notably the provision whereby the Sheikhs would receive a percentage of the shares of the exploitation Company, in the place of a definite royalty based upon output) and they would in any case require to be substantially modified before I should feel justified in advising the Sheikhs to accept them. But quite apart from this consideration, I felt very strongly that the Sheikhs should not be allowed to disregard their solemn undertakings with impunity, nor that 'unapproved' firms, by approaching them direct, should be permitted to secure Concessions the terms of which had not previously been submitted to His Majesty's Government; particularly in view of the fact that the Anglo-Persian Oil Company would thereby have directly suffered as a result of having carefully observed all the conditions required by His Majesty's Government.

A telegram was therefore sent to Colonel Trevor on the 30th June instructing him to inform the Sheikhs that His Majesty's Government would not approve of the Concessions which contrary to their undertakings they had granted to Major Holmes, and that the Eastern and General Syndicate was not a firm approved by His Majesty's Government.

7. By this latter statement it must not be understood that my intention was to give the impression that His Majesty's Government had any objection in principle to the Eastern and General Syndicate except on the grounds of the irregular manner in which their Concessions had been obtained. On the contrary, so far as His Majesty's Government are aware the Eastern and General Syndicate although possessing neither the wide experience of oil production nor the efficient organization and financial strength of the Anglo-Persian Oil Company are a substantial and reputable firm.

8. In view of the attitude taken up by His Majesty's Government the Eastern and General Syndicate have now volunteered to modify their Concessions in such manner as to conform to the wishes of His Majesty's Government. These Concessions have not yet been closely examined in the Colonial Office but if in fact the Syndicate show themselves willing to accept such suggested modifications as may upon examination prove to be necessary, and moreover are able to satisfy me that they are in a position and intend to carry out to the full the obligations which they propose to undertake towards the Sheikhs, I shall no longer, subject to my remarks in the following paragraph, feel any strong objection to their candidature.

9. On the other hand, the Anglo-Persian Oil Company both by their priority of application and by their due observance of the correct procedure, have in my opinion established an undoubted claim to His Majesty's Government's prior support, provided, of course, that they are prepared to offer to the Sheikhs terms at least as favourable as those submitted by the Eastern and General Syndicate. Consequently I am not prepared to recognize the Concessions granted to the latter Company unless and until the Sheikh of Kuwait has set forth his objections to the draft Agreement submitted to him by Sir Arnold Wilson: has explained his statement (reported in the telegram of the 13th July from the Political Agent, Kuwait, a copy of which accompanied your despatch of the 30th July, No. 378) that the draft Agreement, is entirely unacceptable to himself and his people: has satisfied me of the validity of those objections: and has further stated precisely in what manner he wishes that Agreement modified.

10. If when this information has been obtained from the Sheikh, his proposals are not accepted by the Anglo-Persian Oil Company I shall then be prepared to approve of the Kuwait Concession being granted to the Eastern and General Syndicate, provided that the terms thereof are no more favourable to the Company than those rejected by the Anglo-Persian Oil Company, and that the Concession is recast upon approved lines.

11. You should therefore communicate to the Sheikh the revised draft Agreement contained in the later paragraphs of this despatch and inform him that His Majesty's Government consider it a fair basis for discussion with the Company. You should also call upon him to formulate his objections if any to this revised draft Agreement, and to state in the clearest terms in what way he proposes that it should be modified. The Sheikh's proposals should then be communicated to me and also to the local representative of the Anglo-Persian Oil Company, who would then decide, in consultation with his principals, to what extent these proposals may be accepted. I am also prepared, provided the alterations proposed do not involve any large question of principle, and provided that they are acceptable to the Company to authorize you at your own discretion to approve on behalf of His Majesty's Government the grant of the concession by the Sheikh without further reference to myself. In dealing with this matter you should proceed in consultation with the local representatives of the Company as proposed in the last paragraph of your despatch No. 405 of the 24th of July.

12. The procedure which I have set out above as regards Kuwait, should also be followed, *mutatis mutandis* with regard to Bahrein. You should proceed as you propose in your despatch No. 406 of the 24th of July and, when you think that the time is ripe for further action, the Sheikh should be informed that His Majesty's Government have not found it possible, in view of the prior claims of the D'Arcy Exploration Company, to approve the concession granted by the Sheikh, contrary to his predecessor's solemn undertaking, to the Eastern and General Syndicate, and that they desire him first of all to consider the application of the former Company. In the event however of the Company not being prepared to offer the Sheikh terms which are as favourable in the opinion of the Sheikh himself and of His Majesty's Government as those offered by the Syndicate, His Majesty's Government are prepared to withdraw their opposition to the grant of a concession to the latter.

13. I will now proceed to examine the Anglo-Persian Oil Company's draft Agreement, the form of which in my opinion, leaves much to be desired.

I fully sympathise with the Company's desire to regularize their position once and for all, which has led them to provide in one Agreement for the grant of exploration and prospecting licenses and a mining lease, but their draft has in consequence lost in precision and clarity, and appears to charge the Sheikh with large and ill-defined liabilities over a long period of years.

This is illustrated by the wording of Article 1, which at present gives the impression that the exploration license is to remain in force for a period of 60 years. This is presumably not the intention, and in any case directly conflicts with Article 2.

I should therefore not find it possible to inform the Sheikh that I am satisfied that his rights and interests would be adequately guarded by the existing draft Agreement.

The Company can, however, attain their object equally well, and without imparting ambiguity into the Agreement, by adopting the expedient of framing an agreement which provides for the grant of exploration and prospecting rights, and also irrevocably binds the Sheikh—subject to the fulfilment of certain specified conditions—to grant at the request of the Company (on or before the termination of the prospecting period) a mining lease upon terms set forth in detail in a document scheduled to the agreement. The advantage of this method, which is generally employed in similar cases in the Colonies, is that it defines without ambiguity the exact relations between the two parties in all the various stages: moreover it is by this arrangement open to the Company at the expiry of the prospecting period either to withdraw from the whole undertaking without incurring any obligations in respect of exploitation or, if oil is discovered in profitable quantities, to obtain a mining lease upon previously agreed terms.

14. This view should be represented to the Local Representative of the Company, and he should be pressed to substitute for the present draft Agreement a revised draft modified on the following lines.

He should be informed that if the Agreement is revised in this way I shall have no hesitation in commending it to the Sheikh.

Article I should read:

'The Sheikh grants to the Company by these presents an exclusive exploration licence for a period of two years from the date of this agreement, whereby the Company shall be entitled to search for natural gas, petroleum, asphalte and ozokerite throughout the whole extent of the territories under his control and to enjoy the privileges set out in the first schedule to this agreement; and he undertakes on behalf of himself and his successors to grant to the Company the further exclusive licences and privileges set forth below.'

Article II

The Exploration licence referred to in the preceding Article may be extended for a further period of two years . . . and so on as in the existing Article II.

Article III

The Sheikh hereby undertakes on behalf of himself and his successors to grant to the Company on application after the expiry of the above mentioned period . . . and so on as in existing Article III.

Add at the end: 'Under the terms of this licence the Company shall enjoy the privileges set out in the second schedule to this agreement.'

Article IV

The Sheikh hereby undertakes on behalf of himself and his successors to grant to the Company on the expiration of the prospecting licence a Mining Lease over an aggregate area not exceeding 100,000 acres divided into not more than three blocks on the terms and in the form of the Lease attached as the third schedule to this Agreement.

Article V

First sentence as first sentence of Article XIV. Delete 2nd sentence of the latter article and substitute 'They shall not be liable to pay any further sum to the Sheikh unless they receive a Mining Lease from him, but if they receive such a mining lease they shall pay to him the sums provided in the lease but no more.'

Article VI

Present Article XV.

Article VII

Present Article XVI.

Article VIII

Present Article XVII.

FIRST SCHEDULE

Privileges to be enjoyed by the Company under the Exploration Licence.

1. Free access for their agents and servants necessarily employed by the latter, to all parts of the territory under the control of the Sheikh whether private or public property, saving only sacred buildings, shrines and graveyards.
2. Free use of water for the same.
3. Free use of fuel lying on land and property of the Sheikh for the same.
4. The right to purchase fuel and food supplies of every kind for man and beast, being the private property of the Sheikh's subjects, at current market rates, which rates in default of agreement between the parties shall be fixed by the Sheikh or his agents, subject to the right of the Company to appeal to the British Political Agent at Kuwait if they find it necessary.
5. Free importation of all material, apparatus and machinery of every kind necessary for the work carried out by the Company, including the reasonable personal requirements of employees, and office requirements.
6. All protection in the Sheikh's power from theft, highway-robbery, assault, wilful damage and destruction.
7. An extension of the period of the licence if the Company are prevented by causes beyond their control from carrying out the work of exploration, the length of such extension to be fixed by agreement between the parties or failing such agreement by the Resident in the Persian Gulf.

SECOND SCHEDULE

Privileges to be enjoyed by the Company under the Prospecting Licence.

1. 1 of 1st Schedule.
2. The right to carry out in any part of the said territory such works as may be necessary for the

purpose of prospecting. If this work is carried out on uncultivated land the Company shall not be bound to make any payment to the owner of the land but shall restore the land as far as is possible to its previous state unless it is covered by a Mining Lease subsequently granted. If it is carried out on cultivated land the Company shall pay a fair rent for the land occupied. 3, 4, 5, 6, 7, 2, 3, 4, 5, 6 of 1st Schedule.

8. An extension of time on the same lines, *mutatis mutandis* of 7 of 1st Schedule.

9. The right to win up to 100 tons of oil free of payment and further quantities of oil on payment of the royalty per ton provided in the Mining Lease, but on condition that the Company shall apply for a mining lease in respect of each area in which work is proceeding, as soon as more than 100 tons of oil are won from one single bore-hole within it.

THIRD SCHEDULE
DRAFT MINING LEASE

Article I

The Sheikh hereby grants to the Company on behalf of himself and his successors the exclusive right for a period of fifty-five years to prospect and drill for, extract, treat, refine, manufacture, transport and deal with petroleum products, naphtha, natural gases, tar asphalt, ozokerite, and other bituminous materials within the area or areas described in words in the first Schedule to this lease and delineated upon the map attached as the second Schedule hereto. Such right, however, shall not include the exclusive right to sell such products within the leased area. He also grants the Company the exclusive right to construct and operate refineries and storage tanks within this area and also the right, but not the exclusive right to construct, erect and operate pipe lines, refineries and storage-tanks, railways, wharves and jetties, tramways, roads, buildings, machinery and telegraph apparatus of all kinds in any part of the territory under the control of the Sheikh so far as may be necessary for the purpose of their business.

The Company shall have the right to prospect for, collect and use free anywhere within the leased area, but not to export, or sell stone, gypsum, salt, sulphur, clay, wood and water, whether from rights or springs, for the purpose of their work.

The selection of routes and sites for such works and the course of their pipeline shall rest with the Company.

Article II

As Article VII of the original agreement, but add at the end:

It is however understood that save within the leased area the Company shall be entitled to acquire only such land as is necessary for their pipe-lines, refineries, offices and other works and that save in so far as it is necessary in order to enable their pipe-lines and railways to pass freely to their destination they shall not occupy any land outside the leased area which might otherwise be leased for the production of oil to a third party, unless they can satisfy the Sheikh that no other land suitable for their purpose is available.

Article III

As Article VIII of the original agreement, *mutatis mutandis.*

Article IV

As Article IX of the original agreement but for the third sentence substitute:

'If the Company's work is delayed by such events as are mentioned in this article then the period of such delay shall be added to the period for which this lease is granted and to the period for which the royalty mentioned in Article VII hereof shall hold good and the fixed rent of Rs. 30,000 shall not be payable in respect of such delay.'

Articles V, VI, VII and VIII. As Articles X–XIII of the original agreement, subject to correction of the numeration of the Articles quoted therein.

Article IX

The Company shall carry on work within the leased area with all proper zeal and diligence and shall maintain at least two rigs in continuous operation, except in so far as they may be prevented by causes beyond their own control. In the event of the Company failing within 5 years of the commencement of this lease to declare that oil has been found in commercially payable quantities the Sheikh shall have the power to call upon the Company either to give forthwith the undertaking mentioned in the preceding article or to abandon the lease.

Article X

In the event of the Company failing within six months of the end of any calendar year to pay to the Sheikh the royalties due in respect of that calendar year or failing, save for causes beyond their own control, to carry out their other obligations under this lease the Sheikh shall have the power to terminate the lease in which case the provisions of the last sentence but one of Article II (i.e. the last sentence of Article VII of the original Agreement) shall apply.

Article XI

The rights conveyed by this lease shall not be conveyed to a third party without the consent of the Sheikh acting with the advice of the Resident in the Persian Gulf.

Article XII

A repetition of Article VIII of the new Agreement (i.e. Article XVII of the original agreement).

The lease would be accompanied by two schedules the first containing a description in words of the leased area and the second a map of the same.

15. I have communicated a copy of this revised draft Agreement to the Company in this country and have informed them that I am prepared to commend its terms to the Sheikh for acceptance, but that I shall not be able to press the Sheikh to sign an agreement in these or any revised terms which you may propose if he is in a position to satisfy me that he has received an offer which is definitely more favourable from the Eastern and General Syndicate.

I have the honour to be,
Sir,
Your most obedient
humble servant,
(Signed) Devonshire.

---

24. *Extract from the* Baghdad Times, *13th May 1924:*

Among the visitors to Baghdad at present is Major Frank Holmes, the consulting engineer who was primarily responsible for the granting of the Hassa oil concession, to a British syndicate, by Sultan Ibn Saud. We understand that, with the consent of the British Colonial Office, Major Holmes has just concluded a further agreement with the Sultan of Nejd and the Shaikh of Kuwait for oil rights over an area of 2000 square miles known as 'the Neutral Zone' and situated between the provinces of Hassa and the territory of the Shaikh of Kuwait. This Neutral Zone was created by Sir Percy Cox at the Ojair Boundary Conference in December, 1922. The Eastern and General Syndicate has now commenced operations on both the Hassa and the Neutral Zone concession. A geological survey is being made under the direction of Dr. Heim, a noted Swiss oil geologist. Work was begun in the middle of April by Dr. Heim and his three assistants. It is hoped that, at the end of the hot weather, several test bore holes will be sunk in the Neutral Zone and in the Northern part of the Hassa Concession. Arrangements have already been made for this work to be done. It is reported that Major Holmes is now taking an active interest in the proposed development of the Iraq oil-fields, and that in this connection he will shortly submit a definite project to the Government of Iraq. In the opinion of many people, however, the Turkish Petroleum Syndicate is regarded as having a prior claim to oil rights in Northern Iraq.

---

25. *Dr Heim's geological survey, dated 5th September 1924. The maps, plates, etc. referred to are not reproduced here, nor the detailed material on El Hasa and Bahrein.*

Geological Report No. 1.
THE QUESTION OF PETROLEUM IN EASTERN ARABIA
(KOWEIT, HASA, BAHREIN)
with 18 plates, maps, sections, and photographs,
by
Dr. Arnold Heim
Docent at the University of Zurich

Content

NOTE 25

INTRODUCTION

The observations for this report have been made between April 20th and July first 1924, commencing and terminating at Koweit. The photographs pl. VIII–XVIII with explanations are roughly in a chronological order and may give some information concerning the manner of travelling. The caravan route is indicated on the map pl. I.

On account of the excellent preparations which were made already before the writers arrival at Koweit, the expedition through the Province El Hasa proceeded without special difficulties. We are especially indebted to His Majesty the Sultan Ibn Saud of Nejd, who sent a guard of 25 men to the disposition of the expedition, to His secretary Dr. A. S. Damlooji and to Major Frank Holmes.

Partly on account of the heat, partly caused by hesitation of His Excellency Sheikh Ahmad of Koweit, not all the excursions in Eastern Arabia could be executed as desired. About a fortnight was lost at Koweit, with the final success however that Djahra and the Oil seepage of El Bohara were reached. The gradual increase of the heat made the continuation of field work nearly impossible. The last trips on foot, camel or donkey were made during strong winds with sand blows up to 113°F. It is true that the Beduins are used to travel in summer time, but only at night, which would be useless for making geological observations.

For a better judgement of the oil possibilities it would have been necessary not only to continue the field work in the above named countries, but also outside of them, as it was originally suggested: to study for instance the formations underlying the Bahrein dome we need to go to the interior of Nejd on one side and to the Persian mountains on the other, and to judge on the oil possibilities of Koweit, a comparison with the Persian field and with Iraq (especially Hit) would be necessary. This report thus is not considered as a final one although the observations and deductions will be given as far as they can be made at present.

Geological Literature

containing special data on Eastern Arabia:

Lit. 1. 1908 *Pilgrim, G. E.*, The geology of the Persian Gulf and the adjoining portions of Persia and Arabia. *Mem. geol. surv. of India*, Vol. XXXIV, part 4. On Bahrein, pl. 12, 13, pag. 112–122, 152–154. Kubbar, Qaru, Farsi islands, p. 143.

Lit. 2. 1909 *H. Höfer*, in Engler-Höfer, Das Erdöl, Bd. II.

Lit. 3. 1914 *Blanckenhorn, Max*, Syrien, Arabien und Mesopotamien. Handbuch der Regionalen Geol. Bd. V. Abt. 4. Verlage Carl Winter, Heidelberg.

Lit. 4. 1919 *Schweer, Walter*, Die türkisch-persischen Erdölvorkommen. Abh. Hamburg. Kolonialinstitut Bd. XXXX. Friederischsen & Co., Hamburg.

Lit. 5. 1921 Geology of Mesopotamia and its Borderlands, compiled by the Geogr. section of the Naval Intelligence Division, Naval Staff, Admiralty, London (out of print).

Lit. 6. 1922 *Pascoe, E. H.*, Geological Notes on Mesopotamia with special Reference to Occurrences of Petroleum. *Mem. geol. surv. of India*, Vol. XLVIII, Calcutta.

Bahrein excepted, on which Pilgrim gives a good description, almost nothing is published on the geology of the region El Hasa–Koweit. The latest geological map, contained in Lit. 5, leaves all of Hasa and further South in plain white.

KOWEIT AND NEUTRAL ZONE.

(Pl. I, II Fig. 2, III, IV, VIII, IX Fig. 1 and 2, XI Fig. 1, XIII Fig. 1.)

Shore of Koweit

The greatest part of the country South of Koweit Bay is a flat desert of sand and gravel. On a few points only along the shore outcrops of rocks are visible: The American Mission at the SW-side of the town of Koweit is situated on a sandy hill, the base of which is formed of a hard, gray, *coarse sandstone* with layers of lumachelle (broken shells) and conglomerate. This sandstone rises about 4 metres above the sea. No indication of oil was found, but it is said that oil was encountered in the foundation of the Mission Buildings.

Apparently the same coarse sandstone forms the shore of Ras al Ajuza at the side of Sheikh Ahmad's Palace, 5 kilometres further NE. There, the dip is 5 to 10° towards NNW (1 in pl. IV Fig. 2). It seems however, that this dip is only apparent, caused by oblique stratification, as already noted at the Mission.

On the latter locality, the sandstone is overlain by a sandy looking light coloured *oolite* of fine grain, which later was found even as far south as Bahrein, and which we propose to call *Jubail Formation*. According to its fossils, amongst which are Gastropods and Pelecypods of recent type, it seems to be subrecent or Pleistocene. The hard low-lying platform on which Sheikh Ahmad's Castle is situated is entirely formed of this oolite.

More important for the natives is another oolitic limestone which is used as the chief *building stone* of Koweit: greenish, dirty-coloured rock full of unregular holes and of white shells of recent type, the most characteristic one being a large Conus. According to the Natives, this rock is brought from the sea at Ras Asheridj W of Koweit.

Djahra

The little town of Djahra with Sheikh Ahmad's Palace is situated at the end of Koweit Bay on flat sandy desert, according to the map of Asia 1:1.000.000 about 30 metres above sea level.

Some 6 km NW of it is the Southwestern end (Dj. Mutla) of the 50 km long escarpment called Djal ez Zor which surrounds the North side of the Bay (Pl. IX Fig. 4). The highest part is situated N of Djahra and is called by the natives Dj. Umm Hasa. The elevation is indicated as 410 feet on the Nautical map, and 127 m on the new Map of Asia. The latter is conform with the writers aneroid measurements.

From the top of Dj. Umm Hasa it was plainly seen that *Djal ez Zor is not a mountain range*, but only the escarpment of a wide table land which gradually, although not quite regularly, dips towards E or NE, ending at Sabiya. The according dip of the strata seems to be a few per mille as a whole, i.e. a little more than the dip of the surface. Regarding the question of oil it may be stated that the stratification is *unfolded* and practically horizontal. On a clear day this is seen already with the telescope from Koweit (Pl. III Fig. 2).

Dj. Umm Hasa is formed of the following strata (numbers according to Pl. 4 Fig. 1):

1. White soft calcareous sandstone.
2. Gray calcareous sandstone with calcareous concretions (nodules of limestone).
3. White sandy limestone.
4. Coarse sandstone with pebbles up to ½ inch, bedded in white tuffaceous clay.
5. 1,5 metres of sandy conglomerate with pebbles of Quartz, red and green siliceous shale etc. up to egg-size.
6. Coarse whitish sandstone.
7. Surface pebbles of Basalt, Porphyry, Quartzite, schists etc. originating from Central Arabia, up to 6″.

Thus, this mountain is formed of a series of over 100 metres of more or less calcareous, light-coloured, soft and partly porous sandstone and conglomerate of horizontal stratification. No trace of oil impregnation was found.

The Oil Seepage El Bohara

Much difficulty was encountered to find this place, as it is known only of a few men in the whole country.

The first attempt, going on land from a sailboat North of Koweit, on an extended flat of sticky mud, was a complete failure. The two old Beduin guides who pretended to know the place were lost, and nearly died from heat and thirst.

The second attempt, after a loss of three days, was successful, the guide being *Mesfer* of Sabiya, the only man who exploits the tar and brings it to the market of Koweit. It is said to be used for sail boats.

The landing place of the sail boat is indicated with + on Pl. II Fig. 2. We arrived there on day-break of June 27th. Until noon Mesfer had brought donkeys from Sabiya. We left at 3^h.p.m. with a strong North wind of 113°F., walking first 2 km over wet salty and slippery mud, thence riding the donkeys (Pl. XVIII Fig. 1). On the foot of the escarpment, the water wells of Atubaj and Mugheira are passed. The first rock seen from a distance is a brick-coloured, *red sandstone* of about 8 metres thickness. At Atubaj, numerous fossils were found, especially a round shaped marine Pelecypod, the shells of which are mostly crushed from crabs. Unfortunately, they are very badly conserved, and do not allow stratigraphic conclusions. *Lucina* sp. ind. is abundant, the snail *Xenophora* rare.

The bed of red sandstone is overlain by 1 metre of greenish clayey sandstone which is covered with a bed of brown to yellowish hard sandstone with cross-bedding. No trace of oil impregnation was noted. The red sandstone seems to dip very gently with some per mille towards E and to disappear below the surface before Sabiya (Pl. III Fig. 2).

A few days before this excursion, our guide Mesfer was robbed out at Sabiya. He thus advised us not to lose any time before reaching the seepage, where we were hidden in a shallow depression between some scrub and had to leave again before day-break. The writer thus was unable to pursue the stratigraphic studies.

The seepage El Bohara was reached shortly before sunset. It is situated in a wide sandy flat, and according to Mesfer just reached and flooded on exceptionally high spring tide. All the way South of the seepage for miles is formed of salty mud and swamp, so that it can hardly be reached on the straight way coming from Koweit. (Pl. II Fig. 1 and Pl. XVIII with explanations.)

The location as indicated on the map Pl. II has been made as accurate as possible. The difficulty was that the air was not clear enough to see Koweit from El Bohara, nor can El Bohara be located from Koweit with the compass. The following directions were taken; indicated without correcting the declination:

Koweit town to Dj. Mugheira	N 6½ E
Koweit town to Dj. Roddha	N 3½ W
Koweit town to Dj. Nuheiden	N 7 W
Seepage Bohara to Dj. Mugheira	E 28 N
Seepage Bohara to Dj. Roddha	N 6 W
Seepage Bohara to Dj. Nuheiden (middle) ...	N 28 W

If it is added yet that the seepage is situated a little more than 1 km South of the telegraph line Koweit–Basra, these indications will be sufficient to find the place again without a special guide. Seen from Koweit, it seems that the seepage is lying directly in front of Dj. Roddha, as indicated on Pl. III Fig. 2.

The main seepage consists of a roughly oval hole of about 8 by 5 metres, dug out of the sandy muddy ground to a depth of some 2½ metres below the surface, the salt water being at 1 metre below it. The pool is filled of concentrated yellowish salt water, the salt of which is cristallized in beautifully shaped white crusts formed of large cristal tablets. This salt is used by Mesfer for his household. From the ground, oil is rising at more or less regular intervals in the form of black drops which rapidly spread on the surface, and become surrounded by thin skins of iris colours made of the lighter parts of the oil. They chiefly emanate from the SW part of the hole, and drive away to the SE-corner, where the lighter parts entirely evaporate and black tar is accumulated. This tar is gathered by Mesfer every 6 months with greased hands, and brought to Koweit. The output is 10 oil tins of 4 gallons each for about every 6 months. This would make 1 Liter daily, or 0,6 cub. cm per minute.

The oil flow seems to be accompanied by salt water. No gas has been observed.

The salt water level of the seepage seems to coincide about with the normal high water of the sea.

Two other smaller holes have been dug out in the direct vicinity of the main seepage: one about 100 feet towards E, another about 200 feet N. In both, earth impregnated with tar was found, but the flow of oil was too small for output.

The tar of El Bohara is very heavy and viscous, and of a deep black. A *chemical test* made at the Federal Institute for Research of Combustibles of Zurich has given the following result.

Density at 15°C. . . . 1,054
Asphalt insoluble in normal benzine . . . high content
Asphalt insoluble in ether-alcohol . . . still higher cont.
Distillation: boiling begins at 270°C.
  from 270 to 300°C. . . . 21,8% of weight of brown, dull liquid oil.
  from 300 to 320°C. . . . 40,3% of weight of dark brown, a little thicker oil.
  Residue and loss of distillation 37,9%, pitch-like.
  The fraction of 300–320 can be cooled to −20°C. without getting completely solid; it thus cannot contain much solid paraffine.
  From 300 to 320° decomposition beginning. H_2S.
Heat of combustion . . . 9175 W E per kilogram.
Content of sulphur . . . 6,3%.

The oiltar of Bohara thus may be characterized as a strongly asphaltized oil with *high content of sulphur*, which does not contain fractions boiling below 250°C to be mentioned. Through distillation about *50% of thin liquid oils for combustion durable in coldness* can be obtained.

The oil of El Bohara thus is of an *asphalt base*, not of a paraffine base. The volatile contents being evaporated, it is impossible to say if the oil at the depth is of high grade or not.

Region South of Koweit

On the caravan route, the following hills were encountered (see pl. I and pl. III fig. 1):

Umm Ridjm. Very gentle sandy hill with some signs of coarse sandstone, probably the same as at Koweit.

Madaniyat (means Sulphur). Three very small hills with indications of coarse weathered sandstone, flint, and white kaoline-like substance. The natives claim to have found sulphur in digging.

Wara. This is the most prominent hill of Koweit region, a round table mountain of perhaps 200 feet elevation above the sea, resembling a little volcanoe from far distance, but exclusively formed of sediments (Pl. III fig. 1, and pl. XI fig. 1, pl. XIII fig. 1). The top is formed of a brown weathered, inside white, coarse sandstone with conglomeratic layers and quartz pebbles, similar to that of Koweit-town. The stratification is horizontal (Pl. IV fig. 3).

However, the underlying beds (g = thin beds of gypsum, m = greenish marl with violet and brown streaks, mostly covered by boulders) are dipping 10 to 15° towards North. It seems thus that the sandstone overlies the clayey base with an angular *unconformity*.

Burgan. This place was considered as of the greatest importance. Amongst the Arabs it is generally believed that an oil seepage is existing, and there was even talk of a dozen of such seepages. But notwithstanding special attention, we were unable to find any trace of oil. It is true that we could spend not even a full day at the place. Our guide Emir Mohammed refused to let us camp at Burgan, claiming that we might be attacked by Beduin tribes. We thus studied first the stratification of Burgan Hill. Therefrom the writer made a rapid trip to the wells of *Subahiya*, where also oil was thought to occur, while Mr. J. L. Popham walked around the Burgan Hills. Subahiya is situated 5 miles S 52° W of the highest point of Burgan, from where it is seen with the telescope as an extremely flat basin of brown colour in the yellow desert. About 10 wells have been dug, the water level of which was 2–3 metres below the surface (April 30, 1924). Much dirt from sheep and camels is washed in, but no trace of oil nor any outcrop of rocks was found.

The hill of Burgan is composed as follows (pl. IV fig. 4):

1. 10 metres of clayey and sandy beds with gypsum
2. 8 metres of coarse, soft sandstone
3. White, tuffaceous and clayey layer with quartz grains.
4. 15–20 metres of coarse white and brown sandstone with cross-bedding, apparently the same as Wara (pl. III Fig. 1)

Seen from the South, it seems that the lower beds are dipping about 5° towards West.

One month after this inspection of Burgan, Dr. A. S. Damlooji and Mr. Alajaji of Hofuf mentioned to have seen samples of tar brought from a place somewhere between Burgan and Subahiya, dug out from the sand at about 2 feet depth. We have no reason not to believe this, but without a special guide who knows exactly the spot weeks might be lost in seeking this seepage. And at the end, being in the desert flat, we would not know much more, all the sedimentary strata being hidden below the surface sand. The oil must rise from a certain depth, since the outcrops of Burgan do not show traces of impregnation.

Grain. About 18 km South of Burgan follows another low hill called Grain (Pl. IV fig. 5). It is composed as follows, commencing with the lower beds:

1. 10 metres or more of light-coloured, uncoherent and porous sandstone
2. 5–6 metres of greenish sandy marls with small cristals of gypsum, and with limey nodules at the base passing to
3. 1 metre of limestone in beds of 5 to 10 cm each, with marly intercalations, containing irregular, peculiarly shaped concretions of flint, of which the slopes of the hill are covered. Observed with a strong lense, the limestone shows some small Foraminifera (Globigerina). It is thus purely marine. These strata are lying perfectly horizontally.

The next following outcrops are presented by low tables which are called by the natives *Athaimi*. They rise about 15 metres above the low plain, and probably not more than 20 metres above the sea. They are composed of greenish sandy marls with calcareous nodules, resembling completely those of the Grain, and apparently belonging to the same subdivision.

Ain el Abd. Travelling again over flat, sandy country with low sand hills, which here and there show the above named marls as their substratum, we reach Ain el Abd at the end of a 20 km wide salty marsh which extends to the sea shore. This 'spring of the Slave' is a round pool of 15 to 20 metres in diameter, in which *warm salt water* with *strong gas bubbles* is coming to the surface (Pl. XIII fig. 2). The smell of H_2S is noticed from a distance. The salt water is of small quantity and flows off to the sea. Along the rim of the basin are found black Algae, apparently incrusted with $Fe\ S_2$. This black substance seems to have been regarded as tar by some natives.

Stratigraphic Conclusions

The very few and isolated outcrops and the lack of determinable fossils make it very difficult to obtain an idea of the stratigraphic succession on far distances. If we parallize the coarse sandstone of the Koweit-Mission with that of Wara and the top of Burgan, we come to the conclusion that the strata are gently rising towards South, as indicated with . . . on Pl. III fig. 1.

The red sandstone of Djal ez Zor (Mugheira) probably belongs to the Miocene, especially to the Red beds of Pascoe's *Kurd Series*, lower subdivision.

The lower part with marls and gypsum beds then might be considered as belonging to the *Fars Series.*

If these coordinations (based on insufficient observations) are right, we would have the following succession: 1. Marls with beds of limestone, gypsum, sandstone (Athaimi–Grain) 2. Coarse sandstone (Burgan–Wara–Koweit, and probably Djahra) 3. Red sandstone (Mugheira) 4. Oolites, Pleistocene to subrecent.

The oil probably derives from the lower part of the above series (Fars), as is the case of the Anglo-Persian oil field and of Iraq, rising in cracks to the surface.

However, it must not be forgotten to mention that as a whole we have not yet found in Koweit and El Hasa the typical Fars series with its thick masses of gypsum as they are described by Pilgrim and Pascoe. The above conceptions thus are only preliminary.

Oil Seepages in the Gulf

In most of the books treating on oil in the Persian Gulf region, seepages are mentioned on different islands: *Qishm*, *Salak*, *Halul*, and further North on *Farsi*, *Karu* and *Kubbar*, the latter being close to the Arabian Coast.

The weather was too rough, and the time insufficient to visit Karu Island, as expected. But good information was obtained from the Captain of His Excellency Sheikh Ahmad of Koweit, called Yuzef Bin Nazralla. Some other data are given by Pilgrim.

According to these observers, Kubbar and Karu are small low-lying rocks, littoral and subrecent. No oil seepages occur on these islands, but on the *bottom of the sea*; they thus can only be observed during exceptionally calm weather. The oil spring of Karu (which means tar) is said to be situated about 6 to 8 miles a little West of North from Karu, thus between Karu and Kubbar, the oil rising from a sea bottom of 16 fathoms depth. (Pl. I).

We thus understand wherefrom the tar on the coast of Jubail and Bahrein is coming. *The prevailing Northerly winds drive the oil to these shores, where the light parts evaporate and oxydizing is taking place.*

The first indication of true folding having been found on Bahrein Island, it is possible that the oil seepages Halul–Farsi–Karu derive from anticlines hidden below the sea.

Stratigraphic Compilation

After having travelled through the Arabian Coastal region from Koweit to Hofuf, and studied Bahrein, we now may try to order the formations encountered in a preliminary way. Some new names will be necessary:

Recent	Sand and gravel of desert. Mud along the coast with Potamides.
Subrecent to Pleistocene	Koweit Building Stone, greenish, oolitic. Jubail Formation: white to brownish oolite with cross-bedding.
... Unconformity ...	
	Sandstone with conglomerate: Koweit, Ez Zor, Burgan. *Zor Formation.* Possibly similar with Pascoe's Kurd series of Iraq.
Miocene?	
... Unconformity? ...	
	Marls with gypsum and layers of limestone with flint (Grain, Nadjira). *Nadjira Formation* (possibly similar) with Fars series of Pilgrim and Pascoe)
	Limestones of undetermined age (Dj. Muddra–Dhahran–Hofuf)
Eocene	Nummulitic limestone, limestones with flint and lumachelle. Bahrein.
Upp. Cret.:	Limestone of Djebel Dukhan and chalk rock

For distribution of these divisions see Pl. I.

GENERAL CONCLUSIONS

Although the observations for a proper judgment are insufficient on account of lack of time, some preliminary conclusions can be made.

Koweit

To Koweit belongs *the only important seepage* encountered on the writers voyage: *El Bohara.* We may add the seepage of Burgan although it was not found by the expedition.

The North side of Koweit-Bay (Djal ez Zor) is formed of *unfolded strata* of probably Miocene age, the same division to which belong the oil horizons of Persia and Iraq. If this conception is correct, it must be mentioned however that the facies of strata is remarkably different from the NE-side of Iraq Valley. Especially the big bodies of gypsum mentioned from there in the literature (Fars Series) have not been encountered.

Although very remarkable, the seepage El Bohara is small in comparison with those of Hit, which produce as much as 2500 tons of oil and tar per year, or *30,000 times as much*!

In the southern part of Koweit we have found two places where the lower strata (Nadjira Formation) show a dip of 5 to 15°, while the top sandstone (Zor Formation) is horizontal (Wara, Burgan). The question thus arises if these dips are caused only locally by sliding or weathering, or if they are caused by gentle folding with subsequent unconformity and transgressions of the sandstone. In the latter case, we might parallise the lower part with the Fars, and the upper with the Kurd series. It would thus be *possible that gentle anticlines of the Fars would occur at the depth,* totally hidden in the North by the unfolded Kurd series, and to the greatest part hidden in the South by desert sand and gravel. We will regard the two cases:

I. Supposed that the Miocene strata are lying *conformably* upon each other down to the Eocene, we would conclude that there is *no reason for a* large accumulation of oil below the surface. Oil might be found by drilling at El Bohara, but not in paying quantity. (In regard to the extraordinary difficulty of landing the quantity ought to be large for a paying proposition.)

It is true that big oil fields have been developed on practically horizontal and unfolded formations, like the American Mid-Continent Fields (Oklahoma). The detailed studies however have clearly shown that there are gentle dips forming waves and terraces, on which the oil is accumulated, although these dips are only $\frac{1}{2}$ to 2° as a rule. If this is the case of paleozoic regions, it is not the same for the Tertiary. Indeed, *no unfolded Tertiary country yet is known of which oil is produced by boring in commercial quantities.* This fact may be explained as follows: Within 200 million years, a dip of one degree has been sufficient to separate the salt water from the oil of the porous strata, to let the oil migrate to the higher parts of the strata, while the time of 1 to 2 million years (Miocene) was insufficient for such an accumulation.

It may be objected that the Patagonian Field is situated on Miocene and Oligocene Strata, the dip of which is only measured by tenths of percents! The oil however does not derive from this Tertiary, but from the Upper Cretaceous beds, which underlie the Tertiary with an unconformity, and are *gently folded.* The oil is found on the anticlines of the Cretaceous series hidden below the unfolded Tertiary.

II. In the case of a folded Fars Series below the unfolded Kurd series, the oil could be accumulated at the depth, and it would be *possible to develop an oil field of commercial importance.* The difficulty however would be to locate the hidden anticlines otherwise than by numerous test bores. Such a case was encountered not only in Argentine, but especially in the Caspian Region (Guriew), where the Shell Co. with much difficulty has succeeded to develop an oil field based on hundreds of test bores, and on a complete geological study of them.

We thus see that *Koweit is a country of some possibility, but not of high promise.* Also it should not be forgotten that the cost of boring would be considerable owing to the great difficulties of landing on the muddy coast. The first location would have to be placed not far to the land-side of the seepage El Bohara.

Before taking the chances for very expensive test bores, the writer would advise to continue the less expensive geological observations at the surface, as far as they can clear up the situation. This would consist especially in studying the questionable unconformity inside and outside of Koweit.

NOTE 26

Neutral Zone

In the Neutral Zone, the few existing Tertiary outcrops all show to be unfolded. No oil seepage is known. If we further consider the difficulty of access, the lack of water, the lack of a harbour and the lack of any permanent settlement, we find, with the actual knowledge, *no reason to recommend this concession for oil.*

El Hasa

In the entire region of El Hasa we have found neither a seepage nor any proper folding of the strata. Should Dj. Dhahran, and the dune culmination El Ala between Ojair and Djisha prove to be anticlinal folds, their dips would scarcely amount to 1° and not account much as of accumulative value.

On the other hand, it seems that the limestones of Dj. Muddra–Dj. Dhahran and of Hofuf are lower than Miocene. The chief oil horizon of Persia and Iraq (Miocene) would thus already be weathered away. This would explain the absence of seepages.

We thus, with the actual knowledge of the country, have *no reason to recommend the concession of El Hasa for drilling on oil.*

Bahrein

Bahrein forms a large and very gentle *anticlinal dome*, but unfortunately, the oil bearing Miocene is completely removed.

For the case of a mesozoic or paleozoic oil horizon the structure would be magnificent. But there is no reason to suppose such an oil horizon at the depth below Dj. Dukhan. It is true that the Cretaceous and Eocene Formations of Persia and of Syria and Palestine show numerous scattered oil seepages and asphalt outcrops in the limestones, as is the case on Bahrein, *but none yet has proved to produce oil in paying quantities.* To drill on the dome of Dj. Dukhan not only would be extremely expensive, but also a pure gamble.

The countries of Eastern Arabia thus rapidly traversed by the writer *do not present any decided promise for drilling on oil.*

If we compare what Pascoe in his excellent study on Mesopotamia considers as 'areas of first class importance', 'less certain but of decided promise', 'uncertain' and 'gamble', boring on the Arabian coast would have to be classified as a pure gamble.

However, the studies as given above are insufficient in several directions, especially for Koweit, and need to be pursued before a definite decision is taken.

Zurich, Sept. 5th, 1924 sig. Dr. Arnold Heim

---

26. *Haji Abdullah Fadhil Williamson (1872–1958) entered the service of APOC at Basrah in 1924 after an extraordinarily varied and adventurous career (his biography* Arabian Adventurer *by Stanton Hope, edited and with a Foreword by the writer of this Record, was published in London by Robert Hale Ltd. in 1951). Born William Richard Williamson, of British parents in Bristol, England, in 1872, he left there in 1885 as an apprentice merchant seaman in a sailing ship bound for Australia and California. He deserted the ship in California to become a gold miner and subsequently, after various adventures whaling in the Arctic and in the South Seas, enlisted in the British police-force in Aden in 1891. He then became interested in Islam, learnt Arabic, studied the Koran, and became a Moslem in 1893 on the sponsorship of one of his Aden Arab friends. He wished to leave the British police and travel in Arabia, but this was objected to by both the British and Turkish authorities (Arabia was then under Turkish control), and in 1893 he was transferred to Bombay, discharged from the British police, and forbidden to return to Arabia. Having made friends with some Kuwaiti horse-traders in Bombay, he eluded the vigilance of the British authorities and travelled with them late in 1893, disguised as an Arab groom, on the mail boat to Kuwait where, and in Basrah, he lived for two years under his Moslem name Abdullah Fadhil Zobeiri. To avoid arrest by the Turkish authorities, who had been advised of his escape from Bombay, he went in 1895 with Beduin friends into the interior of Arabia where, with occasional visits to Basrah, Kuwait and other Gulf ports, for the next fourteen years he lived as an Arab trader, contracting several marriages and becoming widely known as a devout and learned Moslem (he twice did the pilgrimage to Mecca in 1895 and 1898), with a great reputation for physical endurance, personal bravery, and skill in camel and horse breeding. In 1909 he settled in Kuwait as a shipowner and pearl merchant.*

Soon after the outbreak of the 1914–18 War, he offered his services to the British

Army in Mesopotamia, who gladly availed themselves until 1920 of his intimate local knowledge both there and in Kuwait and elsewhere on the Arab side of the Gulf. At that time he resumed his former English surname, adding it to his Arab name of Haji Abdullah Fadhil (this last name was that of the Adeni friend who had sponsored him as a Moslem).

On leaving the British Army in 1919, Williamson first became an Assistant Collector of Customs at Basrah, but in 1924 was engaged by APOC (as inspector of Gulf agencies) who were then enlarging their oil-sales organisation on the Arab side of the Gulf where he had acquired close business and personal contacts, including marriage connections, from Kuwait to Muscat. In addition to this work, being already well acquainted in Kuwait since 1893 with the ruling Sabah family as well as with the Alghanim family and other prominent Kuwaitis, he was also occasionally employed there by APOC in connection with its concession negotiations with Sheikh Ahmad from 1923 onwards.

In August 1932, when Mr Chisholm took up residence in Kuwait as APOC's negotiator with the Sheikh, Williamson accompanied him as general assistant and interpreter, proving invaluable in both capacities owing to his perfect Arabic, his close and long acquaintance with Kuwait and its personalities, and the respect which his devout character and established reputation commanded. Having worked with Mr Chisholm in Kuwait during 1932 and 1933, Williamson was transferred for APOC duties elsewhere in 1934 (during the last six months of that year's negotiations Mr Chisholm's interpreters were Mr Juma and Mr Helmi). The reason for this was that Williamson's antipathy to Major Holmes's interpreter-assistant Muhammad Yatim, and to a lesser degree to Major Holmes himself, had by then become so strong that it became impossible to retain him with Mr Chisholm while the latter was a joint negotiator with Holmes, who was dependent on Yatim's interpreting and assistance. This antipathy felt by Williamson, which was reciprocated by both Yatim and Holmes, was due so far as Holmes was concerned to the bitter business antagonism which had existed between them since as far back as 1922; and this, so far as Yatim was concerned, was exacerbated by his convivial habits which, while they were appreciated by Shaikh Ahmad (see Note 17), were anathema to Williamson's rigidly conventional character and devout Moslem beliefs.

In 1937, at the age of sixty-five, Williamson retired on pension from APOC and for the rest of his life lived at Basrah, with his last wife (a Bedu lady) and sons, thus never returning (or wishing to return) to England which he had last seen as a boy of thirteen. During his retirement, in which he again went on the pilgrimage to Mecca, Williamson re-adopted his former Arab dress and style of life, devoting his time mainly to reading, religious studies, and conversation with old friends of whom a large number from Arabia and elsewhere visited him in Basrah; including the writer of this Record who continued the friendship, begun in 1932–33 when in close companionship with this most remarkable and agreeable man, until Williamson's death in 1958.

---

27. *The Alghanim family is one of the oldest and most distinguished in Kuwait since the eighteenth century.*

When APOC was developing its oil-sales organisation in the Gulf in 1925 Ahmad Muhammad Alghanim, then head of the family, was recommended to them by Shaikh Ahmad as the best person to represent them in Kuwait, and he subsequently included their agency among his business interests.

By 1932, when the Company's intensive negotiations for the Kuwait oil concession began, his eldest son Yusuf Ahmad Alghanim (now head of the family) was in charge of the APOC oil-sales agency. From then onwards (see Foreword) his support, with that of many influential Kuwaitis who because of the high reputation of the Alghanim family and business followed his lead, greatly assisted the APOC and later the KOC in Kuwait.

---

28. *See Note 32 for EGS–APOC correspondence March–April 1926.*

---

29. *In 1925 the pre-war Turkish Petroleum Company's 1914 oil concession in Mesopotamia (Iraq) was confirmed by the new Iraq Government to its post-war successor (renamed Iraq Petroleum Company in 1929), owned, except for a five per cent Gulbenkian interest, jointly by APOC, Shell, and a French oil company. These were joined in 1928* *by a group of American oil companies, following strong and successful pressure exerted since 1920 by the United States Government on the British and French Governments for an 'Open Door' oil policy in Iraq. This controversy and first appearance of American oil companies in the Middle East attracted much attention there, especially in Kuwait (see Note 34).*

---

30. *Exchange of APOC London departmental memos, October 1925:*

From Mr. Nichols — To Sir John Cadman

PERSIAN GULF CONCESSIONS

The question of concessions for Kuwait and Bahrain remains in suspense.

In July last the Colonial Office refused to accept the conditions of the Muscat Concession as applicable to the concessions under negotiation with the Shaikhs of Kuwait and Bahrain.

Under Colonial Office instructions all negotiations have, in principle at least, to be conducted with the Shaikhs through the medium of the Political Agents, who would, at the very least, strongly dissuade the Shaikhs from accepting conditions disapproved by the Colonial Office.

The Colonial Office now refuse even to allow direct negotiations between us and the Shaikhs on the question of permanent Customs immunity for Company goods (as obtained at Muscat) and are disposed to continue their insistence on the previously prescribed conditions of immunity for 10 years and thereafter liability on a scale which though known now may be altered at any time.

The royalty accepted in 1922 of 3½ rupees per ton, subject to revision after 10 years, still remains.

I feel that the acceptance in the Gulf of any liability to Custom duties on Company goods (as distinct from those of our staff) is a most dangerous precedent in connection with the D'Arcy Concession.

I should be glad to know what degree of importance you attach to these Concessions, having regard to:

1) their prospects on geological grounds.
2) The customs liabilities and royalties at present attached to them, as well as the limitation of their total area to 100,000 acres, to be chosen by the concessionaire in not more than 3 blocks.
3) my information from Mr. Mackie that the Shaikhs are unlikely to be attracted by any potential increase of their revenues which are already ample for their needs. In Kuwait, moreover, the Shaikh, though favouring the Company beyond other possible competitors, is faced with strong opposition from the leading merchants, who fear that with constant work available they will lose their present absolute economic control over the labour engaged in pearling.
4) the apparent complete failure of the Nejd concession to materialise in the hands of the Eastern & General, and, to the best of our knowledge, the absence at present of competition for Kuwait and Bahrain.

Sir Arnold Wilson telegraphed recently that he understood that Kuwait might be obtained with the assistance of a mediator, and I replied that our position remained unchanged, viz., that we would not take up the Concession without complete immunity from Customs on Company goods, which the Colonial Office were unwilling to concede.

1st October 1925. — H. E. Nichols

From Sir John Cadman — To Mr. H. E. Nichols

PERSIAN GULF CONCESSIONS

My personal opinion is that we should let the concession question, referred to in your note, drop for the time being, but meantime I am ascertaining the views of our technical people regarding the possibilities of Kuwait and Bahrain.

2nd October, 1925. — John Cadman

---

31. *APOC geological report on Kuwait, February 1926, and 1931 addendum:*

I. INTRODUCTION

(1) Mr. A. G. H. Mayhew and the writer were deputed to make an examination of the territory of Koweit, and were thus employed from January 7th to February 7th of this year.

(2) The object of the expedition was primarily to collect general geologic information with a view to solving certain problems of a regional character which have been brought to the front by the researches of Prof. de Böckh and his colleagues. In particular, it was hoped that the place of the Koweit Series in the Stratigraphy of the Persian Gulf region might be definitely determined by tracing this series to its contact with older formations or by a further study of its character.

(3) Permission to carry out this work was obtained from Sheikh Ahmed of Koweit through Major J. C. More D.S.O., British Political Agent; and the writer was instructed to follow Major More's advice in every respect.

(4) A route programme was prepared which would allow of examination of all the principal features indicated on the maps of Koweit territory, including the Jal Ez Zor, El Ya, Abatih and Hamar ridges; Jebel Sanam; the Batin valley as far as Rigai; the Shagg valley at its northern end; and the isolated hills south west and south of Koweit, including especially Jebel Warah, Burgan, and Mishrif (Qasr es Sirrah).
(5) Major More was of the opinion that the Batin valley section would be impassible by motor car. A camel caravan was therefore engaged for this part of the area. Sirrah, Warah, and Burgan were visited by car; and another car journey was made to Mudairah on the north shore of Koweit Bay.
(6) It was found inadvisable to attempt to go far into territory under Ibn Saud's control. Major More was emphatically opposed to it; and Sheikh Ahmed, naturally, would not send his men outside his own boundaries. For this reason Jebel Qrain could not be revisited; and the limits of the Koweit series were nowhere reached, nor, apparently, approached.
(7) Major More very kindly made all arrangements for escort, guides, and camel transport; and, together with Mrs. More, accompanied the writer on the motor journeys. For this, and for advice and information on innumerable subjects, and for hospitality during a large part of the time, it is desired to express the warmest possible gratitude to Major and Mrs. More.

II. GEOGRAPHY

(8) The region examined is a fairly even plain rising gradually westwards from Sea level to about 800 feet at the furthest point reached. The surface is never absolutely flat, but is always more or less etched by erosion into slight hollows and ridges. The hollows are sparsely covered with a growth of one or more species of scrub and grass, while the ridges are generally bare and covered with loose gravel.
(9) The major negative features of the country are the Batin and the Shagg, which are river-cut trenches; and Koweit Bay, which is regarded as the work of marine erosion.
(10) *The Batin* is a remarkably straight feature as represented by the maps, and it was at first suspected that it might be a long, narrow rift-valley. Field evidence, however, is against this. The Batin has been hollowed out by swiftly flowing water. The heavy rainfall giving rise to this must have occurred at a distance,—presumably in the central Arabian highlands,—for the main channel has been deepened more than the tributaries in the section examined, so that the latter tend to hang over the former in many cases. The deepening of the main channel shows at least three stages, as the banks, especially near Rigai, show traces of two gravel capped terraces, about 25 and 50 feet above the present bottom. The depth of the trench is perhaps 80 ft. at Rigai, and diminishes gradually northwards.
(11) *The Shagg*, where examined, is a much shallower channel than the Batin; and, on the ground, failed to give the impression of a straight trench that the maps convey. It appeared rather like a bundle of hollows, trending more or less north-north-west. The maps shows it terminating in a broad shallow depression, without outlet to the Sea.
(12) A number of closed shallow depressions of this nature were seen. Their origin is not explained, but it is suspected that they are subsidences due to removal of underlying soluble materials by percolating rain water. The existence of a soluble formation underground is, however, not proved by observation.
(13) *The Jal El Ya* is a low narrow elongated ridge, rising perhaps 25 feet above the plain. The regions called El Abatih and El Hamar are characterised by a succession of shallow wadis, draining mostly to East and North East, but in one section to South West. The Wadi bottoms tend to be sandy and scrub-covered, while the intervening ridges are gravelly and bare.
(14) Jebel Sanam (or the Hill of the Camel Hump) is just outside the Koweit border, but deserves mention because it is an unique feature in this region. It is a landmark visible for 30 or 40 miles around, and must have been so for ages past, as it is a pinnacle of old rocks projecting above the sheet of Young Sediments which covers the whole of the surface of Koweit territory.
(15) The Jal Ez Zor is the most prominent geographic feature in the neighbourhood of Koweit. It is a sharp escarpment, precipitous in many places, rising to over 400 feet above Sea level north of Jahrah. North-eastwards the altitude decreases gradually, until the ridge is cut off entirely by the Khor Es Sabiyah. South Westwards, there is actually a rise to over 500 feet, but the escarpment dies out at this point. The cliff face shows marked terraces at three levels opposite Jahrah, at about 50, 100, and 250 feet above sea level. The terraces are determined by bands of slightly more resistant rock, but it seems clear that they have been cut by the sea. Traces of marine deposits can occasionally be seen on them.

On top of the Jal Ez Zor there is a narrow strip of flat land strewn with pebbles. Beyond this, the land surface sinks gradually, and is much dissected by Wadis draining to north-east.
(16) South of Koweit, the land rises in the form of a low flat ridge, which at Jebel Warah is over 200 feet above Sea level. Jebel Warah is an isolated block of indurated rock rising abruptly some 80 ft. above the surrounding country. The Burgan hills are of similar composition, but are lower and less abrupt. Mishrif (Qasr Es Sirrah), and numerous low flat-topped hills in the same neighbourhood, are also determined by the presence of an indurated stratum. They show, in places, relics of a capping of shelly grit, containing a species of Conus which is very abundant in the recent littoral formations. There is thus, here also, direct evidence that sea level in this region stood in recent times some 180 feet, and probably more, above the present level. Jebel Warah and the Burgan hills are therefore relics of marine erosion.

III. STRATIGRAPHY

(17) There is a slight development of shelly sand (littoral concrete) on the present shore line at and near Koweit, and on former shore lines. A muddy shell rock is found in Koweit Bay and on the Islands outside it (Failajah etc.). These are of recent formation. Relatively recent also are the 'Iraq gravels', which cap many of the eminences throughout the northern part of Koweit territory, and the terraces of the Batin Valley. With the one exception of Jebel Sanam, all the remainder of the surface examined is covered by the Koweit Series. This series is exposed to a depth of some 450 feet in the Jal Ez Zor Escarpment north of Jahrah. It outcrops all along the Batin as far as Rigai, where it has been penetrated by old wells to a depth of over 100 ft. in some places (Hulaiba and Qulban El Obeid); and it seems probable, from the description by Philby, that the wells at Kafar are also sunk in it. Mishrif, Warah and Burgan present a superficially altered form of the same formation.

(18) *Koweit Series*

These beds are typically formed of quartz grit, or gritty and pebbly sand, often felspathic, and very often with a matrix of white argillaceous substance that seems to resemble Kaolin. The quartz is translucent, and the grains are relatively angular. The materials have clearly been derived from the waste of a mass of decomposed medium grained granite, and have been washed down and deposited rapidly by flood waters, without much sorting, as grit particles and small pebbles are found mixed indiscriminately through finer materials. The Kaolinic substance which commonly occurs as a matrix is found as a thin bed at Burgan. Elsewhere argillaceous bands are rare, but frequently a streak of reddish color appears on outcrops, and is found to be due to the presence of a ferruginous argillaceous matrix in the grits. The most striking development of the sort is seen on the south side of the Burgan hills, where there is a section showing some 50 ft. of variegated beds, comprising not only the normal grey and white Kaolinic grits with a thin Kaolin bed, but also beds of pale green, yellow, red, purple and brown color. The colored argillaceous beds frequently contain thin plates and veinlets of selenite; and sometimes show a salt content by the presence of an efflorescent crust of sodium chloride.

(19) *Calcareous Cementation*

The Koweit grits are very often found in a loose friable condition. But even in such cases, they are characterised by an abundance of small or large concretions, which are more or less round masses of the grit cemented into a coherent rock by carbonate of lime. In most sections this calcareous cementation has also produced ledges of hard rock at intervals. The Zor terraces are an instance of such selective cementation on a large scale. The rocks supporting the terraces are coherent calcareous grits, while the intervening beds are incoherent.

(20) The Calcareous content occasionally increases still more, producing a rock which can be called a sandy limestone. The sandy limestones in the Koweit series are white or grey, less often pink, and are sometimes rather coarsely crystalline. Sand grains can, however, generally be seen even in the most crystalline varieties, and there is a gradation from limestone to calcareous grit. The limestones proper were never observed to attain a thickness greater than one or two feet. But beds of coherent calcareous grit up to 20 ft. thick can be seen.

(20a) Cross bedding is occasionally seen in the Koweit grits, for example in the massive cliff of Jebel Warah. The true bedding gives everywhere the impression of horizontality, though the long section of the Jal Ez Zor escarpment seems to indicate a gradual dip north-eastwards. This is the direction in which the sediments were originally carried, and no tectonic significance can be attached to the observation.

(21) It may be remarked at this stage that the assemblage of features described is rather puzzling. Detrital materials from a predominantly granitic source have been washed down rapidly and deposited in an unsorted condition; and though there are only rare traces of fossil shells, the resulting sediments have a high content of calcium carbonate, with subordinate but distinct traces of gypsum and sodium chloride, the latter—as is natural—associated with the more argillaceous members of the series. It could be expected that such a source of detritus would supply Alkali Carbonate, and not lime carbonate to any extent. One is tempted, therefore to suppose that the sediments were piled up in an enclosed sea, free from currents. Such a sea would have a high salt content, and little or no lime secreting fauna. It would, however, probably have a high calcium sulfate content. Calcium carbonate would thus be immediately precipitated from the incoming flood waters, if, as supposed, the latter contained alkali carbonate.

$$CaSo_4 + Na_2Co_3 = CaCo_3 + Na_2So_4$$

It is possible to imagine that precipitated $CaCo_3$ might occasionally be segregated to some extent and deposited with only a slight admixture of detrital grains, giving sandy limestone, while ordinarily it settled in a diffused condition among the detritus and was later redistributed by the usual processes of diagenesis.

On the other hand, both the sandy limestone and the Calcareous grits occasionally show tubular structures that seem to be casts of worm burrows. This points to the conclusion that the basin of deposition was shallow, and that the water was not concentrated enough to inhibit all forms of life, though shell bearing organisms were rare.

NOTE 31

(21a) *Note:—Decomposition of Felspar*

$$\left\{\begin{array}{l} Na_2O + Co_2 = Na_2Co_3 \text{ sod. Carbonate.} \\ \;\;62 \qquad\;\; 44 \\ \left.\begin{array}{l} Al_2O_3 \\ \;\;102 \\ \\ 2\,SiO_2 \\ \;\;120 \end{array}\right\} + \underset{36}{2\,H_2O} = \underset{258}{H_4Al_2\,Si_2O_9} \text{ Kaolinite} \\ 4\,SiO_2 \text{ --------- } \underset{240}{4\,Si_2O} \end{array}\right.$$

If all the felspar is albite, and is completely kaolinised, the product will be, very roughly, 16% sodium carbonate, 40% Kaolinite, and 40% Silica.

The presence of so much kaolinic matter in the Koweit grits implies that a great quantity of alkali carbonate was liberated at the same time—something approaching 2/7ths of the weight of Kaolin.

Some $CaCo_3$ would, of course, also be formed, as the felspar would probably not be pure albite. Again, some of the felspar would be orthoclase, which would give a higher relative weight of carbonate than the Soda mineral.

In the conditions imagined, it is hardly too much to suppose that these decomposition products would have accumulated to some extent.

(22) *Silicification*

As already recorded by Mr. Lister James, the Koweit grits at Burgan and Warah show a rather remarkable silicification. This phenomenon affects only the superficial beds on the hill tops, where they are more or less indurated, and occasionally altered into dense splintery quartzite, in which there is no longer any trace of calcareous cement. The thickness of beds affected is seldom more than about ten feet. Underneath, the ordinary friable grits, with calcareous nodules and ledges, are found. Silicification has clearly been a downward process. One specimen collected from below the hard quartzitic crust shows calcareous grit containing a geode lined with quartz crystals. Even the Kaolinic bands are indurated when they are found on or near a hill top, producing a kind of porcellanite. The surface with which the silicification process was associated lay above the level defined by the hilltops of Warah and Burgan, which are relics of denudation.

(23) Perhaps more remarkable still is the silicification seen near Koweit, at Mishrif (Qasr Es Sirrah), and on a number of small hills between there and the Jahrah gate of Koweit. The rock affected here is a sandy limestone, which is quite often pink in color, but commonly grey, and highly crystalline. Geodes lined with calcite crystals are sometimes abundant. The sand content varies a good deal. But even in the purest limestone, tubular masses of sand are frequent. Small gasteropod shells are visible in one specimen. This rock nearly everywhere shows incipient replacement by brown or black chert, and replacement is sometimes complete over large masses. The sand tubes are, however, frequently preserved where the surrounding limestone has been replaced. This indicates that the replacing medium expelled Calcium Carbonate with relative ease, but did not affect quartz grains—at any rate with the same facility. It was also observed that a small gasteropod shell was silicified, while the sandy limestone in which it was imbedded remained untouched. This would indicate probably a preference for aragonite on the part of the silica solution. Underneath the hard cap formed by the chert–limestone bed, the Koweit grits are again found in the usual incoherent condition, with calcareous nodules. Alteration to Chert is entirely a superficial phenomenon, and is confined to rocks rich enough in Calcium Carbonate to be called limestones. It seems clear that the same siliceous agent which changed limestone to chert, changed the less calcareous grit into a quartzite.

(24) Mr. Lister James' description of Jebel Qrain leaves no room for doubt that the Koweit series is found there also with a cap of sandy limestone partly replaced by chert. The silicification surface thus appears as a plane, rising from Koweit, where it is found at 50 ft. above sea level, or even less, to Mishrif (180 ft.), over Jebel Warah and the Burgan hills, which must be about 300 ft. to Jebel Qrain (perhaps 400 ft.). The nature of the silicification process is not clear. C. K. Leith has recently (*Econ. Geol*, XX, 6, 1925, pp. 513–523) remarked on the frequent association of erosion surfaces and silicification; and silicification of surface rocks has been described from the Kalahari desert of Southwest Africa. Silica appears, under appropriate conditions, to be a relatively soluble substance. But the best known condition for the solution of silica is the presence of Alkalis. It is just conceivable that sodium carbonate waters coming from the granites of the interior of Arabia may have been brought into action on this elongated, and comparatively narrow strip of the Koweit series, by some accident of drainage after deposition had ceased. This is, of course, entirely a speculative idea, but it explains the apparent absence of chert and quartzite elsewhere in the Koweit series. The superficial position now occupied by the Mishrif chert and the Burgan–Warah quartzite would be due, on this line of reasoning, to the ability of these hard rocks to arrest denudation for a time. The silicification process may have been active elsewhere in the Koweit series, but erosion has not revealed it, because it has taken place underground when surface waters containing silica in solution have been held up by a relatively impermeable limestone or calcareous grit.

(25) *Gypsum*

It has already been mentioned that some of the more argillaceous beds of the Koweit series show gypsum in the form of thin plates and veinlets of Selenite. An attempt has been made to indicate how this gypsum, and the salt which accompanies it, may have been derived from saline waters into which the Koweit sediments were poured. But this is the least conspicuous occurrence of gypsum in Koweit. Almost universally, it can be found within a few inches of the surface in the form of a loose mass of crystalline particles, enclosing grains of sand, grit, small pebbles, or lumps of disintegrated rock from the local formation. The thickness of this surface gypseous layer is variable. It may be only a few inches, but as much as 10 feet has been seen. Occasionally it has the appearance of a solid bed, but closer inspection reveals impurities of various kinds, dependent on the local materials. The torn up ground around Koweit town is due to exploitation of this gypsum, which occurs as a cake 2–3 ft. thick, cementing the surface layers of Koweit grit. On the top of the Jal Ez Zor, and in numerous places in the Batin, gypsum crust could be found by digging a few inches into the loose grits of the surface. Its origin was well displayed in some Wadi sections near Shiquiya (Batin), where a gypseous salty argillaceous bed was overlain by ordinary Calcareous grits. Thin gypsum veins appeared in the argillaceous bed, and became more numerous towards the surface, until the 3–4 ft. of rock immediately under the surface was found to be seamed in all directions and thoroughly permeated with gypsum. Possibly some of the lime cement had been removed in this process, for the material was no longer coherent, but could easily be powdered into a loose mass of which gypsum crystals were the most obvious constituent.

(26) The indurated rocks of Mishrif, Warah, and Burgan show traces of a similar concentration of gypsum near the surface. The porous rocks underneath the hard cap are often found to be permeated with fine crystals, while the hard rocks are veined along joints. On a small hill about 1½ miles south of the Jahrah Gate, Koweit, the sandy limestone, where not placed by chert, is strongly gypseous. Calcite-lined druses have fine gypsum crystals deposited in the space between the Calcite crystals. The whole rock, in places, appears to have been partially disintegrated, and it is difficult to avoid the conclusion that some of the $CaCo_3$ has been actually changed to gypsum, since lumps of limestone coated and permeated with gypsum can be found embedded in a powdery mass of gypsum. Sometimes, in fact, it appears as if only the silicified part of the original limestone bed had survived the attack of the gypsum as a kind of skeleton.

(27) In parenthesis, it may be here noted that the Kharag Island limestone shows a very similar development of gypsum. The quarry operations have shown that the surface rock to a depth of 2–3 ft. is often quite rotten, and contains abundant powdery gypsum crystals. The remainder of the rock is relatively hard and compact, but frequently shows a coating of gypsum along cracks. In one case, a highly coralline layer in the limestone, about 6 inches thick, is reduced to a powder by growth of gypsum. Here again replacement of $CaCo_3$ by $CaSo_4.2\ H_2O$ is an observed fact. The coralline structure is in some cases totally obliterated.

Mr. G. M. Lees and the writer have observed a similar replacement of coral by gypsum in a Pleistocene raised-beach at Titaweb, Sudan coast. There, as at Kharag, the beds which underlie the coral rock are highly gypseous.

(28) There can be little doubt that the surface concentration of gypsum in the Koweit Series is due to capillary rise and evaporation of sulfate-bearing solutions. The source of the sulfate must in most cases be the gypsum disseminated through the underlying strata. But in low-lying regions adjacent to the sea, if the ground is porous, sulfate may be derived directly from sea water. Mr. Lees and the writer observed this process of gypsum formation on the Red Sea coast at Port Sudan. The beach material was composed of porous shelly sand, chiefly coral powder and foraminifera with some quartz sand. Thin layers of gypsum crystals were encountered at depths of about 10 inches, in places where the beach was flat for some distance away from the sea margin. Similarly, a gypsum cake was found at about 12 to 16 inches from the surface in the low sandy flat a mile south of Jahrah gate, Koweit. Concentration of sea water takes place in such circumstances more slowly than in a closed lagoon. But given time, and repeated wetting by sea water, quite considerable gypsum masses may be formed. In addition to direct precipitation of the gypsum contained in sea water, it is possible that gypsum may also be formed at the expense of any $CaCo_3$ present by interaction with the sodium and magnesium sulfates also contained in sea water. Thus $MgSo_4$ plus $CaCo_3 = MgCo_3$ plus $CaSo_4$. This reaction is known to take place with a saturated solution of $MgSo_4$; and under appropriate conditions, the gypsum may remain as a solid.

(29) The 'rotten sandy gypsum' mentioned by Mr. Lister James in the neighbourhood of Tel Mughayir (Ur) and Jalib Saadun may indicate that the Koweit series extends to that point. There is certainly no topographic reason for expecting a change of strata in that direction.

(30) *Jebel Sanam*

The writer has no hesitation in confirming Mr. Lister James' identification of his group 4 at Jebel Sanam with the Koweit Series.

Jebel Sanam was only casually visited, but some observations were made which it is desirable to record.

(31) *Mr. James's Group 3*: 'Consists of a limestone breccia, sometimes sandy, containing very angular to sub-angular fragments of limestone, indurated fine grained sandstone (approaching

quartzite), also of agglomerate and volcanic rocks. In parts . . . great masses of dark blue shattered limestone are found at all angles. These are frequently many tons in weight.' The writer found only quartzite and black foetid dolomite and dolomite shale in this formation. Igneous rocks were noted, but as they were rounded boulders they were referred to the Iraq gravels.

The rocks (specimens of which were brought to Mohammerah) correspond exactly with Prof. de Böckh's description of the black foetid dolomite which is almost universally found overlying the salt plugs of the Persian Gulf. So characteristic is this rock that Professor de Bockh and his associates have, on the strength of its occurrence alone, assigned the salt of Kamarij and some of the gypsum at Ahram to the Hormuz series, though these are not otherwise incongruous with the Fars beds with which they are in contact.

If this black foetid dolomite is so sure a criterion, then there can be no doubt whatever that the whole of the pre Koweit rocks exposed at Jebel Sanam i.e. Mr. James's Groups 1, 2, & 3, belong to the Hormuz Series and that the gypsum seen there has nothing to do with the Fars period.

Specimens of the black dolomite from Jebel Sanam show brecciation both incipient and fully developed. In the writer's opinion, the brecciation can be explained only by intense crushing.

The descriptions of the structure and strata exposed in this hill, especially when read immediately after Prof. de Bockh's account of the Hormuz rocks and Salt plugs, point directly to the conclusion that the whole hill, except for its girdle of Koweit grits, and a few detrital elements belonging to the Iraq gravels, is a tectonic breccia, thrust upward by a salt intrusion, and possibly further confused by collapse due to leaching away of the upper part of the Salt plug.

(32) *Iraq Gravels*

The Koweit series contains many beds which are pebbly, but these pebbles are invariably small. In no instance can it be recalled that pebbles more than 2 inches in diameter were seen actually in the Koweit series.

On the other hand, the surface of the desert over wide areas, especially the tops of all the low ridges such as the Jal Ez Zor, Jal El Ya, and the plateau surface on either side of the Batin as far as it was traversed, is strewn with a cover of distinctly coarse pebbles and even cobbles, ranging up to 8 or 10 inches in diameter.

(33) The Batin in its upper part shows three distinct pebble terraces, of which the upper one is the general level of the country, with surface etched by wind and rain into a network of shallow hollows and low ridges. The second terrace is some 20 to 30 feet lower, and is marked by only a few isolated gravel capped hillocks, of which Tel El Obeid (The Little Nigger) is the most conspicuous. The third terrace is some 25 feet above the Batin floor and is fairly well preserved as a chain of gravel capped ridges. The present floor of the Batin is mostly scrub covered, and shows very little accumulation of gravel, except on some flat ridges which apparently formed low Islands separating a network of channels when the Batin last flowed.

(34) There is thus revealed a rather lengthy history of denudation, extending back from the remote time when the Central Arabian highland last had a climate moist enough to send water down the Batin, and beyond that to the time when great floods were pouring granitic debris into an enclosed sea in this region. Geographic data available permits, however, only a glimpse of the events of this period.

(35) The pebbles of the Iraq gravels may be classified roughly in order of diminishing frequency as follows:—

Vein Quartz, white or pink.
Acid Intrusive rocks—rhyolite, quartz porphyry, porphyry and felsite.
Rhyolitic agglomerate.
Quartzite and quartzitic sandstone (not unlike some Nubian sandstones).
Grey compact unfossiliferous limestone.
Chert and Jasper.
Granite, rather coarse grained, with pink sub-porphyritic felspars—evidently the parent-rock of the Koweit grits.
Pegmatite containing garnets.
Pisolite (1 specimen only found).

(36) All these rocks with the exception of the limestone are siliceous, and very hard. They show an extraordinary development of wind-facetting, a large proportion of the smaller pebbles being cut into the triangular prism or orange-segment-form typical of a desert with regularly alternating winds.

(37) *Age of the Koweit Series*

Fossil evidence on this question has not been found. There is a shelly bed on the terraced slopes of the Jal Ez Zor, above Mudaira. But the shells are small and badly preserved and it is doubtful if any could be identified. Moreover, it is uncertain whether this is a bed intercalated in the Koweit sequence, or merely a patch of shelly concrete deposited on a terrace at a recent date.

(38) The diagenetic state of these beds varies so much, from extreme mineralisation to the opposite end of the scale, that no exact judgment can be based on this feature. Nevertheless, if it is correctly surmised, that most of the $CaCo_3$ in the series was deposited along with the detrital elements, and

not introduced subsequently, the normal degree of consolidation due to diagenesis must be rather low.

(39) The abnormal silicification of a restricted part of the formation is due to some special cause and has no bearing on the question of age.

(40) The nature of the sediments points to a period when a rainy climate prevailed in Central Arabia, and when there was an enclosed sea flanking the Arabian continent in this vicinity. It may be noted that the higher contours of the Arabian mass (1000–3000 feet, see Bartholomews' map) show a very pronounced reentrant south west from Basrah. The effect of this would be, given a rainy climate in the highlands, to concentrate the run off along a line roughly bisecting the reentrant—that is, towards Koweit.

(41) A branch of the Fars Sea would fulfil the second condition, if, as has been suggested, $CaCo_3$ really takes the place of what in normal Fars Strata has been deposited as gypsum.

(42) On the other hand, it is not impossible that an inland sea was initiated in this region, by rift faulting, at a much later period. The period when this might naturally be expected would be the late Pliocene or early Pleistocene, when the Dead Sea–Jordan Rift was definitely formed and when the Red Sea trough finally sank so far that the Indian Ocean and the Mediterranean were joined.

(43) The latter supposition would accord more naturally with the first requisite, namely, a humid climate in central Arabia. It is known that the Glacial conditions which prevailed in northern Europe in Pleistocene times produced a change of climate even so far south as Egypt and the Red Sea. The Nile then carried down gravels, which must have been discharged into it by Wadis which are now permanently dry. The deep gorges cut far back into the Red Sea Hills can scarcely be the work of a climate so relatively dry as the present one. There is, of course, the direct evidence of the Batin trench and its gravels, which proves that there has been a pluvial age even in Central Arabia. But the question is now whether the capacity of this pluvial time can be extended so as to include first a period of great sedimentation into a closed Koweit Sea, then a breaking up of the Koweit barrier and a rise of these Koweit sediments into the sphere of erosion.

(44) As regards the length of time during which pluvial conditions prevailed, there seems to be no impossibility in these assumptions. The Ice Age in Europe has been found to cover a very long interval, and to comprise four main periods of glacial advance, separated by retreat or interglacial period. In the very attractive tables of Professor Koppen, the oldest glacial period is plausibly dated 550–600,000 years ago; while the youngest falls only 70–120,000 years ago. It is presumed that each glacial period had a similar effect in pressing southward the limit of the wet belt.

(45) With regard to the Batin and its gravels, the suggestion may be made that perhaps the climate of Arabia is not so arid on the average as present dry conditions indicate. It may be that once in a while—perhaps once in several hundreds, or even thousands of years—a deluge such as Noah experienced may take place. The Batin terraces may be steps cut by successive single deluges at long intervals. One knows (to quote analogy in lieu of evidence—of which there is not enough) that the arid northern region of Peru is affected by a real rainfall only once in about 33 years. But this occasional rain is such as to effect tremendous changes in the landscape, excavating gorges and piling up debris which appear inexplicable in the intervening rainless years. The well known tanks at Aden were evidently built with a knowledge of a considerable potential rainfall such as Aden has never known. But it may be that, at long intervals, an adequate fall of rain does take place.

(46) Sufficient evidence is not available to settle the age of the Koweit series. Speculative support can be found for any age from Miocene to Pleistocene. The writer would prefer, on the whole, to accept the latter, but cannot pretend to decide on first hand evidence.

IV. PETROLEUM

(47) Nothing whatever can be added to observations already made, on the bitumen occurrences of Koweit, by Mr. Lister James and Mr. G. W. Halse.

(48) There can be no justification for regarding these as indigenous to the Koweit series in which they appear. But the nature of the underlying beds from which they emanate remains unknown.

(49) It seems possible that the petroliferous beds, which are overlain unconformably by the Koweit series, may be a development of bituminous Asmari Limestone similar to that at Hit. But this would not make Koweit an attractive field for wild-cat drilling.

(50) Professor de Böckh includes the whole of the Koweit territory in the 'continental' region, which means that sedimentary rocks will be only thinly developed; folding very gentle or none; and conditions therefore unfavourable for good accumulation of oil.

(51) The view that Koweit belongs to the Continental region is unavoidable, if the Δ_9 curve produced by the geophysical party last season is correctly interpreted as showing a gradual rise of old rocks towards the surface, and their emergence at Jebel Sanam.

(52) Further geophysical work—a traverse from Jebel Sanam towards Koweit, for example—might provide more data. But it is very improbable that, in the total absence of direct geological evidence, a convincing case could ever be made out by this means for exploring the unknown depths of Koweit territory with the drill.

V. CONCLUSION

(53) The Koweit Series extends inland, northwestwards and probably southwards far beyond the limits which it was permitted to explore.

(54) The nature of the series, the conditions of its deposition, and its age are considered. Speculative answers are put forward to some of the questions raised.
(55) A Pleistocene age is preferred, though certainty is not attainable on first hand—nor indeed on any evidence.
(56) Reasons, backed by specimens, are presented for regarding all the rocks, except the insignificant outer girdle of Koweit series at Jebel Sanam, as belonging to the Hormuz series; and it is asserted that the evidence for salt-upthrust here is as strong as the absence of visible salt allows it to be.
(57) The petroleum question is not affected by any fresh evidence obtained. The unfavourable view of prospects in Koweit that is deducible from Professor de Bockh's synthesis of all previous work, cannot be gainsaid.

Mohammerah,
13th February, 1926. Sd—B. K. N. Wylie

KOWEIT

Report B.K.N.W. 33, A.G.H.M. 4 gives a full and detailed account of the surface geological evidence obtainable in Koweit, and Mr. Wyllie's Note of 1st January 1931 summarizes this evidence and the regional geological considerations bearing on the prospects of the Territory.

The absence of geological structure suitable for the accumulation of oil in commercial quantity shows that there is no justification for drilling anywhere in Koweit. This conclusion is supported by the absence of evidence of the existence of a rich oil series in the area and also by general geological considerations. There is no reason to doubt the soundness of the general geology depicted in Figures 2 and 39 of the Preliminary Report of the 1924–25 Field Season's geological survey. These figures show that Koweit forms part of a stable Continental region which was only temporarily submerged beneath the sea and is therefore composed of old rocks which are probably covered by a relatively thin skin of Cretaceous and Tertiary sediments. The lithology of the Koweit Series and the absence of folding in the area fits in with the above account of the general geology. In addition it may be permissible to mention that the Signer has for some years suspected on very general grounds that the rich oil areas of Persia lie outside the Territory under review.

In view of the agreement between the local evidence and the general regional geology and in the absence of any alternative hypothesis the Signer is of opinion that the Company need not take any further interest in Koweit.

London,
6th January, 1931. (Sgd) T. Dewhurst

---

32. *Correspondence between APOC and Eastern General Syndicate Ltd., March–April 1926:*

Messrs. Eastern & General Syndicate, Ltd.,
19, St. Swithin's Lane,
E.C.4.

A.P.O.C.
Britannic House, E.C.2
9th March, 1926.

Dear Sirs,

We have now had time to consider the proposals put forward by Mr. E. W. Janson on 23rd February, together with the further explanations contained in your letter of 24th February. Subject to the conditions specified below, we are disposed to consider accepting the transfer simultaneously and collectively of the four concessions in question, viz. those which you at present claim to hold, covering Bahrain, Hasa and the 'Neutral Zone', together with the concession for the Principality of Kuwait, when obtained, for such sum as may have been expended by you in order to secure the same, but not exceeding £42,000, plus such further sum not exceeding £10,000 as in agreement with us it is necessary to spend in validating the first three concessions mentioned and in obtaining the Kuwait concession:

(a) In all cases we should require documentary evidence that the Chiefs concerned consent to the transfer to this Company of the respective concessions without any liability on the part of this Company for any breaches of contract which may have occurred up to the date of the transfer of the concessions.
(b) The concessions to be handed over to us free of any arrears in respect of rent or of arrears in respect of other obligations.
(c) Actual signed text of each concession and original ancillary documents, if any, to be produced for our inspection.
(d) Certain clauses in the Hasa concession appear to us to be loosely drawn up, and these should be supplemented by an exchange of letters with Ibn Saud stating specifically in each case and to our satisfaction what is presumably the intention of the clauses in question.
(e) The 'Neutral Zone' and the Kuwait concessions to be on lines not less favourable than the Bahrain concession as at present drafted.

(f) In the case of the Kuwait concession freedom of import duties in respect of machinery, etc. imported to extend for at least twenty years and not for ten years as in the Bahrain concession and that thereafter the duty shall not exceed a maximum of 10%.

(g) It is to be understood that, of the clauses numbered 6 and 12 in the copy of the Hasa Agreement as sent to us by you on the 15th February, the first of the two in each case is the clause as actually appearing in the signed Agreement.

(h) In the event of the above conditions being acceptable to you in principle you undertake not to negotiate for the transfer of the concessions to any other individual or Company for a period of 12 months, unless we shall have indicated in the meantime that we do not wish to continue negotiations.

Yours faithfully,
A. T. Wilson
For ANGLO-PERSIAN OIL COMPANY, Ltd.,
for Director.

11th March 1926.

Dear Sirs,

We duly received your letter of the 9th instant (2-FRG) but regret that your suggestions are not agreeable to us.

We can only deal with the matter on the following lines, namely:—

We will endeavour to get the Kuwait Concession completed on similar lines to the Bahrein Agreement, to procure the validity of the Hasa and Neutral Zone Concessions to be established, and to secure, if possible, such modifications to the concessions as you desire. This, it is estimated, would cost us from £12,000 to £13,000.

You on your part to pay us immediately, the sum of £13,000 for an option for one year from this date to acquire the rights we now have, with such modifications and further rights as we may for the time being have succeeded in acquiring in all or any of the four Concessions, namely, Bahrein, Kuwait, Hasa and Neutral Zone, at the definite price of £55,000, the £13,000 to form part of the purchase consideration in the event of the option being exercised.

Should you not exercise the option, you to have a $\frac{13}{55}$ interest in the four Concessions or such of them as have had their validity established.

The Neutral Zone Concession is actually signed, as was made clear to you by Mr. Janson at his interview on the 23rd ultimo, and follows generally the lines of the Hasa Concession. To amend it as suggested by para. (e) of your letter under reply would involve an entirely new agreement being negotiated, the cost of which is not allowed for in the above figure of £13,000; and which negotiations, if not successful, might jeopardise the completion of the signed concession.

It is necessary for us to know whether these terms are acceptable to you by Thursday next, the 18th instant, and we should be glad to have your reply before that date, if possible.

We are, dear Sirs,
Yours faithfully,
For EASTERN AND GENERAL SYNDICATE LIMITED
(Sgd) H. T. Adams
Secretary.

18th March, 1926.

Dear Sirs,

We beg to acknowledge receipt of your letter of 11th March.

We confirm the writer's conversation with Mr. Janson in which he stated that, subject to our receiving an option for two years from this date to acquire the rights you now have, with such modifications and further rights (on the lines already indicated by us) as you may during this period have succeeded in acquiring, in all or any of the four concessions referred to, we are prepared to reimburse you for expenditure incurred in this connection, as and when incurred, up to a total not exceeding £13,000. We are to have the option on the expiry of the two years period of acquiring a 100% interest in the above concessions, at a sum represented by £42,000 + x, (x representing the £13,000 or lesser sum above referred to). Should we not exercise our option, we are to have the right to a free interest of $\frac{x}{£42{,}000 + x}$, in the four concessions, or such of them as have had their validity established.

It is further understood that should we be approached on the subject by the Colonial Office we shall intimate that we do not propose to press our claim to a concession at Kuwait.

As regards Qatar, it is understood that you will not take any steps at present towards obtaining a concession for this Principality, but if at any time during the next two years it is considered desirable by us that such a concession should be obtained through your intervention, you will undertake to use your best endeavours to obtain it, we reimbursing such expenditure as you may incur on the understanding that the total sum of £13,000 is not thereby exceeded.

We agree that Major Holmes should take an early opportunity of visiting Ibn Saud with a view to regularising the concession you now hold.

You emphasized in your conversation with the writer the importance of sending a geological party to Hasa in the near future in order to demonstrate to Ibn Saud that active work is proceeding. We will arrange to do this during the next cold weather, and Major Holmes may inform Ibn Saud accordingly. The geologists that we propose to send will enter and work in Hasa ostensibly under the aegis of your Company.

As regards the proposed formation of an exploration company to undertake work in Hasa, we will give this matter further consideration in due course, but for the present we think it preferable that any work that may be done should be in the name of your Company.

Yours faithfully,
for D'ARCY EXPLORATION COMPANY LIMITED,
A. T. Wilson
Managing Director.

24th March 1926.

Dear Sirs,

We duly received your letter of the 18th instant (50-FRG).

As we pointed out in our letter of the 11th instant to the Anglo-Persian Oil Company Limited, it is estimated that from £12,000 to £13,000 would be required for expenditure in procuring the establishing of the validity of the Hasa and Neutral Zone Concessions and in endeavouring to secure such modifications in those Concessions as your Company has indicated, and in our endeavours to obtain the Kuwait Concession. This was on the footing of your Company having an option for one year to acquire all four Concessions at the price of £42,000 plus such expenditure. If that option is to be extended to two years, there will require to be added a further £7,500 for rentals payable under such Concessions in addition to other probable expenses during the additional year.

My Directors are prepared to fall in with your wish to extend the option to two years on the terms of your letter, but only on the condition that all additional expenditure incurred during such two years beyond the estimate of £13,000 is to be borne by your Company and that this Company incurs no further expense on the four concessions in question.

Kindly Confirm this arrangement and oblige,

Yours faithfully,
For EASTERN AND GENERAL SYNDICATE LIMITED,
(Sgd) H. T. Adams
Secretary.

7th April, 1926.

Dear Sirs,

We beg to acknowledge receipt of your letter of March 24th.

We regret that we are unable to accept your suggested modifications of our original proposals. We have since been in telegraphic communication with Sir John Cadman, who, in view of the opinion of our geologists on the spot, is now unfavourable to the proposals. In the circumstances, therefore, it will be better to consider our letter of 18th ultimo (50-FRG.) as being hereby withdrawn. On Sir John Cadman's return to this country it will be open to you, should you so desire, to approach him afresh on the subject.

Yours faithfully,
for D'ARCY EXPLORATION COMPANY LIMITED,
A. T. Wilson
Managing Director.

---

33. *Mr Thomas E. Ward Sr. (d. 1967) was President of Oilfield Equipment Co., Inc. of New York in 1926 (he had founded the company a few years previously and remained its President until his retirement in 1958) when he became EGS's representative in the U.S.A. In that capacity he negotiated the sale of their Bahrain, El Hasa, Neutral Zone and Kuwait concession interests in 1927 to Gulf Oil Corporation (and subsequent resale in 1928 of the Bahrain concession to Standard Oil Co. of California), whereafter he continued to be closely concerned with EGS Gulf developments in Kuwait until the end of 1933 when EGS's interest in Kuwait terminated.*

Ward had been a pioneer in 1910 in the Trinidad oilfields (where he first met the future Chairman of APOC, Lord Cadman) and later had extensive experience of the oil industry in Mexico and South America as well as the United States before founding his Oilfield Equipment Company, through which he was to become acquainted with most of the leaders of the world petroleum industry.

Following his retirement in 1958 (he was succeeded as President by his son T. E. Ward Jr.), Ward issued, for private circulation only, 'The Story of the Negotiations for Oil Concessions in Bahrein, El Hasa (Saudi Arabia), the Neutral Zone, Qatar, and Kuwait'. In this 300-page monograph he described in detail his part in those negotiations, including much correspondence between EGS and Gulf Oil Corporation executives in connection with the Kuwait negotiations between 1926 and 1933. In 1965 he issued a second edition of this monograph, again for private circulation only, with various corrections and amendments to its Kuwait chapters supplied by the writer of this Record, who from 1948 onwards enjoyed Ward's acquaintance and appreciated his intimate knowledge of and keen interest in the Kuwait concession negotiations.

---

34. *The various companies comprised in the Turkish Petroleum Company (which was to be renamed Iraq Petroleum Company in 1929; see Note 29) had agreed in 1914 not to interest themselves in oil concessions anywhere in the then Ottoman Empire otherwise than through that Company. It is noteworthy that, although since 1899 (see Note 1) Kuwait was generally regarded as independent from Turkish rule, it was felt necessary in the relevant clause of their 1914 agreement to make specific mention of Kuwait's exclusion from it. The clause, as quoted in a letter dated 1st December 1921 from APOC in London (who were partners in the Turkish Petroleum Company) to Sir Arnold Wilson in Persia authorising him to open concession negotiations with Shaikh Ahmad, was as follows:*

The three groups participating in the Turkish Petroleum Company shall give undertakings on their own behalf and on behalf of the companies associated with them not to be interested directly or indirectly in the production or manufacture of crude oil in the Ottoman Empire in Europe and in Asia, except in that part which is under the administration of the Egyptian Government or of the Sheikh of Kuwait, or in the 'transferred territories' on the Turco-Persian frontier, otherwise than through the Turkish Petroleum Company.

This undertaking was reaffirmed in 1928, when a group of American oil companies (see Note 29) joined in the Turkish Petroleum Company, in an agreement dated 31st July 1928 by all its partners that that Company should have the sole right to seek for or obtain oil concessions within 'the defined area' outlined on a map attached to the agreement (which was to become known as 'The Red Line Agreement', because the outline had been drawn on the map in red by Mr Gulbenkian as the TPC acknowledged expert on Turkish territory). This 'defined area' excluded Kuwait, but included Bahrain. Consequently the Gulf Oil Corporation, being a member of the American group in the Turkish Petroleum Company, in 1928 had to offer its Bahrain rights under its option agreement with EGS of November 1927 to that Company; as the offer was declined, Gulf eventually transferred its Bahrain interests on 21st December 1928 to Standard Oil Company of California.

---

35. *The text of Holmes/EGS draft concession terms of July 1928 was as follows:*

In the Name of God The Merciful.
THIS AGREEMENT made the day of 1928.
corresponding to day of 1347.
At KUWAIT between HIS EXCELLENCY SHEIKH AHMED BIN JABIR AL SUBAH C.I.E. SHEIKH OF KUWAIT in Arabia (hereinafter called THE SHEIKH which expression where the context so admits shall include HIS HEIRS, SUCCESSORS, ASSIGNS and SUBJECTS) of the First Part, and FRANK HOLMES of 18, St. Swithin's Lane, E.C.4. London, England, the true and Lawful Attorney of THE EASTERN AND GENERAL SYNDICATE LIMITED whose Registered Office is at 19, St. Swithin's Lane, E.C.4. London, England, (hereinafter called THE COMPANY which expression where the context so admits includes its ASSIGNS and SUCCESSORS) of the Other Part.

WHEREAS THE SHEIKH is desirous of developing the Oil and Petroleum Resources of His Territory, he has for that purpose, agreed to grant unto THE COMPANY the Concession hereinafter contained.

(1) In consideration of the rights, covenants and royalties hereinafter reserved and contained, THE SHEIKH in exercise of his powers, as Ruler and Sovereign of His Dominions for Himself, Heirs, Assigns, Successors and Subjects, hereby grants unto THE COMPANY exclusively:

The exclusive right whereby THE COMPANY shall be entitled throughout the whole of the territories of THE SHEIKH to explore and search the surface of such territories for natural gas, petroleum and all products of oil. THE SHEIKH grants free access to all the agents, successors and servants of THE COMPANY necessarily employed by the latter to all parts of the Kuwait territory under the control of THE SHEIKH, saving only sacred buildings, shrines, graveyards and the area within the existing town wall of Kuwait.

THE SHEIKH grants unto THE COMPANY the right, for a period of Two (2) calendar years to count from the date hereof, to select from the whole of the Kuwait territories under the control of THE SHEIKH, saving only sacred buildings, shrines, graveyards and the area within the existing town wall of Kuwait, an aggregate area not exceeding six hundred forty (640) square miles, divided and located as THE COMPANY may decide, into five (5) or more blocks, which blocks selected by THE COMPANY shall be hereinafter called the 'CONCEDED TERRITORY', and THE SHEIKH hereby grants the following rights and privileges as stated herein, together with the easements and rights and privileges over the Kuwait territory to be exercised in connection with the rights and operations of THE COMPANY in the said CONCEDED TERRITORY, including surface rights over the Kuwait territory necessary for the buildings, camps, storage facilities for water and oil, pipe lines, water and steam lines, roads, railways, telegraph and telephone lines and other ways of communication. In the event that THE COMPANY might desire to occupy lands covered by village areas, gardens or private water wells included within the CONCEDED TERRITORY, then THE COMPANY shall not be permitted to enter into possession thereof without first making arrangements to do so with THE SHEIKH and through THE SHEIKH with the owners.

Within the above mentioned period of two (2) calendar years THE COMPANY will advise THE SHEIKH of the CONCEDED TERRITORY which THE COMPANY may select under this clause by delivering to THE SHEIKH for attachment to His copy of this agreement a signed map showing in detail the boundaries of the said CONCEDED TERRITORY.

The term for which this concession covering the CONCEDED TERRITORY is granted unto THE COMPANY is of seventy (70) calendar years from the date of the execution of these presents, THE COMPANY yielding and paying therefore the fees, payments, royalties, privileges and rights to THE SHEIKH and subject to the provisions hereinafter enumerated.

If THE COMPANY, unless prevented by the Act of God or from war, fire, flood or lightning or some other thing beyond human control, shall not have commenced its exploratory work of the surface of the Kuwait territory through examination and investigation by its geologists and engineers within a period of nine (9) calendar months from the date hereof, then the provision of these presents shall lapse and this deed shall be null and void and neither party shall have any claim against the other in consequence thereof and no moneys already paid to THE SHEIKH shall be returnable.

(2) THE COMPANY shall have the exclusive right during the term of the concession hereby granted to explore the CONCEDED TERRITORY, search for, carry away, export and sell Petroleum, Natural Gas, Asphalt, Ozokerite, Oil and its Products where the same may be found in, on or under the CONCEDED TERRITORY and for that purpose and in connection therewith, exercise in, over and upon the said lands any or all of the following things:

(a) To drill, sink, make, erect, set and construct Wells and Pits, Waterways, Pipelines, Engines, Machinery, Furnaces, Brick-Kilns, Cement Ovens, Workmen's Cottages, construct Railways, Bridges, Tramways and other Ways of Communication, Canals, Wharves, Dams, Erections and other Works, to build Dwelling Houses for THE COMPANY's agents and workmen and set up Stations thereto, to install Telephone and Telegraph lines, and to do generally whatever THE COMPANY may deem expedient for the proper exploitation of the CONCEDED TERRITORY provided that does not harm private and general interest.

(b) To exclusively erect Oil Refineries, Oil and Water Tanks, outside the existing town wall of Kuwait, wherever THE COMPANY shall deem suitable whether in proximity to discover wells or otherwise, provided such action does not in anyway harm the private and general interest.

(c) To have and use for any purpose connected with the working of the said CONCEDED TERRITORY any water within the territory of THE SHEIKH and with the assistance of THE SHEIKH make and construct Water-Courses, Reservoirs and Ponds for collecting such water, provided no harm to the general public or to individuals is apparent.

THE COMPANY shall, in peace time throughout the period of THE CONCESSION, accept and transmit, on its telegraph lines, THE SHEIKH'S GOVERNMENT's telegrams whether in cypher or in clear and likewise allow him the use of its telephone lines, and also THE SHEIKH may use the railways, on special personal services throughout the period of THE CONCESSION during peace time, and have the full use thereof when His Country is at war.

(3) The port and buildings which THE COMPANY require shall be erected outside the town wall of Kuwait, THE COMPANY shall have the power within or without the CONCEDED TERRITORY to construct and develop the harbours along the coast of Kuwait territory and to erect and construct Wharves, Cranes, employ Dredgers, lay down Buoys and erect Lighthouses and do whatever may be necessary to make the harbours safe for the navigation of ships and barges, the unloading of machinery and other goods belonging to or sent to THE COMPANY, THE SHEIKH granting to THE COMPANY the necessary and proper surface rights in connection with such harbours. The Customs Administration of the ports developed by THE COMPANY shall be under THE SHEIKH's local Customs Officials and THE COMPANY undertakes to erect a conveniently large building for a Custom House at each such point and a suitable residence for THE SHEIKH's Official Representative. Should it be necessary also to maintain a guard for the protection of THE COMPANY's works (wells, etc.) inland or along the pipelines or other communications to the sea, THE COMPANY shall build suitable buildings for such guards at its own expense.

THE SHEIKH's Flag and no other shall be used within the CONCEDED TERRITORY.

(4) THE COMPANY shall be free to construct in the Kuwait territory roads, tramways, railroads and other ways of communication and telegraph and telephone lines and one or more pipelines for the purpose of carrying oil, gas and kindred substances, also water and steam, connecting THE COMPANY'S fields and works with each other and with the harbour or harbours

which THE COMPANY may establish on the coast of Kuwait and shall have the right to establish and maintain one or more coaling or oil stations along the coast of Kuwait, the use of such facilities being vested entirely in THE COMPANY. THE SHEIKH retains the right to grant permission to others besides THE COMPANY to import oil and coal and lay pipelines for these purposes.

(5) THE COMPANY shall be free and at liberty to export, sell and dispose to any place or people or country it may wish to and in any manner it may desire the oil and its products won from the CONCEDED TERRITORY and THE SHEIKH and those acting under Him shall not interfere with the internal management of THE COMPANY, but THE SHEIKH shall have the right to keep a general eye over the doings of THE COMPANY. And He shall have the right to levy, and THE COMPANY undertakes to pay Him, on all the oil and its products exported, a custom duty of one per cent (1%). In calculating such one per cent (1%) on the oil exported, the value of the oil at the wells producing same shall be used.

(6) THE COMPANY shall have the right to import all its machinery, equipment, plant, timber, utensils, iron work, building materials and everything belonging to or consigned to it, including medicines and food supplies for the use of THE COMPANY and its employees but not for resale to others, free of customs import duty and taxes, but it shall pay on all personal goods, clothing and general merchandise imported by THE COMPANY for the personal use of its employees the ordinary duty in vogue in the Kuwait territory computed on the values shown in original invoices plus expenses.

(7) THE COMPANY shall be exempt and free, during the period of THE CONCESSION, from all harbor duties and taxes of all kind, tolls and land surface rent of whatever nature, it being understood that THE COMPANY has no right to lease any building to any but its employees and agents. Should ships other than those engaged and used solely for THE COMPANY's business hereunder, either bringing or taking away, make use of the harbor, THE SHEIKH has the right to collect the usual harbor dues and taxes from such ships, and not from THE COMPANY, it being understood that the wharves erected by THE COMPANY are solely vested in THE COMPANY during the period of THE CONCESSION hereby granted and can only during such period, be used by ships on other than THE COMPANY's business with the written permission of the COMPANY.

(8) The ownership of this CONCESSION may not be transferred to or the rights sold to any other company or companies, whether British or otherwise, except to one or more British companies nominated by THE EASTERN AND GENERAL SYNDICATE LIMITED and THE SHEIKH undertakes to sanction such nominated transfer when it becomes necessary provided always that the rights, privileges and interests accruing to THE SHEIKH shall not thereby be prejudiced.

And if by any other ways or means THE CONCESSION is transferred or sold to a THIRD PARTY, this CONCESSION will then become null and void, and THE COMPANY shall leave all the immovable property and wells intact and they will be the property of THE SHEIKH.

(9) THE COMPANY's representatives in the Kuwait territory shall be immune from local interference except with the leave of THE COMPANY and (in matters concerning themselves but not where the Subjects of THE SHEIKH are concerned) shall be responsible for their conduct to THE COMPANY's Board of directors.

(10) If after commencing its operations in THE CONCEDED TERRITORY, THE COMPANY for any reason other than THE ACT OF GOD, or from War, Fire, Flood or Lightning or some other thing beyond Human Control, should discontinue the same for a continuous period of TWO CALENDAR YEARS (2 years) THE SHEIKH shall have the right to cancel the AGREEMENT and no responsibility shall attach to either party.

(11) THE COMPANY or Companies or their employees shall not interfere in any manner or way with the Politics of THE SHEIKH's Dominions or with His Subjects.

(12) THE COMPANY shall employ only native labor (i.e. countrymen) under the supervision of THE COMPANY'S European or other appointed officials and THE SHEIKH agrees to assist with the help of His Amirs and other local agents to procure and provide for THE COMPANY such native labor as THE COMPANY may require, and THE COMPANY on its part undertakes to make the fullest use of the local unskilled labor, in its judgment capable of performing work, to the extent of the suitable supply for its requirements, but THE COMPANY has the right to import unskilled labor should the local supply prove insufficient or unsuitable to the extent of its requirements. THE COMPANY has the right at all times to import skilled workmen of every kind.

(13) THE COMPANY shall pay to the native workmen it employs a fair wage, such wage to be decided and stated by THE COMPANY'S representative at the time the workman is engaged. THE COMPANY shall provide where possible medical attention and medicines free of charge, to its native workmen during the time they are in employ of THE COMPANY.

(14) THE SHEIKH shall always afford the officials and employees of THE COMPANY every facility and assistance in carrying out their plans and projects as far as lies in His Power, and shall allow them to excavate, dig, quarry or drill the Soil in THE CONCEDED TERRITORY wherever they shall have reasonable prospects of discovering and winning petroleum or kindred products, and THE COMPANY by its officials shall be at liberty to abandon any excavation pit or well wherever and whenever they shall seem it expedient so to do, provided always that nothing in this article shall be presumed to give to THE COMPANY or its Assigns or Agents right of entry into,

or on, to private properties without prior sanction of THE SHEIKH or His duly appointed representative.

(15) Within SIXTY DAYS (60) days from the signature of this AGREEMENT, THE COMPANY in consideration of THE SHEIKH'S granting this CONCESSION and the assistance to be afforded to their employees, shall pay to THE SHEIKH the sum of RUPEES THIRTY THOUSAND (Rs. 30,000). But if the payment of this sum of Rs. 30,000 is not made by THE COMPANY within 60 days specified, then this AGREEMENT will become null and void.

And after first payment of Rs. 30,000 as specified above in this article on each anniversary of the date of the signature of this AGREEMENT, THE COMPANY shall pay to THE SHEIKH the sum of RUPEES TWENTY THOUSAND (Rs. 20,000).

The yearly payment of RUPEES TWENTY THOUSAND (Rs. 20,000) shall continue without fail whether THE COMPANY is working or not, until THE COMPANY should declare that oil has been found on THE CONCEDED TERRITORY in commercially exploitable quantities, in which event it agreed that this yearly payment shall cease after the expiry of the then current year for which the rental of RUPEES TWENTY THOUSAND (Rs. 20,000) has been paid.

(16) Should THE COMPANY succeed in finding oil in commercially exploitable quantities, it agrees to pay to THE SHEIKH in lieu of the annual payment of RUPEES TWENTY THOUSAND (Rs. 20,000) provided for in Article (15) FIFTEEN, a royalty of RUPEES THREE AND ANNAS EIGHT ONLY (Rs. 3/8/-) per English ton of net crude oil got and saved (i.e. after deducting water and foreign substances, and oil required for the customary operations of THE COMPANY's installations in the SHEIKH'S Territories).

(17) THE COMPANY hereby undertakes that the amount received by THE SHEIKH in respect of royalties shall not be less than RUPEES SEVENTY THOUSAND (Rs. 70,000) in any complete calendar year in which THE COMPANY continues work, such calendar years to begin at the end of the last day of the year for which the annual rental of RUPEES TWENTY THOUSAND (Rs. 20,000) has been paid, and it is only on the oil won after said day that THE SHEIKH's royalty begins to accrue. In the event of THE SHEIKH, in consultation with the Political Resident in the Persian Gulf, disputing THE COMPANY's decision as to the commercial exploitation, THE COMPANY hereby undertakes its readiness to submit the matter to arbitration as provided in Article 21 TWENTY-ONE below.

(18) In the event of THE COMPANY failing within SIX CALENDAR MONTHS (6 months) of the end of any calendar year or failing, save for causes beyond its control, to carry out its obligations under This AGREEMENT, THE SHEIKH shall have the Power to terminate THE CONCESSION, in which case the provisions of Article (19) NINETEEN shall apply.

(19) THE COMPANY for itself, Successors and Assigns hereby covenants with THE SHEIKH in a manner following:—

(a) To pay the fees and payments required by this DEED at the time and in a manner appointed and also to observe the provisions herein contained.

(b) At the termination of THE CONCESSION whether by the expiration of its period of SEVENTY CALENDAR YEARS (70 years) stipulated, or before such expiration under Article (10) TEN of This AGREEMENT, but after the lapse of THIRTY-FIVE (35) calendar years from the date hereof, deliver to THE SHEIKH all buildings and erections of brick, stone or other materials whatsoever, the railways, telegraphs and telephones and other things standing and being on THE CONCEDED TERRITORY, and all pits, wells, mines, waterways, pipelines, refineries, oil and water tanks, and all such other works and other things belonging to any of the mines and wells, machinery, plants, railways and their cabins and wagons, telegraph and telephone lines and port appurtenances belonging to THE COMPANY, and to leave the ports and harbors as they are, and to relinquish all rights vested in it under Article SEVEN (7) of this AGREEMENT, leaving also buoys and barges, in fact all things belonging to it which are on THE CONCEDED TERRITORY. Provided always that if this CONCESSION shall terminate under Article TEN (10) of these presents within a period of THIRTY-FIVE (35) calendar years from the date hereof, THE COMPANY shall have the right to remove from Kuwait any or all its plant, machinery, tools, apparatus and other things belonging to it above mentioned.

(20) THE COMPANY shall do or cause to be done nothing in THE CONCEDED TERRITORY, which, unless expressly authorized by the provisions herein contained, shall be an infringement of or derogatory to the rights, privileges and prerogatives inherent in THE SHEIKH as Ruler of THE CONCEDED TERRITORY, and in case any such infringement shall inadvertently have been committed by any of THE COMPANY'S officials, upon due proof of such infringement being received by THE COMPANY's local representatives, THE COMPANY shall forthwith make such amends, as may seem fair and reasonable and suitable, and in case of dispute, the Local Judge may be asked to arbitrate and in case of further disagreement, it may be referred direct to THE SHEIKH for judgment.

(21) If at any time during or after the currency of this AGREEMENT any doubt, difference or dispute shall arise between THE SHEIKH and THE COMPANY, concerning the interpretation or execution hereof, or anything herein contained, or in connection herewith, or the rights and liabilities of either party hereunder, the same shall, failing any agreement to settle it in any other way, be referred to two arbitrators, each party choosing one of such arbitrators, and an Umpire who shall be the Political Resident in the Persian Gulf, or a person nominated by Him, before proceeding to arbitration. Each party shall nominate its arbitrator within THIRTY (30) DAYS after being requested in writing by the other party to do so. The decision of the Arbitrators, or in

the case of a difference of opinion between them, the decision of the Referee, shall be final. The place of arbitration shall be such as may be agreed by the parties, and in default of agreement shall be Basra.

(22) THE COMPANY shall pay all monies that may become due to THE SHEIKH under this agreement into his account with the

Bank in

The bank receipt for such money shall be a full discharge for THE COMPANY in regard to due payments.

(23) In addition to the aggregate area of six hundred forty (640) square miles, which THE COMPANY is entitled to select under Clause (1) above, THE SHEIKH grants to THE COMPANY the further right to select, from time to time as its exploration work progresses but not later than ten (10) years after THE COMPANY might designate its selections under Clause (1) above, from the whole of the KUWAIT territory under the control of THE SHEIKH remaining after said selections under Clause (1), saving only sacred buildings, shrines, graveyards and the area within the Town Wall of Kuwait, further areas not exceeding a total of one thousand two hundred eighty (1280) square miles, divided into ten (10) or more blocks, which further areas so selected shall also be known as 'CONCEDED TERRITORY' for the purposes hereof, and in which THE COMPANY shall have the same rights, privileges and benefits as granted to it on the 'CONCEDED TERRITORY' referred to in Clause (1) above.

THE COMPANY will advise THE SHEIKH whenever a selection is made under this Clause by delivering to Him a map of the Kuwait Concession with the selected area or areas clearly shown and identified thereon.

For such areas selected under this Clause, THE COMPANY will pay to THE SHEIKH an annual rental payable in advance of THIRTY RUPEES (Rs. 30) per square mile. The first rental payment however will only cover and will only be proportionate to the period of time between the date of the respective selection and the anniversary of the execution of this Agreement.

Should THE COMPANY win oil, in its judgment in commercial quantities, from any of the areas selected by it under this Clause, THE SHEIKH will be entitled to receive on the oil so won a royalty of THREE RUPEES AND EIGHT ANNAS (Rs. 3/8/) only per English ton and on said oil and its products which may be exported THE SHEIKH will be entitled to levy and collect an export tax as provided in Clause (5) above.

In the event oil is produced from any block selected under this Clause, then from and after the expiry of the then current year as to which the annual rental has been paid on such block, no further annual rental shall be payable on the respective block. In lieu thereof THE SHEIKH shall be entitled to receive the above mentioned royalty of THREE RUPEES AND EIGHT ANNAS (Rs. 3/8/) per English ton on the oil therefrom, and in this case THE COMPANY only guarantees that the royalty per year on the oil from the respective block shall never be less than three and one-half ($3\frac{1}{2}$) times the amount of each annual rental which it has been currently paying on the respective block from which the oil is won.

(24) In the event of any discrepancy between the meanings of the English and Arabic versions hereof, the English version shall prevail.

(25) This AGREEMENT, which comprises the preamble and TWENTY-FOUR (24) Articles other than this, is made and signed by the parties hereto in Original, Duplicate and Triplicate, the Duplicate being retained by His EXCELLENCY THE SHEIKH and both the Original and Triplicate by THE COMPANY.

This AGREEMENT extends over pages all of which are signed by the parties at foot.

IN WITNESS HEREOF the said parties have hereunto set their hands and sealed the day, month and year shown below their respective signatures, and GOD is GRACIOUS.

---

36. *A letter dated 22nd May 1928 from Mr Wallace (see Note 81) in New York to Major Holmes in Bahrain included the following:*

I have just received yours of 20th April in which you so fully explained the difference between Kuwait and Bahrein with respect to the degree of control and supervision which the British Government exercises over the two territories. Your explanation has been very enlightening as naturally it deals on subjects with which we are but little familiar. In view of your explanation, we have cabled you today that we leave it to your own good judgement as to how vigorously you should press negotiations with the Shiekh of Kuwait. In such case however we are again pointing out that one of the tenets of the religion of the oil people is to take what is obtainable as and when it can be had, because experience has taught that the future is always uncertain, particularly when political questions or viewpoints become involved. Therefore our cable indicates that we feel that more or less constant contact should be maintained with the Sheikh of Kuwait even though in your judgement you deem it inadvisable to press negotiations to a conclusion.

---

37. *On 28th November 1928, EGS in London cabled T. E. Ward in New York: 'Holmes reports by cable he has interviewed Sheikh of Kuwait and discussed revised agreement* *generally further detailed discussion taking place this week stop Holmes reports Shaikh seems amenable and thinks early settlement.'*

---

38. *Conditions in lieu of a British Nationality Clause in the Bahrain concession agreement, as agreed between the Colonial Office and EGS in January 1930:*

The 'conditions' as set out in the Colonial Office letter of September 16, 1929 and modified by the letter of January 3, 1930 were as follows:

(a) That the Company formed to take over the Concession from the Syndicate shall be and remain a British Company registered in Canada but shall maintain a registered office in Great Britain which shall at all times be in charge of a British subject who shall be the recognized channel of communication between the Company and His Majesty's Government in the United Kingdom of Great Britain and Northern Ireland.

(b) That of the five Directors of the Company one Director shall at all times be a British subject who shall be persona grata to His Majesty's Government. His selection for appointment as Director shall be made in consultation with His Majesty's Government, and his salary as Director shall be provided by the Company; it being understood that a reasonable time shall be allowed for the replacement of the British Director in the event of that post falling vacant.

(c) That the Company shall at all times maintain in Bahrein an official to be called the 'Chief Local Representative' of the Company whose appointment shall be approved by His Majesty's Government in the United Kingdom and who shall be the sole representative of the Company empowered to deal direct with the local authorities and population in Bahrein. All communications which that official may desire on behalf of the Company to address to the Sheikh of Bahrein shall be made through the British Political Agent in Bahrein. For the first five years after the Company starts to operate in Bahrein, or for such lesser period as the Company may operate in that territory, their Chief Local Representative shall be Major Frank Holmes, provided the arrangement between the Company and Major Holmes continues to be mutually satisfactory to them during such five years, or such lesser period above mentioned, provided also that any sooner determination of the appointment of Major Holmes shall be subject to the consent of His Majesty's Government, which shall not be unreasonably withheld; it being understood that a reasonable time shall be allowed for the replacement of the Chief Local Representative in the event of that post falling vacant.

(d) That as many of the employees of the Company in Bahrein as is consistent with the efficient carrying on of the undertaking, shall at all times be British subjects or subjects of the Sheikh of Bahrein.

(e) His Majesty's Government also desire to receive an assurance from the Syndicate that neither the Syndicate, nor the Company to which it is proposed that the rights of the Syndicate should be assigned, will take any steps which would prejudice the position of the proposed sites for a landing ground and seaplane station in Bahrein, the precise location of which will be communicated to the Syndicate as soon as possible.

The exploration license was accordingly extended and the original Bahrein Concession was assigned to the Bahrein Petroleum Company, a Canadian subsidiary of the Standard Oil Company of California.

---

39. *Lt-Col H. R. P. Dickson (1881–1959) was British Political Agent in Kuwait from 1929 to 1936. An exceptional Arabic linguist, having spoken the language since childhood (he was born in Beirut, his father being then British Consul in Damascus and later Consul-General in Jerusalem), he had served in Mesopotamia (Iraq) in 1914–15 as a soldier and then as a Political Officer until 1919 when he was appointed Political Agent in Bahrain. In 1921 he returned to Iraq as Political Adviser at Hillah until 1922 when he was sent to Bahrain (when he first met Major Holmes; see Note 15) to arrange the Ojair Conference (see Note 22) and interpret for Sir Percy Cox in his conversations with Ibn Saud.*

This wide experience and personal acquaintance with the leaders and people of the adjoining countries were the background of his very distinguished thirty-year career in Kuwait, where he arrived to take up his appointment as Political Agent in May 1929, after service in India from 1923 to 1927 followed by one year as Secretary to the British Political Resident in the Gulf at Bushire.

Dickson's term of office as Political Agent lasted seven years (during which he was intimately connected with the Kuwait oil-concession negotiations as described in this Record) until his retirement from Government service, at the compulsory retirement age for Political Service officers of fifty-five, in 1936. In that year he was appointed Chief Local Representative of the Kuwait Oil Company by Shaikh Ahmad who, from their first acquaintance in 1929 until his death in 1950, held Dickson in increasing respect and affection; the Shaikh, in accordance with the KOC Concession Agreement of December 1934, had the right to select the Company's first Chief Local Representative in consultation with H.M.G. Dickson continued in this position for twenty-three years until his death in Kuwait in 1959; he and his wife were then still residing there (as his widow does to this day) in the house which they had first occupied in 1929 and of which Shaikh Ahmad had granted them a permanent tenancy in 1936.

Since early in this long career in Arabia, Dickson and his wife (they married in 1920) compiled notes and records of local folklore, habits and events; including tribal customs and traditions in which his intuitive sympathy with the Beduin way of life and mastery of tribal dialects facilitated their researches. They also studied the local flora and fauna, Mrs Dickson being an enthusiastic botanist and zoologist. In 1949 Dickson published these notes and records, together with some of his own adventures and travels, in his encyclopaedic and profusely illustrated (Mrs Dickson being a talented artist) The Arab of the Desert *(664 pp.; Allen & Unwin). Towards the end of his thirty years residence in Kuwait, he followed this in 1956 by* Kuwait and her Neighbours *(627 pp.; Allen & Unwin), an equally comprehensive volume resulting from his unique experiences and extensive journeys in and around Kuwait. These two books were further supplemented by Mrs Dickson in her* Wild Flowers of Kuwait *and* Forty Years in Kuwait *(both published by Allen & Unwin) in 1955 and 1971 respectively.*

---

40. *Correspondence between H. W. Cole of the Petroleum Department and Sir John Cadman of APOC:*

Confidential 18 October, 1930.

Dear Sir John Cadman,

As you no doubt know the Eastern and General Syndicate have for some time been negotiating for Oil Concessions on the Arabian Coast of the Persian Gulf. They have recently applied for permission to negotiate an Oil Concession with the Shaikh of Kuwait and to transfer their rights to a Canadian Company in which the bulk of the capital will be held by the Standard Oil Company of California. The Syndicate have already been granted a Concession in Bahrein, and before recommending the grant of any further areas to American interests, we should like to know definitely whether there is any prospect of British interests being disposed to assist in prospecting for oil in Kuwait?

Do you consider that there is any chance of either the Anglo-Persian Oil Company or the IPC being prepared to undertake exploration for oil in this district or have your geological advisers considered this district and turned it down on the ground that prospects are too unfavourable to make it worth carrying out any test drilling? The American interests referred to appear to be willing to risk some money and we do not like to see any area which offers any promise going entirely into American hands.

Perhaps you could also say whether the Standard Oil Company of California is in any way linked up with the Standard Oil Companies who at present have share-holdings in the I.P.C? As the matter is somewhat urgent I should be grateful for an early reply.

Yours sincerely,
H. W. Cole

21st October, 1930.

Dear Mr. Cole,

I was interested in what you told me in your letter of the 18th inst. on the subject of the intention of the E. & G. Syndicate to renew negotiations for an oil concession in Kuwait. According to our latest information, the representative of this Syndicate (Major Holmes) left Kuwait in 1928, after the breakdown of negotiations with the Sheikh.

As you doubtless know, the A.P.O.C. has very long been interested in the subject of concessions both at Bahrein and Kuwait; the earliest negotiations are more than 10 years old. Very little detailed geological work has been done and reports leave little room for optimism. At the same time, the Company had repeatedly indicated its willingness to conclude a concession on reasonable terms, both with Kuwait and Bahrein. It has always been recognised that the search for oil was a very 'long shot' and that the probability was that expenditure would be nugatory. None the less, in view of the special geographical position of these territories, we were quite willing to take the chance. What was important, however, was that in the event of success, the concessionary obligations should be of reasonable character, and it was on this point that our negotiations broke down. The Colonial Office, acting in a kind of tutelary capacity, was unwilling to recommend modification of the terms such as to suit our views. We considered the royalty obligations too high and the clause relating to Customs Duties unduly onerous.

Negotiations were rendered yet more difficult and protracted by the fact that the E. & G. Syndicate was even then in the field and seemed ready to accept obligations to which we objected, and to outbid us on other points. Their object was not our object: what they wanted were concessions with which to traffic; ours was to obtain a concession for honest development should oil be found to exist. They appear to have got away with the Bahrein concession, and I do not doubt their capacity, provided they bid sufficiently high, to do the same in Kuwait.

Before I can answer specifically your question as to whether the A.P.O.C. is prepared to reconsider its own position in this matter, I should very much like to know the character of the terms which the E. & G. Syndicate is prepared to offer, as I presume these terms will need some form of endorsement on the part of the Colonial Office. I wonder whether you can also let me know the terms on which the concession in Bahrein was actually concluded.

As regards the two other questions which you put to me, the Standard Oil Company of California is not one of the parties constituting the Near East Development Corporation, which, as you know, forms the American Group in the I.P.C. Further, Kuwait is not in the area covered by I.P.C. interests; in fact, it has been specifically excluded.

Yours sincerely,
(Sgd) John Cadman

---

41. *Petroleum Department to A. C. Hearn of APOC:*

20 January, 1931.

Dear Mr. Hearn,

Many thanks for your letter of 13th January addressed to Mr. Cole concerning the possibility of the Anglo-Persian Oil Company undertaking exploration for oil in Koweit. I am sorry to note that the prospects are not regarded as being more favourable, but am obliged to you for looking into the matter. We have informed the Departments concerned of the Anglo-Persian Oil Company's decision.

Yours faithfully,
F. C. Starling

---

43. *Correspondence between APOC and Petroleum Department:*

Anglo-Persian Oil Co. Ltd.
Britannic House,
London, E.C.2.

The Director,
Petroleum Dept.,
(Mines Dept.)
Dean Stanley Street,
MILLBANK, S.W.1.

25th August 1931

Sir,

We have the honour to refer to the correspondence ending with Mr. Starling's letter of the 20th January, P.D./98, and to state for your confidential information that the results of recent test borings on the Persian coast south of Bushire give us some reason to think that the prospects of the existence of petroleum in Koweit territory are perhaps somewhat less remote than have hitherto appeared to be the case. In these circumstances we should be prepared, subject to the approval of the Shaikh of Koweit and the concurrence of H.M's Government, to send a small party of geologists to Koweit in the autumn to make a thorough examination of the surface geology.

We anticipate that a reconnaissance on the above lines would take several months and we assume that the Shaikh of Koweit would be prepared to give us every facility in the matter without requiring any payment from us other than out-of-pocket expenses incurred.

If your Department is prepared to support these proposals we shall be glad to receive a very early intimation of the decision reached by the Departments concerned in order to enable us to lay our plans accordingly.

We have the honour to be,
Sir,
Your obedient Servants,
A. C. Hearn

26th August, 1931.

A. C. Hearn, Esq.,
The Anglo-Persian Oil Company, Limited.
Britannic House,
Finsbury Circus,
London, E.C.2.

Dear Hearn,

Many thanks for your letter of August 25th from which I am glad to note that the Anglo-Persian Oil Company are now prepared to send out a party of geologists to Kuwait. Starling is at present away on leave, but the Departments concerned have been informed of the Anglo-Persian Oil Company's proposal and we hope to let you have an official reply at an early date.

Yours sincerely,
H. P. W. Giffard

42. *Shaikh Ahmad's letter of 2nd July 1931 to Holmes (facsimile of original Arabic text and certified English translation):*

Ahmad Al Jabir Al Sabah
KUWAIT.

احمد الجابر الصباح
كويت

كويت ١٦ صفر ١٣٥٠ موافق ٢ جولي ١٩٣١

حضرت الاجل الافخم حميد الشهم المحب العزيز الميجر فرنك هولمز المحترم دام محروسا
بعد التحية ؛

وصلني كتابكم المؤرخ ٢٨ جون ١٩٣١ وماذكرتم صار معلوم — نسخة الشروط التي قدمتموها لاجل امتياز لعمل الكاز في اراضي الكويت اشرفنا عليها ؛

نحن قد بينا لكم في مذاكرتنا الشفاهيه عن المواد التي بينتها لكم حكومة صاحب الجلاله البريطانيه ان ادماجها لازم في اي مقاوله بعمل الكاز — لاني اعتقد ان حكومة صاحب الجلاله البريطانيه هي حكومتي الصادقه التي لازال باذله عنايتها لمصلحة بلادي وحفظ حقوقي — فينبغي ان لا نترك مانراه مفيدا لنا ولبلادنا ؛

بناء عليه اذا انتم وشركتكم اتفقتم مع حكومة صاحب الجلاله البريطانيه على تلك المواد التي بينتها لكم وسمحت لكم في الغائها — فحينئذ نحن يصير لنا معكم مراجعة اخرى هذا مالزم ودمتم

مخلصكم
احمد الجابر

Kuwait 16 Sefar 1350
2nd July 1931.

To Major Frank Holmes.

After Greetings,

I am in receipt of your letter dated 28th June 1931 contents of which are noted. The draft of the conditions which you have presented for an oil concession in the territories of Kuwait has been considered by us.

We have mentioned to you in the course of our conversations regarding the clauses which have been indicated to you by H. B. M.'s Government, that their incorporation in any oil concession is necessary. Because I believe that H. B. M.'s Government is my own sincere Government which is always devoting its care to the welfare of my country and the safeguard of my rights. We must not ignore what it considers useful to us and to our country.

Therefore if you and your Company agree with H. M.'s Government on the said clauses mentioned to you and it allows you to omit them, then we shall have another opportunity of discussing matters with you.

Yours sincerely,
sd. Ahmad al J.

True translation
J. E. Elkabir
Advocate, Baghdad

44. *The Acting Political Agent in July 1931 during Lt-Col Dickson's absence on local leave was the Agency's medical officer, Dr Alan Greenway, an Indian Medical Service doctor of wide experience long resident in Kuwait. He had not been closely concerned with the concession negotiations or the nationality clause controversy and therefore could not appreciate the implications of Shaikh Ahmad's letter of 2nd July to Holmes.*

---

45. *Extract from Major Holmes's letter from Kuwait to EGS London of 1st November 1931:*

In summing up the position several points are clear to me. The Political people here have been instructed or with the full approval at least of the Colonial Office to endeavour to make the Sheikh of Kuwait retract his letter to me, secondly when that failed their efforts were concentrated on delaying the whole of our Concession business as much as possible.

The only way to overcome this delay is to use every means in our power to force a decision from the Colonial Office. I do not see how I can do anything further here until such a decision is given.

Up to the present the Sheikh of Kuwait has given no sign that he will alter his attitude on the Nationality Clause and he insists that it is a question to be settled between our Company and the Colonial Office.

Therefore it resolves itself as to what course is taken to obtain a definite statement from the Colonial Office as to their intentions. I personally consider that we have been patient and considerate to a fault and that we are now justified in making use of every weapon, diplomatic or otherwise we possess.

It has been proved to the Colonial Office that the Sheikh of Kuwait is willing to grant us the concession without the Nationality Clause if the Colonial Office will permit him.

It is the Colonial Office and not the Sheikh who is the stumbling block. The Sheikh and his people are anxious to have exploration work commenced at the earliest possible.

---

46. *Following State Department's telegram of instructions of 3rd December 1931 (see below) to the American Embassy in London, that they should request British Government action to waive the nationality clause which was holding up Gulf's Kuwait negotiations, the matter was discussed on 4th December with the Foreign Office by Mr Atherton, the Embassy's Counsellor and Charge d'Affaires, in the temporary absence in France of the Ambassador, General Dawes. On 14th and 28th December, Ambassador Dawes continued the discussion in considerable detail with the British Foreign Secretary, Sir John Simon; after which the Embassy was informed by the Foreign Office (see below attached letter of 29th December 1931 from Atherton to State Department) that, although the Colonial Office and other Departments had to be consulted and the precise significance ascertained of Shaikh Ahmad's letter of 2nd July to Holmes, their considered views would be sent as soon as possible.*

Department of State,
Washington,
December 3, 1931.

AMEMBASSY
LONDON (ENGLAND)

336. Please refer to the Department's 61, March 28, 6 p.m., 1929, regarding a petroleum concession in the Bahrein Islands.

This Department is informed that at the time the Gulf Oil Corporation obtained the option on the Bahrein concession it also obtained from the Eastern and General Syndicate an option on a concession for which the Syndicate was negotiating in Kuwait. The Syndicate has continued its negotiations with the Shaikh of Kuwait who seems to be willing to grant a concession contract on terms acceptable to the American company, but the Colonial Office appears to have intervened and to have insisted upon the inclusion in the concession contract of the so-called 'British nationality clause'. As in the case of Bahrein, the insertion of this clause would effectively exclude a company controlled directly or indirectly by American interests from holding or operating the proposed concession on Kuwait. Unless the Colonial Office is willing to substitute for the 'nationality clause' a clause similar to that which was finally agreed upon in connection with the Bahrein concession, the Gulf Oil Corporation will be barred from proceeding with the Kuwait development.

From evidence furnished by the Gulf Corporation it appears that it, as well as the Eastern and General Syndicate, has made sincere efforts during the past two years to adjust this matter, but it has met with one delay after another. Finally, on August 4, 1931, the Syndicate addressed a letter to the Colonial Office inquiring whether it would now be prepared, in view of the favorable attitude which had been shown by the Shaikh, to notify the Political Resident in the Persian Gulf that the British Government had no objection to the Shaikh granting a concession from which the British nationality clause was omitted. No definite answer has been received to this inquiry, but on November 25, 1931, the Colonial Office informed the Syndicate that it would reply 'as soon as practicable'.

In view of the delays which have already occurred in this matter it is desired that you seek an informal interview with the appropriate authorities and express the Department's hope that it will soon be possible for the Colonial Office to give a favorable reply to the American Company through the Eastern and General Syndicate. In this connection you may state that the Department assumes that the Colonial Office has no desire to exclude American interests from participating in the development of any petroleum resources which may exist in Kuwait and that it is hoped there will be no difficulty in omitting from the proposed Kuwait concession the British nationality clause.

You will understand that the principle which the Department wishes to establish in this case is the right of American interests to participate in the development of the petroleum resources of Kuwait upon an equal basis with British interests. The Department does not wish to insist that any particular concession be granted to any particular American company, but it does wish to open the door in Kuwait so that American interests may have an equal opportunity to compete. In this general connection it may be mentioned that the Department is informed that several draft concessions have been submitted to the Shaikh by the Eastern and General Syndicate. Some of these drafts provide for the grant of certain areas for exploitation, such areas to be selected by the Syndicate or its assignees; other drafts provide for an exclusive exploitation concession for the whole of Kuwait. The Department has made it clear to the Gulf Oil Corporation that this Government would not be prepared to ask for any exclusive rights for American interests in Kuwait and that corporation has expressed its willingness to confine its exploitation rights to a reasonable area. In the event that this aspect of the question is raised by the British authorities you may inform them of the attitude of this Government and of the Gulf Oil Corporation.

It is understood that the Ambassador has been informed of the background of this whole situation. In the event that further or more detailed information is needed it is suggested that you consult Major Harry Davis, the London representative of the Gulf Oil Corporation, who is understood to be thoroughly informed in the matter.

Please inform the Department by telegraph of the results of your informal representations.

(STIMSON)

Embassy of the
United States of America
London, December 29, 1931.

No. 2482
The Honorable
The Secretary of State,
Washington, D.C.
Sir:

I have the honor to refer to the Department's telegraphic instruction No. 336, December 3, 5 p.m., regarding the Koweit oil concession, and, as of record for the Department in its discussions with any interested American company, to state, as set forth in the Embassy's telegram No. 453, December 4, 5 p.m., that on that date this matter was discussed by the Chargé d'Affaires with the Assistant Secretary of State. On December 14, upon his return from Paris, the Ambassador discussed this matter with Sir John Simon, and supplemented this conversation by an informal note, dated December 22, to the Foreign Secretary, a copy of which is enclosed.

Shortly after the Ambassador's interview with Sir John Simon, the Counsellor of the Embassy had occasion to see Sir Lancelot Oliphant and again referred to the matter, and on December 22 received a note of reply to his representations, which stated:

> 'In the course of our conversation on the 4th December, you raised the question of the application of the Eastern and General Syndicate to be granted a concession in respect of exploration for oil in Koweit.
>
> 'I find on enquiry that some doubt exists as to the correctness of the interpretation placed by the Syndicate upon the letter addressed to their representative, Major Holmes, by the Sheikh of Koweit on the 2nd July, regarding the inclusion in such a concession, if it were granted, of a clause stipulating that the concession should not be transferred to a non-British company.

'In these circumstances it has been necessary to obtain a report from the Political Resident in the Persian Gulf and the Political Agent at Koweit before any further communication can be made to the Syndicate. This report has only just been received, and a further communication will be sent to you as soon as it has been considered by the various Departments concerned.'

This was followed shortly by an acknowledgment from Sir Lancelot Oliphant, in the absence of Sir John Simon, to the Ambassador's personal letter of December 22, referred to above.

On December 28, in the course of a conversation with Sir John Simon, the Ambassador took occasion again to raise the question, and left with him a memorandum which had been prepared by Major Harry G. Davis, the London representative of the American company, setting forth certain facts for consideration. A copy of this memorandum is attached hereto.

On December 29, Mr. Atherton saw Sir Lancelot Oliphant again and in the course of conversation said that he hoped the Foreign Office would satisfy itself that while this matter of the Koweit concession was under discussion between the Embassy and the Foreign Office no facilities for survey and exploration were being sought through the good offices of British officials for any other British oil company. Sir Lancelot Oliphant promised to raise this point at once with the Colonial Office and to indicate his feeling of the correctness of this attitude. He stated further that the text of a note of the Sheikh of Koweit regarding the non-nationality clause which had been referred to in his note quoted in this despatch was under consideration at the moment by the Colonial Office and that he would again inquire as to the likelihood of an early and fuller reply.

The texts of all notes to and from the Foreign Office in this connected have been shown to Major Davis, the London representative of the American company, and reports of conversations with the Foreign Office have been discussed with him, so that he has been kept fully informed of the Embassy's action in every detail, and he has, according to his reports here, fully informed his company in New York from day to day of the progress of the negotiations.

Respectfully yours,
(For the Ambassador)
Ray Atherton
Counsellor of Embassy.

Enclosures:
General Dawes to Sir John Simon,
December 22, 1931.
Memorandum for the American Ambassador from
Major H. G. Davis, dated December 28, 1931.

The Right Hon.
Sir John Simon, G.C.S.I. London, December 22, 1931

My dear Sir John:

In furtherance of my remarks to you the other day regarding the Koweit oil concession, I venture to set forth below the pertinent section of the United States Mining Lease Act of February 25, 1920, under which you will notice that British subjects receive the same treatment as American citizens:

'(Public-No. 146–66th Congress)
'(S. 2775)

'An Act To promote the mining of coal, phosphate, oil, oil shale, gas, and sodium on the public domain.

'Be It enacted by the Senate and House of Representatives of the United States of America in Congress assembled, That deposits of coal, phosphate, sodium, oil, oil shale, or gas, and lands containing such deposits owned by the United States, including those in national forests, but excluding lands acquired under the Act known as the Appalachian Forest Act, approved March 1, 1911 (Thirty-sixth Statutes, page 961), and those in national parks, and in lands withdrawn or reserved for military or naval uses or purposes, except as hereinafter provided, shall be subject to disposition in the form and manner provided by this Act to citizens of the United States, or to any association of such persons, or to any corporation organized under the laws of the United States, or of any State or Territory thereof, and in the case of coal, oil, oil shale, or gas, to municipalities: Provided, That the United States reserves the right to extract helium from all gas produced from lands permitted, leased, or otherwise granted under the provisions of this Act, under such rules and regulations as shall be prescribed by the Secretary of the Interior: Provided further, That in the extraction of helium from gas produced from such lands, it shall be so extracted as to cause no substantial delay in the delivery of gas produced from the well to the purchaser thereof: And provided further, That citizens of another country, the laws, customs, or regulations of which, deny similar or like privileges to citizens or corporations of this country, shall not by stock ownership, stock holding, or stock control, own any interest in any lease acquired under the provisions of this Act.'

In discussing this with the representative of the interested company here, he informed me that the syndicate concerned has already advised the Colonial Office that the same nationality conditions as incorporated by the Colonial Office in the case of the Bahrein oil concession, which the Embassy discussed with the Foreign Office in 1929 (reference: Foreign Office note No. E 2521/

281/91 of May 29, 1929) are acceptable in the case of the Koweit concession. I understand the Bahrein concession was assigned to the Bahrein Petroleum Company, the Canadian subsidiary of the Standard Oil Company of California. In this connection a statement to me by the syndicate representative may be of interest to you:

> 'The agreement entered into between the Eastern and General Syndicate, Limited, with the Eastern Gulf Oil Company stipulated that the nominee of the Eastern Gulf Oil Company will be a Canadian or English corporation, at the election of the Eastern Gulf Oil Company. A copy of this agreement has been in the possession of the Colonial Office (in its consideration of the Koweit oil concession) since December 28th, 1928, the same having been furnished to the Colonial Office by the Eastern and General Syndicate, Limited.'

As stated to you the other day, the discussions have been so long delayed that I should be most appreciative of an early word of reply.

Yours sincerely,
(Signed) Charles G. Dawes.

Memorandum for the American Ambassador from Major Harry G. Davis.

London,
December 28, 1931.

1. With further reference to the Koweit Oil Concession, I understand that some doubt exists on the part of the Colonial Office as to the correctness of the interpretation placed by the Eastern and General Syndicate Limited upon the letter (dated 2nd July, 1931) addressed by the Sheikh of Koweit to their representative, Major Holmes, regarding the inclusion in such a concession of a clause stipulating that the concession should not be granted to a non-British company.

2. I wish to point out that the concession when granted is to be executed by the Sheikh of Koweit in favor of the Eastern and General Syndicate Limited, a British Company; that the latter, by virtue of their contract with the Eastern Gulf Oil Company, have undertaken, when it is consummated, to transfer and assign said concession to the nominee of the Eastern Gulf Oil Company, which company, in turn, have elected, in accordance with the terms of the aforementioned contract, that the concession is to be transferred and assigned to a Canadian subsidiary of the Eastern Gulf Oil Company. The nominee company would be and would remain a British company registered in Canada. Furthermore, as previously mentioned, the Eastern and General Syndicate Limited have notified the Colonial Office, by letter dated August 4th 1931, that they were prepared to include in the Koweit Concession similar conditions to those incorporated by the Colonial Office, in the Bahrein Concession, thus they repeated what was on several occasions previously conveyed to the Colonial Office by representatives of the Eastern and General Syndicate Limited during the course of various discussions on the subject.

3. The Sheikh of Koweit in his letter of July 2nd 1931 to Major Holmes has expressed his willingness to omit from the Koweit oil concession the 'clauses' (so-called 'British Nationality Clauses') provided the British Government will agree to the omission of same. The Sheikh has recently confirmed his attitude to Major Holmes in Koweit and has stated if the British Government will notify the local Political officials that there is no objection to the omission of the said clause so that their opposition may be withdrawn he is willing to proceed with the consummation of the concession.

4. In the final analysis the interpretation of the letter written by the Sheikh of Koweit to Major Holmes is quite immaterial. The real and sole question now under discussion between the American Government and British Government is whether the British Government does or does not insist upon the inclusion in the Koweit Concession of the so-called 'British Nationality Clause' and whether the British Government is willing to substitute for the said clause similar conditions to those finally incorporated in the Bahrein Concession.

5. A copy of the conditions that were incorporated in the Bahrein Concession is attached hereto.

(Signed) Harry G. Davis.

Enclosure.

CONDITIONS.

A. The Bahrein Petroleum Company Limited as assignee of the Eastern & General Syndicate Limited shall be and remain a British Company registered in Canada but shall maintain an office in Great Britain which shall at all times be in the charge of a British subject who shall be the recognised channel of communication between the Company and His Majesty's Government in the United Kingdom of Great Britain and Northern Ireland.

B. Of the five Directors of the Bahrein Petroleum Company Limited one Director shall at all times be a British subject who shall be persona grata to His Majesty's Government. His selection for appointment as Director shall be made in consultation with His Majesty's Government and his salary as Director shall be provided by the Company. A reasonable time shall be allowed for the replacement of the British Director in the event of this post falling vacant.

C. The Bahrein Petroleum Company Limited shall at all times maintain in Bahrein an official to be called the 'Chief Local Representative' of the Company whose appointment shall be approved by His Majesty's Government in the United Kingdom and who shall be the sole representative of the said Bahrein Petroleum Company Limited empowered to deal direct with the local authorities and population in Bahrein. All communications which that official may desire on behalf of the said Bahrein Petroleum Company Limited to address to the Sheikh of Bahrein shall be made through the British Political Agent in Bahrein. For the first five years after the said Bahrein Petroleum Company Limited starts to operate in Bahrein, or for such lesser period as such Company may operate in that territory, their 'Chief Local Representative' shall be Major Frank Holmes, provided the arrangement between that Company and Major Holmes continues to be mutually satisfactory to them during such five years, or such lesser period above mentioned PROVIDED ALSO that any sooner determination of the appointment of Major Holmes shall be subject to the consent of His Majesty's Government which shall not be unreasonably withheld. A reasonable time shall be allowed for the replacement of the 'Chief Local Representative' in the event of this post falling vacant.

D. As many of the employees of the Bahrein Petroleum Company Limited in Bahrein as is consistent with the efficient carrying on of the undertaking, shall at all times be British subjects or subjects of the Sheikh of Bahrein.

DATED the 12th June 1930

SEAL

Sheikh of Bahrein

Before me,

(sgd) C. C. Prior

British Political Agent

Bahrein

EASTERN & GENERAL SYNDICATE LIMITED

By Frank Holmes

Its Attorney-in-Fact.

---

47. *In February 1932 General Dawes (see Note 46) was succeeded as American Ambassador in London by Mr Andrew Mellon (d. 1937) who, as a senior member of the family who had founded the Gulf Oil Corporation, had substantial interests in it. At the time of his appointment to London, State Department arranged with him that his Embassy's dealings with the British Government regarding Gulf's position in Kuwait would be conducted by his Embassy's Counsellor Mr Atherton (see Note 46), to avoid his being embarrassed by having to make representations on behalf of a concern in which he was so closely interested. Subsequently, Mr Mellon suggested to State Department, at the end of his first month as Ambassador, that his Embassy's representations to the British Government on behalf of Gulf, which Atherton had been pursuing with increasing urgency since January, might be transferred to Washington and carried on exclusively between the British Embassy there and the State Department; a suggestion which, being considered impractical as well as unnecessary, was not adopted (see Note 54).*

It was for that reason that State Department's telegram of instructions of 26th March 1932 (see below) to London for formal action with the British Government were pursued with Foreign Secretary Sir John Simon by Atherton (see telegram of 30th March below), and not by Ambassador Mellon as would normally have been the case in a matter of such importance. And for the same reason Sir John Simon's reply of 9th April (see below) was addressed to Atherton and, in its opening paragraph, refers to the 'representations made by General Dawes (see Note 46) and yourself on this subject'. The British Foreign Office obviously appreciated, and respected, the delicacy of Mr Mellon's position as well as did the State Department.

It was not until some six months later when, at Gulf's request, State Department from September 1932 onwards (see telegram of 1st September below) had been pressing the Foreign Office to accelerate resumption of negotiations in Kuwait and expressing apprehension that APOC was being unduly favoured, that Ambassador Mellon for the first time took part in his Embassy's representations (see page 29). On 6th September Atherton had complained to the Foreign Office that Gulf's concession negotiations were being delayed unnecessarily and to the advantage of APOC, complaints which the Foreign Office's reply of 16th September (see below) described as unjustified, stressing that both applicants for the concession were receiving equal treatment. A month later Ambassador Mellon himself requested the Head of the Foreign Office (Sir Robert, later Lord, Vansittart) on 17th October (see below) to expedite action, following this in November with a detailed memorandum which Sir Robert answered on 11th and 23rd November (see below) in equal detail and justifying such delay as had occurred as unavoidable in all the circumstances. By the second week of December 1932, there being still no news from Kuwait of

H.M.G.'s embargo on the negotiations being lifted although the American Embassy had been informed by the Foreign Office on 11th November (see below) that instructions to that effect had already been sent, Ambassador Mellon informed the State Department on 15th December (see below) that he had left a stiff note on the subject with the Foreign Office. This resulted, when he was visiting Washington a few weeks later (see below, State Department memo of 27th December 1932) in a suggestion that, in view of his personal position, he should 'go easy on this question' and continue to leave its conduct to Mr Atherton as originally arranged. By the time Ambassador Mellon returned to London from his Washington visit the resumption of negotiations on 14th January 1933 (see page 29) had been authorised by the Political Resident in Kuwait; on 20th March his appointment as Ambassador ended and he returned to America on 23rd March.

Department of State,
Washington,
March 26, 1932.

AMEMBASSY
LONDON (ENGLAND)
100. Your 116, March 23, 1 p.m.

After mature consideration, the Department has arrived at the conclusion that no useful purpose would be served by continuing further with the Foreign Office your informal representations regarding the question of American rights in Kuwait.

In view of the undue delay that has already intervened in this matter and the evident necessity of receiving as soon as possible an expression of the British Government's intentions with respect thereto, it is desired that you seek an early interview with the Foreign Secretary and present to him at the same time a communication embodying in appropriate terms the following considerations:

1. This Government recalls the inquiry which it made through the Embassy in 1929 as to the policy of His Majesty's Government in the matter of the holding and operation of petroleum concessions by American nationals in British-protected Arab territories such as Bahrein. His Majesty's Government is aware of the solution subsequently arrived at in the specific case of the Eastern and General Syndicate which on behalf of the Eastern Gulf Oil Company was at that time seeking a modification of the Nationality clause, the inclusion of which in any oil concessions granted by the Sheikh of Bahrein was being insisted upon by the Colonial Office. The arrangement then agreed upon had appeared to this Government only just in view of the extremely liberal treatment accorded in the United States and in its possessions in regard to the operation of petroleum concessions by British controlled companies. This Government had therefore supposed that the policy of His Majesty's Government would be no less liberal in the matter of according open-door rights to American nationals in Kuwait than it had shown itself to be in the almost identical case of Bahrein. This Government sincerely trusts that it has been correct in this assumption and would appreciate an early indication that such is the case.

2. This Government understands that it is the policy of His Majesty's Government to require of companies seeking concessions in Arab States such as Kuwait, that such companies obtain the prior consent of the rulers of such States to the entry and operations of such companies in the territories in question. This Government is informed that contrary to the impression that seems to have prevailed in the Colonial Office the Sheikh of Kuwait is understood to be quite agreeable to the specific entry of the Eastern Gulf Oil Company and to the granting on behalf of that company of an oil concession without the inclusion of the 'Nationality clause'. This Government trusts that in view of the apparent willingness of the Sheikh in this matter, the British Government will see its way clear to taking up in the case of the Kuwait concession no less liberal an attitude than was assumed in the case of the Bahrein concession.

3. This Government understands that despite the fact that the Colonial Office as early as 1925 gave its full and unqualified consent to the negotiations by the Eastern and Central Syndicate of an oil concession with the Sheikh of Kuwait that office later qualified its consent by insisting upon the inclusion of the Nationality clause in any agreement arrived at with the Sheikh for the apparently specific purpose of preventing the entry into that territory of the Eastern Gulf Oil Company which had meantime arrived at an understanding with the Syndicate as to the transfer of any concessions that it might obtain from the Sheikh. The continued insistence of the Colonial Office on this point and its apparent unwillingness to accord to that Syndicate the same treatment as was accorded in the case of Bahrein has seriously handicapped the Syndicate in bringing to a conclusion with the Sheikh the negotiations which that concern was authorized by the Colonial Office to undertake.

The above situation is further complicated by the fact that at the very moment while His Majesty's Government had under consideration the petition of the Syndicate for the elimination or modification of the Nationality clause, permission was granted the Anglo-Persian Oil Company, a rival concern, to send a small party of geologists to Kuwait for the purpose of studying the surface geology of the ground. It will be recalled that the Embassy on repeated occasions requested of the Foreign Office that the Company in question not be permitted to proceed with its operations pending a decision by His Majesty's Government on the question then before it regarding open-door rights for American nationals in Kuwait. Now, this Government has been informed, this study of the surface geology has been followed by a second expedition equipped with drilling machinery and plant. This Government greatly regrets that no effect has been given to the

Embassy's request in this matter but would appreciate being assured by His Majesty's Government that this fact will not be allowed to militate against the position of the Syndicate and its affiliate, the Eastern Gulf Oil Company in the eventual granting of an oil concession in Kuwait.

STIMSON

From LONDON
Dated March 30, 1932
Rec'd 7:54 a.m.

Secretary of State,
Washington.
123, March 30, noon.

I delivered note (Department's telegram 100, March 26, 4 p.m.) to Sir John Simon this morning who asked me to assure you the matter was receiving his attention and memoranda for a draft reply had already been compiled. The British position will be discussed at the next Cabinet meeting which takes place on April 6th and Sir John stated he hoped immediately thereafter to forward me a note of reply.

ATHERTON

Foreign Office, S.W.1.
9th April, 1932.

R. Atherton, Esq.,
United States Chargé d'Affaires.

Sir,

With reference to your Note No. 1696 of the 29th March regarding the application of the Eastern and General Syndicate for an oil concession in Koweit, which they propose, if granted, to transfer to United States interests, I have the honour to inform you that His Majesty's Government have given careful consideration to the representations made by General Dawes and yourself on this subject and I am now in a position to return you a reply.

2. Your Government will appreciate in the first place that the Sheikh of Koweit, though an independent ruler, is in special treaty relations with His Majesty's Government and enjoys their protection. These special relations lead him to seek their advice on important matters of policy, and place His Majesty's Government under an obligation to watch over his interests. Many years ago the predecessor of the present Sheikh gave an undertaking that he would not grant an oil concession in his territories without their consent.

3. In paragraph 2 of your note of the 29th March you mention that your Government are informed that the Sheikh is agreeable to the 'entry of the Eastern Gulf Oil Company and to the granting on behalf of that Company of an oil concession without the inclusion of the "nationality Clause"'. As was explained to you in a semi-official letter of the 22nd December last from my Department His Majesty's Government on learning this, felt some doubt as to the correctness of this interpretation of the Sheikh's attitude, since the Sheikh had consistently expressed himself emphatically to the local British authority as desirous of confining any oil concession to entirely British interests. In your letter of the 30th December you were good enough to transmit for my information a copy and translation of a letter from the Sheikh to Major Holmes, the representative of the Eastern and General Syndicate, on which the American interests apparently based the information on this point given to your Government. His Majesty's Government have caused enquiry to be made of the Sheikh, who replied that he was still averse from receiving in his principality a company other than an entirely British one and that he did not consider himself as in any way committed by his letter to Major Holmes to grant the Eastern and General Syndicate the concession which they seek. It will be observed from a reference to the Sheikh's letter that its final sentence only expresses a readiness to discuss the matter further with Major Holmes after agreement has been reached between the Syndicate and His Majesty's Government.

4. When examining the necessity for the continued insistence on the inclusion in any oil concession in respect of Koweit of a clause confining it to British interests, His Majesty's Government have been concerned not only with their own interests in the matter, but also with their duty to secure the best terms possible for the Sheikh of Koweit, and in particular, have had regard to the possibility that it would be less difficult for the local British authorities to control the activities of a purely British concern and to reconcile them with the Sheikh's interests. On a balance of all the conflicting considerations, His Majesty's Government are, however, now prepared, for their part, not to insist in this case that any concession must contain a clause confining it to British interests, if the Sheikh for his part is willing to grant a concession without such a clause.

5. I wish, however, to make it clear that this decision does not imply agreement in the immediate grant of the proposed concession to the Eastern and General Syndicate, to which the Sheikh, as stated above, considers himself in no way committed. His Majesty's Government indeed do not consider that they could properly advise the Sheikh to give prior or preferential treatment to the Eastern and General Syndicate, but hold it to be necessary that any application for a concession which may be forthcoming from any quarter be examined with a view to decide

which, if any, will best serve the interests of the Sheikh and his principality. I should add that the draft concession submitted to the Colonial Office by the Syndicate would in any case need revision both in respect of the provisos designed to safeguard the interests of His Majesty's Government (Clause 8) and on many points affecting the interests of the Sheikh.

6. In paragraph 4 and 5 of your Note of the 29th March you have referred to the operations now being carried out by the Anglo-Persian Oil Company in Koweit and reminded me of the requests made to my Department that this company should not be permitted to proceed with its operations pending a decision by His Majesty's Government as to the exclusion of all but British interests. I would explain that the Anglo-Persian Oil Company manifested an interest in Koweit oil, and indeed made a formal application for a concession before the Eastern and General Syndicate had even appeared on the scene, though the negotiations were at that time not brought to a conclusion, chiefly because the terms suggested were not satisfactory. Several months before any representations were made by General Dawes or yourself in the matter, the Anglo-Persian Oil Company made a request for permission to carry out a geological survey in Koweit with a view to decide whether to submit an application for an oil concession. In order to ensure that any oil concession which the Sheikh may grant shall embody the best available terms, it is in the view of His Majesty's Government desirable and proper that any interested companies be given every opportunity in advance of satisfying themselves whether or not they wish to submit an offer. His Majesty's Government therefore raised no objection to the grant by the Sheikh of the application of the Anglo-Persian Oil Company. I understand that their present activities in Koweit are confined to such a geological survey.

7. The position therefore is that His Majesty's Government for their part are prepared to agree to the omission from any oil concession, which the Sheikh may be prepared to grant, of a clause confining it to British interests. If therefore the Eastern and General Syndicate desire to renew their application to the Sheikh for a concession, which they would subsequently transfer to the Eastern Gulf Oil Company, His Majesty's Government will raise no objection to the application being taken into consideration together with any other applications for oil concessions which may be forthcoming from other quarters.

I have the honour to be,
with high consideration,
Your obedient Servant,
(Signed) JOHN SIMON.

Department of State,
Washington.
September 1, 1932.

AMEMBASSY,
LONDON.

Your despatch No. 2, April 11, 1932, regarding Kuwait.

The Department is informed by the Gulf Oil Company that the Eastern and General Syndicate has been unable to obtain from the Colonial Office a definitive reply to its letter of June 10, 1932, requesting information as to whether Clause eight of a proposed draft concession with the Shaikh of Kuwait was satisfactory from the point of view of safeguarding the interests of the British Government, and, if not, requesting an indication of wherein it failed to do so.

It is represented to the Department that the failure of the Colonial Office to give a definite indication of its views in this matter prevents the Shaikh from taking a decision on the draft concession which the Syndicate submitted on May 26, 1932, since he is barred from taking action before knowing what safeguards the British Government requires in a concession which is eventually to be assigned to a British incorporated company controlled by American interests. Meanwhile, it is stated, the Anglo-Persian Oil Company has submitted a draft concession which the Shaikh is being urged to grant immediately. Inasmuch as the question of safeguards does not arise in the case of the latter concession, the Shaikh is presumably free to make a decision on it at any time. Thus the Syndicate is placed in a disadvantageous position since it is unable to obtain consideration from the Shaikh for its draft concession because of the failure of the Colonial Office promptly to furnish information requested nearly three months ago. Consequently effect is not being given to the assurance contained in the Foreign Office note of April 9, 1932, in which the British Government agreed that it would raise no objection to the Shaikh taking under consideration any application which the Syndicate might wish to make for a concession.

Please take up this question immediately with the Foreign Office and urge that appropriate steps be taken in order that the American interest involved may be placed in as favorable a position as the Anglo-Persian Oil Company in having its application considered by the Shaikh. The Department leaves to your discretion the determination as to whether this question should be taken up formally or informally.

Please telegraph the results of your representations. Stacostrep.

CASTLE
(Acting)

NOTE 47

From London
Dated September 17, 1932

Secretary of State,
Washington.
272, September 17, 1 p.m.

In formal reply received today in reply to my note of September 6th (see my 255, September 7, 11 a.m.) Foreign Office states the contentions expressed therein appear to be based upon a misunderstanding, and continues 'As you are aware from my note of the 9th April His Majesty's Government decided that they could not advise the Sheik to give preferential treatment to the syndicate, and that any applications which may be forthcoming for a concession, should be compared in order to see which appeared to be the best interests of the Sheik and his principality. The Anglo-Persian Oil Company submitted to the Sheik last month a draft of a concession; but a copy of this draft concession has only very recently been received by His Majesty's Government and some time must elapse before a thorough comparison of its terms with those of the draft concession submitted by the Eastern and General Syndicate can be completed. The results of that examination will then be communicated to the Sheik in order that he may reach a conclusion as to the respective merits of the two offers from the point of his own interests and those of his Sheikdom. In the meantime no further expression of the views of the Sheik on either proposal has yet been received by His Majesty's Government and he is not in a position as suggested in your note under reply to take a decision in favor of the application of the Anglo-Persian Oil Company since, as was stated in my note referred to above, he is bound by an undertaking given by his predecessor to grant no oil concession in his territories without the consent of His Majesty's Government.

'It will thus be seen that the application of the Eastern and General Syndicate is receiving consideration equally and *pari passu* with that of the Anglo-Persian Oil Company and that the question of the provisions designed to safeguard the interests of His Majesty's Government which would be required in any concession not confined to a British company does not arise until the Sheik has compared the two draft concessions in the light of the comments of His Majesty's Government. The American interests concerned suffer no prejudice therefore from their ignorance of the precise nature of these provisions, and His Majesty's Government for their part consider it preferable not to communicate further with either the Anglo-Persian Oil Company or the syndicate regarding the terms of their respective draft concessions, until their own consideration of the draft concessions and consultation with the Sheik is complete.'

Full text by pouch.

ATHERTON

Embassy of the
United States of America
London, September 17, 1932.

No. 365
SUBJECT: Koweit Oil Concession
The Honorable
The Secretary of State,
Washington, D.C.
Sir:

I have the honor to refer to my telegram No. 272, September 17, 1932, regarding the applications of the Eastern and General Syndicate and the Anglo-Persian Oil Company for an oil concession in Koweit, and to forward herewith a copy of the Foreign Office note received in reply to my note of September 6, 1932.

Respectfully yours,
Ray Atherton
Chargé d'Affaires *ad interim.*

Enclosure:
Copy of Foreign Office note
No. E 4582/121/91, September 16, 1932.

Enclosure to despatch No. 365, September 17, 1932, from American Embassy, London.
COPY.
No. E 4582/121/91.
FOREIGN OFFICE, S.W.1.
16th September, 1932.
Ray Atherton, Esq.,
Sir,

I have the honour to inform you that your note No. 231 of the 6th September regarding the applications of the Eastern and General Syndicate and the Anglo-Persian Oil Company for an oil concession in Koweit has been considered by the Departments of His Majesty's Government in the United Kingdom concerned.

2. It appears from that note that the United States company interested in the application made by the Eastern and General Syndicate have represented the situation as being as follows. They state that the Eastern and General Syndicate have not yet learnt from the Colonial Office what alterations would be required in the draft concession submitted to the Sheikh by them, in

order to render it satisfactory from the point of view of safeguarding the interests of His Majesty's Government. They contend that this fact precludes the Sheikh of Koweit from giving consideration to the Syndicate's application, but that there is nothing to prevent the Sheikh from taking a decision upon the application for a concession submitted by the Anglo-Persian Oil Company. They complain that the Syndicate are thus placed at a disadvantage in their efforts to obtain the desired concession. These contentions appear to be based upon a misunderstanding of the present position in the matter.

3. As you are aware from my note No. E 1733/121/91 of the 9th April, His Majesty's Government decided that they could not advise the Sheikh to give preferential treatment to the Syndicate, and that any applications which might be forthcoming for a concession, should be compared in order to see which appeared to be to the best interests of the Sheikh and his principality. The Anglo-Persian Oil Company submitted to the Sheikh last month a draft of a concession; but a copy of this draft concession has only very recently been received by His Majesty's Government and some time must elapse before a thorough comparison of its terms with those of the draft concession submitted by the Eastern and General Syndicate can be completed. The results of that examination will then be communicated to the Sheikh in order that he may reach a conclusion as to the respective merits of the two offers from the point of view of his own interests and those of his Sheikhdom. In the meanwhile, no expression of the views of the Sheikh on either proposal has yet been received by His Majesty's Government, and he is not in a position, as suggested in your note under reply, to take a decision in favour of the application of the Anglo-Persian Oil Company, since, as was stated in my note referred to above, he is bound by an undertaking given by his predecessor to grant no oil concession in his territories without the consent of His Majesty's Government.

4. It will thus be seen that the application of the Eastern and General Syndicate is receiving consideration equally and *pari passu* with that of the Anglo-Persian Oil Company and that the question of the provisions designed to safeguard the interests of His Majesty's Government, which would be required in any concession not confined to a British company, does not arise until the Sheikh has compared the two draft concessions in the light of the comments of His Majesty's Government. The American interests concerned suffer no prejudice therefore from their ignorance of the precise nature of these provisions, and His Majesty's Government for their part consider it preferable not to communicate further with either the Anglo-Persian Oil Company or the Syndicate regarding the terms of their respective draft concessions, until their own consideration of the draft concessions and consultation with the Sheikh is complete.

I have the honour to be,
with high consideration,
Sir,
Your obedient Servant,
(For the Secretary of State)
(Sd.) G. W. RENDEL.

From London
Dated October 18, 1932.
Recd 6:51 a.m.

Secretary of State,
Washington.
298, October 18, 11 a.m.

I presented views outlined in your 258, October 4, 6 p.m., yesterday to Van Sittart and urged early action.

MELLON.

Re Preferential treatment for American interests in Kuwait.

Foreign Office, S.W.1.
11th November 1932.

His Excellency the Honourable
Andrew W. Mellon.
My Dear Ambassador,

I am glad to be able to let you know that the Departments concerned have now completed the comparative examination of the draft concessions for oil exploitation in Koweit, submitted to the Sheikh by the Eastern and General Syndicate and by the Anglo-Persian Oil Company respectively, and that the document embodying the result of this examination is already on its way to the British authorities in the Persian Gulf.

On its receipt by them it will be communicated to the Sheikh of Koweit for his consideration.

Meanwhile I am arranging to have a detailed reply prepared to the various other points raised in the memorandum which you left with me on November 2nd.

Yours very sincerely,
(Signed) ROBERT VANSITTART.

NOTE 47

Foreign Office, S.W.1.
23rd November, 1932.

His Excellency
The Honourable
Andrew W. Mellon.

My dear Ambassador,

Since I wrote to you on November 11th about Koweit oil, I have been considering the question, on which, as you know, I was not in possession of full details at the time of our interview. I have, therefore, come to it with a fresh mind, and one or two points have at once struck me.

2. The memorandum which you left with me on November 2nd might be interpreted as implying that His Majesty's Government have been purposely procrastinating in regard to the participation of American interests in the development of Koweit oil for over four years. But, apart from the fact that the Anglo-Persian Oil Company were in the field in Koweit long before the British concern which is now acting for the United States interests, I wish to make it clear that the decision of His Majesty's Government (which was communicated to the Eastern and General Syndicate in November 1928) that any oil concession which might be granted must contain a clause which would confine it to British interests, was taken on grounds of general policy and before we had heard anything of American participation in the matter. The decision was in fact taken in pursuance of the then existing general policy of His Majesty's Government which had been in force for many years, and also because the Sheikh of Koweit, whose interests they are, of course, under an obligation to protect, expressed himself as unwilling to grant a concession to any company not under British control. It was not until December 19, 1928, that the Syndicate informed the Colonial Office of their agreement with the Eastern Gulf Oil Company, by which the concession, if obtained, was to be transferred to United States Interests. His Majesty's Government did not however feel able to change their decision until, in December 1931, your Embassy first made representations in the matter. Then His Majesty's Government, in their desire to go as far as they could to meet the United States Government, reconsidered the question and decided after much deliberation that, while they could not commit the Sheikh of Koweit, they would, for their part, not insist in this case that any concession granted must contain a clause confining it to British interests, if the Sheikh for his part was willing to grant a concession without such a clause, and we so informed your Embassy in April.

3. Your memorandum also reverts to the representations made in Atherton's official note No. 231 of the 6th September, to the effect that the American interests concerned are labouring under a disadvantage as compared with the Anglo-Persian Oil Company owing to their ignorance of the provisos which His Majesty's Government would require to see embodied in the concessions granted, in order to safeguard their own interests. But surely these representations were satisfactorily answered in Sir John Simon's reply, No. E 4582/121/91 of the 16th September. As I understand it, the 'safeguards' are a matter for discussion *after* the Sheikh of Koweit has made his decision from the point of view of what is to the best advantage of his own State. (I am advised that though no final decision has been taken on the point it is not unlikely that at least some of them would equally have to be embodied in any concession which might be granted to a purely British Oil Company wishing itself to operate in Koweit.) As these safeguards are not primarily the concern of the Sheikh, and will not affect the comparison of the two draft concessions on their merits, they do not in our view affect the matter at the present stage.

4. As you know, that stage is that the latest draft concession submitted by the Eastern and General Syndicate and the draft submitted by the Anglo-Persian Oil Company have been compared in London by the department concerned on the technical side in order that the Sheikh, who is naturally not well versed in such technical matters, may understand what in fact will be the effect of the main provisions of each offer (e.g. the financial side, conditions of working the oil, etc., etc.). The resulting document is now on its way to the Persian Gulf and we must await the result of the Sheikh's examination.

5. The two offers made for the concession are thus being treated concurrently, and that, I feel sure you will appreciate, was the only correct course for His Majesty's Government to take in order to secure the most acceptable terms for the Sheikh. If only in his interest, His Majesty's Government were naturally bound, as Sir John Simon informed Atherton in his note No. E 1733/121/91 of the 9th April, to allow any interested company to consider whether they wanted to apply for a concession, and if so to give them time to do so. The Anglo-Persian Oil Company formally renewed their efforts to obtain a concession in Koweit in August, 1931 (not October as mentioned in your memorandum).

6. I regret that there has been delay in the whole matter; I cannot of course at this stage say exactly when the Sheikh will decide to grant a concession; I do hope, however, in the light of the preliminary information I have now given, that you will be able to assure your Government that there has been no desire on our part to cause them embarrassment by any avoidable delay.

Believe me,
My dear Ambassador,
Yours sincerely,
(Signed) ROBERT VANSITTART

No. 558
SUBJECT: Koweit Oil Concession.

Embassy of the
United States of America
London, December 15, 1932.

The Honorable
The Secretary of State,
Washington, D.C.
Sir,

I have the honor to refer to my despatch No. 483 of November 12, 1932, and subsequent correspondence, with regard to the matter of American interests seeking an oil concession in Koweit, and to state that on December 13 I called upon the Foreign Office and orally presented the considerations set forth in the enclosed memorandum of conversation, based on Sir Robert Vansittart's note of November 11, which went forward to the Department with the despatch above referred to. Sir Robert expressed some surprise at this delay and promised to look into the matter and inform the Embassy as soon as he is able to get the data from the Colonial Office. I told Sir Robert that it was my intention to sail for America very shortly and asked him to communicate with Mr. Atherton in my absence, who would cable me the Foreign Office reply, since I desired to discuss the matter with the Department during my short visit to Washington.

On leaving, I reminded Sir Robert that if Mr. Atherton did not hear from him within the next week or so he would, under my instructions, again be reminding Sir Robert of my desire for an early reply to his promise to expedite the matter.

Respectfully yours,
(For the Ambassador)
Ray Atherton
Counselor of Embassy.

Enclosure:
Memorandum of conversation,
dated December 13, 1932.

Enclosure to despatch No. 558, December 15, 1932, from American Embassy, London.
Memorandum on which the Ambassador will base his talk with Sir Robert Vansittart.

On November 11 of this year Sir Robert Vansittart wrote the Ambassador that the comparative examination of the draft concessions for oil exploitation in Koweit submitted to the Sheikh by the Eastern and General Syndicate and the Anglo-Persian Oil Company, respectively, had been completed and that the document embodying the result of this examination *was already* on its way to the British authority in the Persian Gulf. The Ambassador informed the American interests concerned of the receipt of this information from the British Government. However, he has been informed by the American interests concerned on December 10 that the British Political Resident in the Persian Gulf had stated that week that he had no knowledge of the receipt of this document, and that consequently it had not presumably been presented to the Sheikh.

In view of Sir Robert Vansittart's note of November 11 and the fact that the November bi-weekly air mail only took some six days from London to the Persian Gulf, the Ambassador hesitated to regard this information as accurate, and would be grateful if Sir Robert would inform him as to whether in fact the document had been received by the British authority in the Persian Gulf and had been delivered to the Sheikh.
London, December 13, 1932.

Department of State,
Division of Near Eastern Affairs.
December 27, 1932.

WSM.
HSV:

Apparently there will be no need for us to consider the possibility of asking the Embassy at London to ask the Foreign Office about the reason for the delay in the British comments on the draft oil concessions reaching Kuwait, since this despatch indicates that the Ambassador (not Atherton it is to be observed) has already taken up the matter at the F.O.

Evidently Mr. Mellon expects to take this matter up while he is here in the United States and I do not see that there is any action for us to take until such time as we hear either from London or from Mr. Mellon.

(Incidentally, it is to be noted that the Ambassador is not at all bashful in pressing this matter himself. Possibly Mr. Castle or the Secretary may wish to suggest to him that he go easy on this question.)

48. *Exchange of telegrams between Gulf (London) and Gulf (New York):*

Cable received from Major Davis May 12 1932 9 AM London 4 May 11th my 2 today letter received from Major Holmes by syndicate dated May 4th *quote* Report serious setback in our negotiations with Sheikh of Koweit and the Council of State stop Told them British Government agreed British nationality clause may be omitted from any oil concession that may be granted by Sheikh and mentioned British Government agreed concession may be transferred to Eastern Gulf Oil Co. stop Council of State very much perturbed on hearing of the suggested transfer and proposed we should at once see Sheikh stop Sheikh became extremely angry, gist his remarks as follows: I had wilfully misled him a most unfriendly action he did not expect of me and having obtained letter from him without which British Government would not have considered let alone agreed to omitting British nationality clause, the British nationality clause settled to my satisfaction I proceed using his letter as lever to arrange through London entirely unknown to him and behind his back alterations in other clauses with which neither British Government nor the American have any concern or right to conclude except through himself; that he knew nothing of American law and is prohibited by treaty obligations dealing direct with any nation except British; that he did not wish to be cut off from advice of British Government stop Sheikh worried why British Government agreed transfer concession to American company without first consulting him and whether it indicated disposition on the part of British government modify long time friendly attitude towards him stop Sheikh also pointed out I had shown him draft of concession which I had stated met with full approval of London and New York in which was written concession was to be transferred to company registered in England or Canada and what is the reason for this sudden and secret desire that now it should be otherwise stop Meeting had depressing effect on our supporters on Council of State and broke up in atmosphere from which big slice goodwill had been extracted and replaced by suspicion stop It gives British Political Agent Koweit opportunity to adorn tale at our expense stop Sheikh put all responsibility on me and during his choleric outburst I said nothing stop When he cooled asked him if he does not think British officials intended transfer sentence react as herring across trail and to remain disturbing influence between us when we discuss terms of concession stop Am of opinion British Government agreed include sentence subquote which they would subsequently transfer to Eastern Gulf Oil Co. end subquote with tongue in cheek knowing it would be more difficult persuade Sheikh ratify alteration made in London in the concession agreement without his consent than it had been to obtain his support for omitting British nationality clause unquote fullstop

There is much more verbiage along similar lines which consider unnecessary to cable stop Holmes evidently lost his head and failed present true aspect situation to Sheikh and the Council of State stop

---

From New York, N.Y., May 14, 1932

To London

3 May 14th difficult for us escape conclusion Holmes has failed somewhat at very critical point stop Assuming perturbation of Sheikh real Holmes with background of past negotiations with Colonial Office should have been able make effective and immediate reply to Sheikh stop However if he had any doubt regarding his actual position under wording of British Government note he should have advised Sheikh he would clear matter up by cabling London immediately stop We have been impressed and disturbed over lack of information from Holmes such as his letter May fourth and previously regarding actual situation in Koweit in spite of Syndicates requests that he keep Syndicate fully and currently advised by cable stop Please discuss tenor foregoing with Syndicate and say two steps seem to us immediately imperative.

Firstly that Holmes be supplied with such facts and suggestions as you in London can furnish to enable him reassure Sheikh and reestablish friendly contact and confidence on the assumption that Sheikh really disturbed over question of transfer and not merely resorting to subterfuge stop

Secondly suggest Janson take opportunity offered by Colonial Office letter May tenth to confer with Colonial Office officials and endeavor arrive at prompt agreement exact wording transfer clause which British Government will approve for use in Koweit concession in substitution for clause eight fullstop Suggest also Holmes be requested cable any information he has or can obtain regarding actual attitude Sheikh and any pressure or inducements to which he may have been subjected by British political agents or Anglo Persian Oil Company stop Will have conference at Department of State Monday morning stop Department of State has not yet made any reply to Foreign Office note April 9th

Wallace

---

49. *APOC Geological Report on the oil prospects of Kuwait dated 12th May 1932 and based on a survey carried out from 14th February to 13th April 1932. The illustrations etc. are, of course, not reproduced here.*

NOTE 49

The Anglo-Persian Oil Company Ltd.
Geological Report No. P.T. C.14.
A Report on
The Oil Prospects of Kuwait Territory.
By
P. T. Cox.

Contents.

1. SUMMARY

1. *Results of Drilling*. One bore-hole was drilled at Bahrah (M'dairah) to 87 ft. and another at Burgan to 218 ft. 6 ins. Neither hole penetrated below the Kuwait Series. Gas, traces of oil and some bitumen were found at Bahrah and at Burgan bitumen impregnation with traces of oil occurred throughout the depth drilled.

2. *Stratigraphy.*

(i) Fossiliferous beds in the Kuwait Series are confined to the North and North-Easterly parts of the Jal-es-Zor escarpment. Fossils found are not conclusive but indicate probable Miocene age.

(ii) Beds outcropping at Wara and Burgan previously referred to the Nadjira Formation are probably not older than beds outcropping in the Zor escarpment but the case is not definitely proven.

3. *Seepages.* Seepages of gas at Bahrah and Burgan and the occurrence of bitumen on the coast are recorded in addition to seepages described in Preliminary Report.

4. *Structure.*

(i) There is a minor anticlinal structure about the Bahrah seepages oriented approximately in a N.N.W.–S.S.E. direction.

(ii) There is probably a general regional dip to the E. or E.N.E. which may be of depositional origin only.

(iii) There is possibly a broad area of structural elevation extending roughly parallel with the coast and bounded by a gently synclinal zone lying to the West of the Jal-es-Zor and extending southwards along the Shagg depression.

5. *Relation of Surface Structure to Seepages.* The seepages probably represent escapes from oil-bearing beds at depth either i. at crest maxima of minor folds on the broad anticlinal area lying between the Jal-es-Zor and Shagg and the sea or ii. at points on a monoclinal where cover beds, effective further to the East, become permeable through change to more littoral facies.

6. *Character of the Oil.* It is probably, in view of regional considerations, that only heavy oil with a high Sulphur content is to be expected in Kuwait though the data obtained on the ground, if taken out of its regional context, is not necessarily adverse to the chances of there being high grade oil at depth.

7. *Recommendations.* If decided to test the area further, it is recommended that

(i) a geological surface reconnaissance of the Province of Hasa be undertaken to throw further light on stratigraphical problems but that results of further surface work in Kuwait Territory would not justify the expenditure involved.

(ii) the possibilities of obtaining information on sub-surface structure by geophysical methods should receive every consideration.

(iii) if geophysical methods shown to be unsuitable, the next step should be to drill a well capable of going to 4000 or 5000 feet and of coping with oil under pressure if found.

(iv) such a well would be located to best advantage at Malah, about two miles east of Madanayat.

(v) if possible, data obtained from this well should be awaited before making further test locations.

2. METHODS OF WORK

Acting on instructions from London, field exploration of the oil prospects of Kuwait Territory was recommenced in mid February. The usual methods of surface geological examination were supplemented by a shallow drilling outfit to investigate the seepages and an aerial reconnaissance. Mr. G. M. Shaw and the writer returned to Kuwait on 14th February and Mr. H. H. Green with the drilling outfit arrived in the Territory on 21st February.

Mr. Shaw completed an inspection of the more important outcrops and seepages and an aerial reconnaissance of Kuwait Territory and part of southern Iraq by the end of February when he returned to Persia. His report on the aerial reconnaissance is contained in Abadan–London Mail Memo No. 27022 of 9/3/32.

The drilling outfit was first erected at Bahrah (M'dairah district) and a hole of 87 feet depth was drilled by 13th March. The rig was then moved to Burgan and a hole of 218 feet 6 ins. was drilled by April 13th when instructions were received to discontinue operations and return staff and equipment to Abadan.

From the beginning of March until work was suspended on April 13th detailed surface work was carried out with a view to determining whether any relations existed between the gas, oil and bitumen seepages and structure of the Kuwait series beds as exposed. This involved measurement of sections and partial mapping of outcrops in which attention was given to careful determination of heights. As pointed out in our Preliminary Report, the outcropping Kuwait Series beds lie so nearly horizontally that any more approximate methods of determining dips are inapplicable in the circumstances.

It was necessary to return to Kuwait before completion of laboratory work on the material collected on our preliminary reconnaissance but heavy mineral analyses of these samples and further search for micro-fossils were carried out in the Central Research Laboratory, Masjid-i-Sulaiman, by Messrs. White, Lowson and Taitt; their results are given in an Appendix by Mr. Taitt to our Preliminary Report (P.T.C. 13; A.H.T. 9).

The only macro-fossils found which may throw light on the age of the Kuwait Series beds are two species of Ostrea; specimens of these have been sent for identification to Dr. Douglas whose report we have not yet received.

Mr. Southwell and Dr. Lees visited the area from March 11th to 13th and Mr. Shaw again returned on April 14th to 15th at the close of operations.

3. INFORMATION OBTAINED BY DRILLING

Graphic well logs of the two holes drilled at Bahra (M'dairah) and Burgan respectively are given on Plate 10. In both cases the beds penetrated consist of grits, sandstones, sands, clayey sandstones or clays closely similar in general character to the Kuwait series beds as exposed at outcrop. No fossiliferous beds were encountered in either well and thus no further information on the age of the series has been obtained from the wells.

(1) *Kuwait No. 1* was located on a surface gas seep 1250 yards to the S.S.W. of the main oil seepage at Bahrah (M'dairah). Choice of the site was further influenced by its being the nearest spot to the oil seep where ground is high enough to preclude the danger of flooding; a precaution which experience showed was well taken.

Slight emanation of gas was struck at 25 ft. and continued to 49 ft. where a stronger show was encountered, and, for 24 hours, gas-cut muddy water was blown out of the casing which stood 12 ft. above the derrick floor. The pressure then blew down and only slow bubbling of gas continued until the well was shut down. Water in the well, which showed a definitely acid reaction with litmus, maintained a steady level at 15 ft. from surface except when disturbed by excessive gas. Traces, and at times a film of oil appeared on the surface of the water brought up from 49 feet downwards and sand grains in the samples from below this depth were frequently stained with bitumen. The tools also became coated with thick sticky oil.

Throughout the drilling of this well there were constant difficulties on account of caving sands. The Formation consists of alternating bands of loose sands, and clayey sands or clay with a few thin beds of hard sandstone. Whenever one of these harder beds was struck and progress retarded caving from the beds immediately above it invariably set in.

No red or strongly coloured beds were found in this well; it is probable that gas action has to a large extent reduced the ferric or prevented oxidation of the ferrous compounds present and so bleached the rocks. The red colouration so common in the beds immediately above the initial horizon at outcrop is also, as was shown by pitting, to some extent a surface weathering effect. That the foetid, and slightly acid water in the well is capable of bleaching red sandstone was shown experimentally by immersing a piece of bright red sandstone from outcrop in this water; after 2 days it had changed to a pale yellow colour.

(2) *Kuwait No. 2* was located 3870 ft. to 335½° (mag.) of the highest point of the Burgan Hills.

The beds penetrated, though similar in general character to those drilled through at Bahrah, are more firmly cemented. From cavings and the character of the mud they appear to have, in general, a greater amount of clayey matrix. There was no trouble in this hole with excessive cavings, and it was not necessary to run casing at any stage. Hard sandstones occur frequently and below 170 ft. there are thin alternating beds of hard calcareous sandstones and soft sticky blue marls.

The beds are more or less bituminous throughout; between 86 ft. and 89 ft. and from 121 ft. to 134 ft. there are thick masses of plastic bitumen containing very little sand or extraneous matter. Traces of oil floating in the water were brought up by the pump. The water had an oily but only slightly sulphurous smell, it showed no acid reaction with litmus paper and did not corrode the tools to the extent of the water in Kuwait No. 1.

4. STRATIGRAPHY

1. *Classification.*

In our preliminary Report (P.T.C. 13; A.H.T. 9) we followed Heim's suggestion in regarding the Kuwait Series as composed of two members: an Upper, 'Zor Formation' and a lower, 'Nadjira Formation'. More detailed work has failed to show convincing evidence in support of this conception and though the bases of correlation are slight there is some reason to suppose that the beds at Burgan and Wara referred by both Heim and ourselves to the Nadjira Formation are in fact equivalent to the higher horizons exposed in the Jal-es-Zor. The use of Heim's term 'Nadjira Formation' has therefore been abandoned and the following provisional classification adopted:—

Recent: Blue, shelly, beach concrete (Kuwait building stone), alluvium etc.

Sub-Recent: 'Oolite', calcareous sandstones and shelly sandstones forming raised beaches.

Kuwait Series: (4. Wara Sandstone Group.
(3. Upper Zor Group—unfossiliferous.
(2. Middle Zor Group—fossiliferous.
(1. Lower Zor Group—unfossiliferous.

The subdivisions of the Kuwait Series are applicable only in certain areas. The Upper, Middle and Lower Groups of the Zor Formation are distinguishable in the North Easterly parts of the Jal-es-Zor exposures but the fossiliferous Middle Group passes laterally to the S.W. into unfossiliferous beds. To the S.W. of Al Mutla's the several Zor groups cannot be distinguished. The true position of the Wara sandstone in the stratigraphic sequence is still questionable; reasons for placing it as shown in the table are discussed below.

The Kuwait Series, as a whole, maintains a general uniformity of character throughout the thickness we know and our examination has not brought to light any sound basis for correlation over more than restricted areas. The localized occurrence of fossils makes palaeontological criteria impracticable and the most hopeful line of approach appeared to be through the petrography of the rocks. Heavy mineral analyses of representative specimens collected on the preliminary reconnaissance showed that throughout the thickness exposed there was no marked change in mineral content of the detritus brought down to form these beds and this means of effecting a correlation between the Zor exposures and those lying to the S. of Kuwait Bay was thus shown to be inapplicable. Other possible indices such as the percentage of carbonates present, the presence or absence of red chert pebbles or fragments, rose coloured quartz, porphyritic and basaltic materials, nature of the clayey matrix and relative coarseness of texture of the sediments were tested but found to be at best of only local value.

2. *Conditions of Deposition of the Kuwait Series.*

Shallow water and estuarine conditions with an abundant influx of detritus derived from dominantly granitic sources in the interior of Arabia prevailed throughout the deposition of the Kuwait Series.

Messrs. Wyllie and Mayhew (B.K.M.W. 33; A.G.H.M. 4) have discussed the nature of this sedimentation; they attribute the large proportion of Kaolinitic material present as matrix in the Kuwait Series sandstones to decomposition of felspars and conclude that the transporting waters must have been rich in alkaline carbonates. The lack of fossils throughout most of the Series may well be due to the unsuitability of such alkaline waters to calcareous organisms. The fossil bearing beds of the Middle Zor Group about M'dairah may represent a facies developed in places rather further removed from river mouths and points of entrance of the alkaline waters, places where more normal marine conditions prevailed. The beds of this fossiliferous group are, furthermore, generally of finer grained elastic material, contain less kaolinitic matter and are marked by relatively abundant veins and pockets of Selenite. Mr. Wyllie explains the $CaCO_3$ content of the Kuwait beds as the result of reaction between the incoming alkaline carbonates and $CaSO_4$ originally contained in the sea water. Our microscopic examinations of the more calcareous, non-fossiliferous beds of the series fully support the suggestion that the lime is due to precipitation rather than detrital accumulation or secretion by organism. The Grain limestone and some of the more calcareous beds encountered by the Burgan well may be cited as examples.

There is a very variable $CaCO_3$ content in the various members of the Kuwait Series and our observations do not confirm Mr. Wyllie's general statement that these sediments 'have a high content of calcium carbonate'. The Plate illustrating the Appendix to our Preliminary Report (P.T.C. 13; A.H.T. 9) shows approximate carbon content of a representative group of specimens. It was hoped to use this criterion as a basis for correlation within the series but no general result of much value has emerged, though it is, as a rule, a sufficient index to distinguish between the Kuwait Series beds and the later recent and sub-recent sandstones of similar appearance. The latter are uniformly rich in $CaCO_3$ when compared with the Kuwait Series beds. The Middle Zor group in the M'dairah district is richer in lime than the Lower or, more especially, the Upper Group. Elsewhere it is difficult to generalise. In the Burgan Well, Kt. No. 2, there was a perceptible increase in calcareous beds below 171 ft. and though no fossils were found this may indicate entry into beds equivalent to the Middle Zor Group.

The Wara sandstone and other siliceous sandstones which we have correlated with that outcrop were attributed by Messrs. Wyllie and Mayhew to secondary silicification of an old land surface. This conception would explain most of the phenomena observed but we suggest an alternative view that these sandstones represent a distinct stratigraphic group.

The widespread occurrence of selenite in small veins and cavities throughout the Kuwait Series has been noted by Messrs. Wyllie and Mayhew. This selenite is especially notable in the Middle and Lower Zor Groups where clay beds are developed and immediately below the Wara Sandstone beds at Burgan. In the latter locality, we mentioned in our Preliminary Report a 3 ft. bed of gypsum at this horizon. Closer examination of the occurrence and exposure of the bed by pitting and trenching showed that

(i) much clay and sandy matter is mixed with the gypsum,

(ii) most of the gypsum is in the form of selenite crystals often of large size.

The occurrence is a lenticular development of a great lateral extent. Whether it represents a recrystallized bed of gypsum or a concentration of gypsum from circulating waters is open to question.

3. *Lower Zor Group.*

This group, as it occurs along the base of the Jal-es-Zor escarpment from below Delamaniya to near Kuwaikab, is similar in general aspect to the fossiliferous Middle Zor beds overlying it, but is distinguished by its lack of fossils. The most striking components of the group are red, current-bedded rather coarse grained and often pebbly sandstones which commonly show a characteristic nodular, sugary weathering surface. As exposed at outcrops the group appears to be almost entirely composed of these sandstones varying in colour to yellow and brown from place to place. Pitting revealed a considerable number of interbedded clays which are generally covered by scree on the surface. Some 80 ft. of this group in the M'dairah district were examined at the surface and in pits and Kuwait No. 1 Well penetrated a further 77 ft. Throughout this thickness of 157 feet there is an alternating sequence of clays, clayey sandstones and sandstones or loose sands. Some of the sandstones are firmly cemented and calcareous but the bulk of the rocks of this group show only slight or no effervescence with acid. The clay beds are usually less than 4 feet thick and either red (at the surface) or green. In the well no red beds were encountered at all but this may be due, at least in part, to bleaching by gas action. Some clay and, to a lesser extent, sandstone beds as exposed in pits were light yellow or greenish in colour but when traced to outcrop appeared as red. It is therefore probable that the red colouration of this and the overlying group is largely the result of surface oxidation. The red colour is most pronounced about M'ghatti and M'ghaira (as the Arabic name signifies) and becomes less marked going S.W. towards Kuwaikab. From Kuwaikab to the S.W. the group is largely obscured by scree and owing to the change in facies of the overlying group its upper boundary cannot be traced further in this direction.

4. *Middle Zor Group.*

In the M'dairah district this group is a comparatively distinct subdivision of the sequences. It is composed of alternating red and yellow sandstones, red and green clays and various intermediate clayey sandstone and silty clays. Throughout the group fossils are moderately common though lateral persistence of any given fossiliferous bed for more than one or two miles is rare. The base of the group is marked in most places by a compact, fine-grained, yellow, calcareous sandstone with small Lamellibranchs (usually occurring as hollow casts) similar to the *Clausinellas* which occur in the Lower and Middle Fars beds of Persia. This small Lamellibranch is probably the most abundantly represented fossil in the group and occurs either alone or associated with other fossils at numerous horizons. Other forms found in beds of this group are:—

Dendritina cf. rangii. This form occurs in the uppermost shelly sandstones of the group about Kuwaikab and Khashmqudhai.

Ostrea sp. A species with elongated ligamental groove similar to that of O. gryphioides. This form was recorded in our preliminary report and we await Dr. Douglas' report on its determination and age significance. It occurs about Khashmqudhai and Kuwaikab in the lower half of the group.

Ostrea cf. latimarginata. Occurs from near M'dairah Wells to Khashmqudhai. In the latter locality it is in the same beds with Ostrea sp. mentioned above. It closely resembles O. latimarginata from the Lower Fars beds of Persia and is probably referable to that species.

Turritella sp. Not common but several poorly preserved specimens were found at Khashmqudhai.

Lucina sp. Abundant in localized patches at various horizons.

Balanus sp. Common in certain beds throughout the group.

Worm tubes. Occur in great abundance in the lower parts of the group throughout its exposure. They vary in size from about $\frac{1}{2}$ c.m. to 3 c.m. or more in diameter.

The assemblage is not extensive and in many of the shell beds it is not possible to make out what forms are present, the remains being entirely fragmentary and often leached. It is not uncommon to find druses in rocks of this group, especially near shell beds lined with well formed calcite crystals.

The assemblage is however quite distinct from the recent and sub-recent fossil faunae. There are certain prominent forms characteristic of the latter and which still flourish on the present day beaches e.g. Conus sp., Solen sp., Turbo sp., etc. but none of these were found in the Zor beds. It may be noted that many of the sub-recent shelly sandstones outcropping on raised beaches along the coast are lithologically similar to Kuwait Series beds but can be distinguished by the presence of this recent fauna.

On tracing this group from the M'dairah district south westwards it was found that it maintained an unexpectedly uniform thickness of 110–130 ft. until the neighbourhood of Kuwaikab. There is a gradual lessening of its red colour to the S.W. of M'dairah wells—a variation which makes its distinction from the overlying group less obvious than in the M'daira–M'ghaira district. About opposite to Kuwaikab the fossiliferous beds, in all but the uppermost 20–30 ft. of the group, peter out and pass laterally into unfossiliferous sandstones, clayey sandstones, clays, etc. which are indistinguishable from the Lower (or Upper) groups. At the base of the group a conglomerate composed of well-rounded quartzite, rhyolite, basalt etc. pebbles occurs at the passage from the lowest shelly horizon to a non-fossiliferous facies. The conglomerate differs from any of the pebbly sandstones so common in parts of the Kuwait Series in being almost entirely composed of pebbles with only a small amount of arenaceous cementing material.

From Kuwaikab to the S.W. a fossiliferous horizon persists at the top of this group until it also passes over to an unfossiliferous facies opposite Kadhama. From this point to the S.W. it was not found possible to differentiate the Zor formation into distinct groups.

5. *Upper Zor Group.*

In the M'dairah district this group is readily distinct from the underlying group by its colour, lack of fossils and the presence of coarser-detrital material. It is composed of white, weathering buff, pebbly sandstones with matrix of white kaolin-like clay. Pebbles of acid volcanics schists and a few dark, fine-grained basalts are common though quartz and quartzites make up the bulk of the detritus. The group maintains these characteristics throughout the length of its exposure. A few red or reddish beds occur in places but are always subordinate members. From M'dairah wells to the S.W. the distinction of this group from the Middle Group becomes less obvious owing to change in the latter which, in its higher parts, approximates more in general aspect to the Upper Group. The distinction between the two groups can be made as far as Kadhama by the criterion of the presence or absence of fossils.

6. *Kuwait Series Outcrops to the South of Kuwait Bay.*

Exposures along the Jal-es-Zor to the S. and S.W. of Al Mutla'a consist chiefly of grey and yellow clayey sandstones and pebbly sandstones. As the land surface rises southwards from Jahrah the thickness of beds exposed is correspondingly reduced. Red beds persist some distance but either die out or pass below the exposed level. As far as could be observed without good marker horizons, there appears to be a slight pitch southwards from Al Mutla'a, as shown in Plate 5.

On top of some of the small hillocks about Al Atraf there are the weathered relics of a brown coloured, quartzose sandstone which in contrast to the underlying beds has little clayey matrix. As the isolated but fairly numerous exposures are traced south eastwards from Al Atraf one finds a tolerably well marked ridge rising to over 500 feet above sea level to the N.E. of the Shagg depression in which light coloured, yellow or pale green, clayey sandstones occur, overlain in places by brown weathering, more siliceous sandstones. Umm Ruus, Umm Twenith, Arhaiyah, Thallat eth Thaba Manaigish, Jerib al Arbid, Umm Agadair and Fowaris are points where such outcrops occur along the North-easterly side of this ridge. The brown weathering sandstone varies in hardness and degree of silicification but is usually preserved as remnants on the higher knolls. Going eastwards from Jerib al Arbid or Umm Agadair exposures are fewer but of similar character —light coloured clayey sandstones with occasional patches of harder, brown, siliceous sandstones on the higher points—until Wara and Burgan are reached. Here also there is, in effect, the same succession and on account of this chain of outcrops of similar character a tentative correlation is made between the brown weathering, siliceous sandstones which in each case form the higher member of the sequence; taking the Wara outcrop as a typical exposure all these brown weathering, more or less silicified sandstones are included in the Wara Sandstone Group.

Wara Sandstone Group. Besides the sandstones outcropping in the area outlined above, other siliceous sandstones overlying yellow and greenish clayey sandstones occur as small, isolated patches capping hillocks such as Umm Khasna to the E. of Wara and Burgan, round the W. flank of the Th'ar ridge, at Madanayat, Mishrif and a number of other small hillocks near Kuwait Town. These siliceous sandstones are all tentatively included in this group. The greatest thickness observed is at Burgan where there are some 40 ft. exposed.

There is considerable variation in character of these sandstones from place to place. The outcrop at Wara is of intermediate type and consists of sugary, clean quartz grains, in parts pebbly, firmly cemented together but with practically no clayey matrix. The individual grains are clearly distinct and there is no chert developed. In Burgan and most of the smaller outcrops to the East and West of Wara and Burgan there is usually some white Kaolinitic matrix but the distinction is always clear between these beds and the underlying clayey sandstones. On the other hand the outcrops nearer Kuwait Town, of which Mishrif is a typical example, are much more siliceous and vary from hard sandstones in which the individual grains are clearly marked to almost homogeneous cherts. In Burgan, Fowaris and some of the smaller outcrops thin fine grained white or cream chert beds occur interbedded with the normal sandstones; these are distinct from the brown cherts of Mishrif. At Grain and Aqunah there are also brown chert fragments lying on top of the limestone. It is possible to distinguish more than one type amongst these cherts and they may be of different origins though they may equally well represent the relics of an overlying sheet of silicified Wara sandstone.

At Wara, Burgan and Fowaris these sandstones lie on an uneven surface of the lower clayey beds which, combined with a sharp change in rock type, suggests some break in sedimentation.

The correlation of all these outcrops with one another is clearly open to criticism and can at best be regarded as only an alternative view to that given by Messrs. Wyllie and Mayhew. The chief objection to regarding these sandstones as relics of a silicified old land surface lies in the sharp break between them and the underlying clayey sandstones in such exposures as Wara, Burgan and Fowaris. This break is sharp for the area under consideration though in essential petrological composition the two types are not far different. At Mishrif and adjacent outcrops no such break is clearly exposed and it is possible that a combination of the two conceptions may be a more correct explanation.

Clayey Sandstones. The light coloured clayey sandstones underlying the Wara sandstone show little variation from place to place. In any one outcrop, apart from those between Al Atraf and Al Mutla'a there is rarely more than 30 or 40 feet exposed. The only continuous section through them of a greater thickness is that obtained by Kuwait No. 2 Well at Burgan. This entered the group some 25 ft. below the base of the Wara sandstones and penetrated 218 ft. 6 ins. In the first 170 feet the beds encountered differed in character from the exposed outcrops chiefly in being more pebbly and somewhat more consolidated. Below 170 feet they were definitely more calcareous. Their colour was masked to some extent by bituminous impregnation but no red beds were encountered.

To the North of Wara and further to the North of Madanayat there are isolated exposures of red sandstones similar to some of those occurring in the Lower and Middle Zor groups about M'dairah. They appear to be interbedded in the normal yellow or white clayey sandstones of the surrounding outcrops.

At Wara, Burgan and Fowaris there is discolouration of the upper few feet of this group, due probably to local secondary action.

The outcrops of the Tha'ar ridge are chiefly of this group with occasional patches of Wara sandstones about the westerly flanks but not on top. This suggests that at time of deposition of the Wara sandstone the ridge had already taken shape and may have been partly emergent.

The Grain limestone was probably due to precipitation rather than normal detrital or organic origin. Its exact position in the sequence is doubtful. A pit sunk on top of the hill and trenches about its flanks showed that the underlying beds are essentially the same as the light coloured clayey sandstones outcropping throughout the surrounding area. The limestone itself, as exposed in the pit, occurs as a jumble of flat, angular slabs set in a mass of soft gypseous material 2 ft. thick. This soft material appears to be a weathering product from the limestone heavily impregnated with gypsum of 'caliche' type. A similar limestone also covered with loose cherts occurs at Zelat Aquanah es Sa'al, three miles to the N.N.W. of Grain but no other outcrop resembling this bed was found. If these two occurrences represent the same bed, there is a general dip to the N.N.W. in this locality of about 1 in 500 but to suppose that this dip continues as far as Burgan and that the limestone should there occur about 25 ft. below the initial horizon of the well is to expect a much greater regularity than is justifiable. One can only suppose that the Grain limestone is a localized development in the Kuwait Series possibly due, as indicated above, to precipitation by alkaline waters from a sea of comparatively high salinity.

7. *Age of the Kuwait Series.*

The only additional evidence as to the age of the Kuwait Series beds, collected since writing the Preliminary Report, is that afforded by finding *Dendritina* cf. *rangii* and *Ostrea* cf. *latimarginata*. Neither fossil is conclusive but both point to the probability of a Miocene age for these beds. The occurrence of *O. latimarginata* in beds, similar to the Kuwait Series in character, immediately above the Euphrates Limestone at Hit is perhaps significant and may indicate that the total thickness of the Kuwait beds is less than was formerly supposed. Whether or not an Oligo-Lower Miocene Euphrates limestone facies is developed in Kuwait is not known and beds of this age may well be indistinguishable from the Kuwait Series as exposed. The occurrence of at least 1500 ft. of Eocene beds in Bahrain in a marl and limestone facies makes it less probable that beds of this age will also be of the same character and therefore liable to inclusion within the Kuwait Series.

5. OIL, GAS AND BITUMEN SEEPAGES

1. *Bahrah* (*M'dairah*).

(i) Subsequent examination revealed the existence of gas seepages covering an area of about 2 acres and lying some ¾ mile to the south of the oil seepages in this locality. There is no oil or bitumen visible at the surface near the gas vents which show the usual discolouration, sulphur impregnation, acid taste and acrid smell associated with gas seepages in the Kuwait Series. Kuwait No. 1 Well drilled within the area of these seeps showed that

(a) even at the shallow depth of 49 ft. there was a small accumulation of gas under slight pressure. The pressure was probably not more than 10 lbs. per sq. in. and the accumulation evidently due to trapping in a pocket beneath a clay bed;

(b) the sands below 29 ft. in this locality contain some oil and bitumen.

(ii) With rise in temperature as warmer weather approached the rate of seeping of oil into Pit No. 2 (see Preliminary Report) greatly increased. On 28th March it was estimated that 150–200 gals. of oil remained in the pit after taking a 42 gallon sample.

2. *Madanayat.*

What probably represents a dead gas seep was found on a small hill 1500 yards to the North of the hillocks near the main Madanayat seep. Discoloured, clayey sandstones of similar appearance to those near the active gas seeps occur here and there is some sulphur impregnation but no perceptible smell or acid taste was observed. A pit dug on the outcrop exposed 4 ft. of white and yellow sandstones showing neither taste nor smell. On the surface there was a small patch of black bituminous sand but no impregnation was found below the surface; it is possible that the bitumen on the surface was carried to the spot from Burgan.

3. *Burgan.*

Further examination of this locality showed that

(a) Sandstones with bituminous impregnation extend to the west and some distance to the East of the Central depression and pass up to the top of the lower clayey sandstone group. The Upper, siliceous (Wara type) sandstones appear to be everywhere free from bitumen;

(b) A small gas seepage occurs about 900 yards to the west of Pit No. 6 and a well half a mile further to the West contains strongly sulphurous water.

(c) The brightly coloured (red, orange etc.) beds below the siliceous sandstones at the S.E. and S. end of the group of hills are confined to this vicinity and appear to be due to local secondary action. The 'burnt' appearance and some porcellanite like rocks in this neighbourhood suggest a possible dead gas seep which may have been on fire. There is no smell or taste indicative of an active gas seepage. A similar discolouration may have been produced by the action of acid water associated with seeping gas. It is probable that the similar discoloured beds at Wara and Fowaris are also of local occurrence only and may represent dead gas seeps.

(d) Mr. Southwell drew attention to similarity between the light, fine-grained, brown-coloured, inflammable substance occurring on top of the bitumen in Pit No. 5 and a substance which occurs about the margin of the pitch in the Trinidad pitch lake. This substance at Burgan shows well developed columnar structure; we had (Preliminary Report) attributed it to partial combustion of the bitumen and taken it as evidence that it had been on fire at some time in the past. Mr. Southwell informed us that in Trinidad the substance is regarded as a mixture of finely divided (colloidal) particles of clay and bitumen. We have not found any published information on the matter but when time is available a chemical and microscopical examination of the substance may yield some useful information on the behaviour of bitumen. It is noteworthy in this connection that analysis showed that bitumen from immediately below this substance differs markedly from that occurring in Pit No. 3 at Burgan. It has a lower Sulphur content (7·60% as compared with 13·6%) and gives a lesser yield on steam distillation up to 350°C. (12·2% as compared with 55·3%). This may indicate that burning did take place and much of the Sulphur-bearing compounds and/or free Sulphur and the lighter fractions were lost through the action of heat. It would be of interest to know whether a similar origin could be attributed to the brown inflammable substance in Trinidad where Day records a fine, bluish clay locally changed to porcellanite through combustion in the neighbourhood of the Pitch Lake.

Bitumen Occurrence on the Coast. About the middle of March it was reported that bitumen had been found washed up along the coast near Kuwait Town and at Ras al Ardh. On a former examination of these localities no bitumen had been noted but a second visit showed a considerable amount occurring as small flakes of soft pitch both at Ras al Ardh and Ras Ajuza which had evidently been washed up during recent southerly storms. Similar occurrences were subsequently found near Fantas and Abu Halaifa and at Qasr al Abid.

This bitumen sank readily in sea water but its density was somewhat increased by the presence of a few sand grains imbedded in it. It is most probable that it was derived from some relatively near submarine seepage—possibly in part or all from that reported near Qaru Island. During transportation it may have been sufficiently light to float, the greater density of the samples tested being due to accretion of sand grains on the beach.

6. STRUCTURE

Evidence of the Zor Escarpment.

From about Qudhai Wells to the N.E. and E. as far as Sabiya an outline map—of which the upper part of Plate II is a reduction—was made on a 3″ to 1 mile scale and the top and bottom of the Middle Zor Group traced round the escarpment. Dips were taken on well exposed beds by measurement with alidade and levelling staff but it soon became evident that such observations were of no great value since determination of the direction of dip could only be determined by taking two apparent dips at a wide angle and individual beds were rarely exposed over a sufficient distance in two such directions to give reliable results. Mapping of horizons with careful height determinations from place to place gave a more reliable indication of the behaviour of the beds.

To the E. of Bahrah (M'dairah) there is a fairly regular dip between M'ghaira and the Saba's wells of about 1 in 40 to E.N.E. From M'ghatti to the west along the escarpment the beds continue to rise until they reach a crest maximum near Nehedin. From there down to M'dairah Wells there is a contrary dip of about 1 in 70 to the W.S.W. There is thus a very flat anticlinal structure at this end of the escarpment with an axis passing slightly to the W. of Nehedin. The precise direction of the axis cannot be determined with but a narrow strip of exposures along the escarpment but it appears to pass near the seepages at Bahrah. It is also impossible to trace the length of the structure since to the S. it passes under Kuwait Bay and to the N. below the cover of Upper Zor beds, Iraq gravels and younger desert alluvium. On Plate I a dotted line shows a possible position of the 50 ft. a.s.l. contour of the base of the Middle Zor Group.

From M'dairah Wells to the S.W. the beds rise again and it is significant that these wells (which are famed amongst the Arabs for having water that is slightly less brackish than any others along the foot of the escarpment) lie in the syncline. From Khashmqudhai mapping was continued on a 1″ to 1 miles scale to Al Atraf and the beds show a general rise in this direction with a few minor undulations as shown in Plate 5. In this stretch of the escarpment the exposures in an E–W-direction are too narrow to give an accurate measure of dip in this sense. There is reason to suppose that the section shown in Plate 5 is not at right angles to the general strike but that there is a slight general dip to the N.W. Such dips as it was possible to take in this direction supported this supposition and the topography further suggests it.

General Structure of the Easterly parts of Kuwait Territory

To the South of Kuwait Bay the absence of distinctive horizons in the exposed rocks makes it impossible to draw definite conclusions as to structure. In the Tha'ar ridge topographic forms and such observations as it was possible to make of disposition of the surface rocks suggest a general easterly dip. The most definite evidence of this was seen at Funtas in a pit 45 ft. deep and 50 ft. × 30 ft. in plan,* where in spite of somewhat irregular bedding there appears to be a seaward dip of somewhat less than 1°.

With the very slight direct evidence available little more than speculations can be put forward as to general structure of the area. The following suggestions are given as such:—

1. that there is a regional dip at very low angle from the interior of Arabia E.N.E.'wards towards the Persian Gulf.
2. subsidiary to this general dip there is a broad anticlinal area bounded to the west by the Jal-es-Zor and the Shagg depression and to the east by the Tha'ar ridge and the continuation of the Jal-es-Zor between M'ghaira and Sabiya.

The section given on Pl. 8 illustrates the latter suggestion. Such undulations as are determinable along the escarpment are thus regarded as minor warpings on the general flat anticlinal. Such a structure would to some extent explain the shape of Kuwait Bay but two topographic features which are not adequately explained by it are the remarkable straightness of the Jal-es-Zor escarpment from Nehedin to Al Atraf and the straightness of the Shagg depression.

Structure of the pre-Kuwait Series Beds.

Where so little is decipherable of the structure of the Kuwait Series beds exposed at the surface an even greater uncertainty must remain as to sub-surface structure. Some consideration of the possibilities is, however, essential to a discussion of the oil prospects of the Territory.

Whether or not the Kuwait Beds are markedly unconformable on Lower Miocene, or older rocks must depend largely on the age of the upper group. On regional grounds it is unlikely that marked folding occurred here between Eocene, Oligocene or Lower Miocene (Asmari) times and Middle Miocene. The climax of Tertiary movements in the Persian mountain system came at the end of Miocene or in Pliocene time and it is improbable that anything older than Pleistocene or late Pliocene should lie almost horizontally over very much more strongly folded beds of Eocene, Oligocene or Lower Miocene age. Though the evidence is not conclusive it is most probably, as discussed above, that the exposed Kuwait Series beds are no younger than Miocene. The probabilities are therefore in favour of the structure of at least Upper cretaceous and young beds conforming approximately with the structure of the Kuwait Series.

In this connection it must be emphasized that no director junction of the Kuwait beds with any older group has been observed to the south and at Bahrein, which lies against the Arabian mainland, there is a clearly defined structure developed in the Eocene and older beds. The possibility remains that the Kuwait Series before removal by erosion did overlie this structure unconformably and that the Oligocene and Lower Miocene beds are absent or represented in the Kuwait Series itself. Such a condition involves the assumption that the late Tertiary movements had little effect on this side of the Gulf but the folding movements were effective here in post Eocene and pre Miocene time.

The evidence to the North from Hit according to Halse indicates a depositional break between the Oligo-Lower Miocene limestone and the Kuwait Series but no marked angular unconformity.

7. RELATION OF SURFACE STRUCTURE TO SEEPAGES

The three seepage localities of the Territory of Kuwait lie within the axial area of the broad anticlinal postulated above. At Bahrah the seepages also lie on the axis of a lesser anticlinal fold. At Fowaris in the Neutral Zone the gas seeps appear to lie towards the westerly side of the supposed area of elevation.

There appear to be two more probable cases worth considering with regard to relation of the seepages to sub-surface structure:—

(i) Assuming a broad anticlinal area to exist as outlined above the several seepages may represent escapes from crest maxima of minor structures on it. This infers that the flat structures probably existing have been sufficient to control accumulation. A well known case of a comparably flat structure having such an effect is that of the Seminole field of Oklahoma but there the oil is Palaeozoic and most probably of lighter gravity and greater mobility than anything that may be expected in Kuwait. One does not know of any Tertiary field showing similar structural conditions.

(ii) If the dominating structure of the area is a gentle Easterly dip, whether of depositional character only or partly accentuated by tectonic movements, it is possible to account for the chain of seepages by supposing a facies change to occur along this line in these beds, overlying the oil bearing rocks, which further to the east have acted as an effective cover. The line of seeps may thus reflect an older shore line where a marl and gypsum facies of the Lower Fars, for example, changes over to a sandy and pebbly facies. A gentle monoclinal structure would allow of migration up from below the Gulf and, where the cover group ceased to be effective, the oil would seep out into the overlying Kuwait Series beds and so to the surface. The seepage reported near Qaru Island may in this case be due to escape from the crest of a structure lying to the East of or on the monocline.

It is also possible to explain the origin of the Kuwait seepages by a combination of both these cases.

In the former case it is reasonable to expect an available accumulation of oil at the top of the main anticlinal area; in the latter case the prospects are less promising and some sealing of the outlet would be necessary for the accumulation of a profitable pool. Such a seal may occur through the agency of bitumen but there is little reason to suppose that this would become effective before approaching the surface.

If the second case represents in general principle the structural conditions prevailing it is more probable that any oil that has come up the monocline would have been dissipated into the Kuwait Series beds. That the seepages are still active may be due to the heavy and viscous nature of the oil which has rendered migration extremely slow. The chief difficulty in accepting this hypothesis is to account satisfactorily from the extensive and active gas seeps. Any gas present might well be expected to have made its way out during the long period available since deposition of the Kuwait beds.

That oil accumulation and consequently the seepages in Kuwait have been controlled by buried faults in the pre-Kuwait Series beds cannot be considered very probable on regional grounds but the scantiness of our knowledge of movements on this part of the Arabian foreland must be borne in mind. No direct evidence of faulting in the Kuwait beds themselves has been found though certain topographic forms (the Jal-es-Zor escarpment and the Batin depression) suggest the influence of fracturing.

8. CHARACTER OF THE OIL

The question of quality of any oil that may be available for exploitation in Kuwait is one of paramount importance in an estimation of the oil prospects of the Territory. The seepages about Hit, in Kuwait and Bahrein of thick viscous oil or bitumen have until lately been regarded as residues derived from a lighter oil by evaporation and weathering at or near the surface. The evidence of wells at Qaiyarah and Khanuka and also at Kuh-i-Mund have, however, indicated that this may not be the case and it is possible that the seepage oil at Bahrah may represent a fair sample of the type of oil to be expected in an accumulation at depth.

The only seepage of liquid oil in quantity in Kuwait Territory occurs at Bahrah (M'dairah). A sample (Bah. 1) from the surface seepage where the oil had been exposed to maximum weathering conditions and another sample (Bah. 4) from a pit dug in the same neighbourhood where the oil had been to some extent protected by a covering of hard bituminous sand, gave the following results on analysis (See Fds. Laboratory Departmental Report No. 456 of 1st March 1932):—

	Bah (1).	*Bah (4).*
Specific Gravity at 60°F*	1·0532	1·0220
Sulphur content on ashless, water-free sample	6·82%	6·42%
% distillate by steam distillation up to 350°C from ashless, water-free sample. Weight-weight	38·9%	25%
Engler on Distillate		
Specific Gravity at 60°F	0·924	0·903
Initial Boiling Point	210°C.	166°C.
% Distillate to 175°C	—	1·5%
,, 200	—	7·5%
,, 225	drops	13·5%
,, 250	1·0%	25·0%
,, 275	9·5%	33·5%
,, 300	21·0%	48·0%
,, 325	—	63·0%
,, 350	—	77·5%
,, 375	—	87·5%

*The Specific Gravity determinations of the sample as collected, in this and the following analyses, are not of much significance since extraneous mineral matter is included. In the above two samples this mineral matter is less than 2% but in Bur (2) and Bur (86) (see below) it is 73·76% and 32·52% respectively.

Bah (1) may fairly be regarded as the weathered product of an oil with the characters of Bah (4). It shows an increase in Sp. Gr., a slightly higher sulphur content and, rather surprisingly, a much greater yield from steam distillation.

If it be supposed that the bitumen at Burgan is the residue from an oil of similar character to that which gave rise to the thick oils at Bahrah, i.e. a stage further in the process of losing its lighter constituents by evaporation and dissipation at or near the surface, one might expect it to show further increase in Sp. Gr., still higher sulphur content and even greater yield from distillation under the same conditions. The sample Bur (2) is more representative of the Burgan bitumen than Bur (4), which has possibly been subjected to abnormal heat by burning at the surface, and this progression in characters is clearly shown by its analysis as follows:—

Specific Gravity	{ 1·8271 1·7662
Sulphur content on ashless water-free sample	13·6%
Matter soluble in CS_2	20·6%
% distillate from ashless, water-free sample. Weight-weight	53·3%
Engler on Distillate	
Specific Gravity at 60°F	0·912
Initial Boiling point	112°C.
% Distillate to 175°C	3·75%
,, 200	7·0%
,, 225	10·0%
,, 250	17·5%
,, 275	23·75%
,, 300	33·75%

In so far as this data goes, it tends to support the supposition that the Kuwait seepages are derived from a lighter oil with lower sulphur content at depth. Mr. L. C. Cowen of Fields Chemical Laboratory is reporting on the analysis stated '. . . it will be seen that a considerable amount of oil was obtained on steam distillation, all except Bah (1) having an initial boiling point and a small fraction of the distillate well within the upper distillate range of motor spirits (F.B.P. about 180°C). The presence of these comparatively low boiling point fractions most definitely suggests that they originally contained much lighter fractions which have weathered off'. The question of importance is that of how much has weathered off.

The presence of extensive gas seeps in the area has also some bearing on this topic. If retained in a reservoir at depth in contact with the oil and at high pressure this gas may reasonably be supposed to be in solution and to impart greater fluidity to the oil. Kuh-i-Mund may be quoted as a case where gas is present in a reservoir in contact with only heavy bituminous oil but in this instance pressure conditions must be taken into account. Kuh-i-Mund conditions might more accurately be compared to those of the Kuwait Series beds after the oil and gas has escaped from its reservoir below and pressure has been much reduced. It is to be noted that at Bahrah where gas and oil seeps occur in close proximity the oil is definitely fluid. At Burgan there is but one small gas show and there are only traces of liquid oil. At Madanayat and Fowaris where there is much gas we have no evidence of the character of associated oil. The most convincing evidence against regarding the presence of gas as a favourable indication is that of Khanuka where dry gas under pressure occurs below a heavy oil accumulation and that of Injanah where viscous oil and gas occur together.

The information derived from well No. 2 at Burgan showed that with increasing depth down to 218 ft. there was no very marked increase in fluidity of the bituminous matter present. The beds and pockets of almost pure bitumen encountered about 86 ft. and between 121–134 ft. are slightly more plastic than the surface bitumen but the difference is relatively slight.

The fact that 218 ft. were penetrated without showing marked change in fluidity of the bituminous matter at least shows that

(i) any great loss of lighter constituents takes place well below the surface,
or (ii) seeping of the oil was in progress during deposition of the beds penetrated,
or (iii) the oil at depth has not contained any appreciable quantity of lighter constituents.

The following analysis was made of the sample of bitumen (Bur. 86) obtained at 86 ft. (ref. *Fds. Lab. Rept.* No. 118 dated 18th April, 1932):—

Specific Gravity at 60°F	1·3562
Sulphur content on ashless water-free sample	9·22%
Matter soluble in CS_2	66·40%
% distillate from ashless water-free sample. Weight-weight	48·3%

Engler on Distillate

Specific Gravity at 60°F		0·940
Initial boiling point		126°C.
% Distillate to	175°C	—
,,	200	—
,,	225	2·5%
,,	250	7·5%
,,	275	12·5%
,,	300	20·0%

In viscosity, Specific Gravity, Sulphur content and percentage distillate up to 350°C (steam distillation) this sample is intermediate between the surface specimen (Bur. 2) and the more weathered oil from the Bahrah surface seep (Bah. 1).

Analyses of samples from 121–134 ft. depth in the Burgan well are not yet available.

An obvious analogy to apply to the Burgan occurrence is that of the Pitch Lake and underlying field in Trinidad. From the rather scanty data available it appears that the bitumen of the lake has a Sp. Gr. at 77°F of 1·40–1·42, contains 56–57% matter soluble in Cs_2 and has a sulphur content of 6·40%. We have not found an analysis of the oil obtained from below the pitch lake but such generalized figures for 'Trinidad' oil as are available show generally high Sp. Gr. and sulphur content of only 1·81% and less. At least there appears to be a considerable concentration of sulphur in the derived bitumen and if the Kuwait seepages do represent residues from an originally light oil it is not unreasonable to suppose that it has a much lower sulphur content than the above analyses show from the seepage products.

As already mentioned, almost all the seepages occurring along the eastern margin of the Arabian massif are of bitumens and oil of a heavy viscous character; the same applies to parts of Dashti and Tangistan where overlying sediments are also comparatively thin but it is worth recalling the seepage of a light, mobile oil at Nafatah near Ramadi as an exception to this generalisation. The well now being drilled in Bahrein by the Standard Oil Co. of California may give valuable data in this question, but, since the structure there is exposed down to the Eocene, the absence of lighter oils does not entirely preclude the possibility that they were originally present but have escaped at the surface. On the other hand if oil of lighter character is found in Bahrein the prospects of Kuwait are greatly enhanced.

Our information in this respect on the results of this well to date is

1. Bitumen or bituminous rocks were found at various depths down to about 1,000 ft.

2. A bottle containing about 100 c.c. of dark brown to black oil with fluidity about equal to that of Fields Crude and a strong smell of petrol was sent by Major Holmes to the Sheikh of Kuwait as a sample of oil from the Bahrein well. It is however clearly unwise to stress this evidence

in favour of the presence of light oil in the westerly belt of the Persian–Iraq basin of sedimentation; a product in colour, viscosity and smell closely similar can be obtained by solution of any of the Kuwait bitumens or seepage oil in refined petrol.

The balance of evidence, as we now see it, undoubtedly indicates the probability of obtaining only heavy grade oil, with high Sulphur content in Kuwait. The bulk of the evidence favouring this view is in the nature of regional considerations; the data on the ground in Kuwait is not in itself, if taken out of its regional context, adverse to the chances of there being a high-grade oil at depth.

9. OIL PROSPECTS

1. Lack of data on both structural conditions and the nature and horizon of oil bearing beds at depth place the Territory of Kuwait as an oil proposition definitely in the 'wild-cat' class. The circumstances in favour of and against the oil prospects of the area may be enumerated as follows:—

PRO. (i) The existence of large seepages indicating that oil has occurred at depth and that much may still remain there.

(ii) The presence of active gas seepages lends some support to the contention that pressure conditions will be satisfactory. The case of Kuh-i-Mund on the other hand may be cited as an example of unsatisfactory pressure conditions where there is a small surface gas seepage.

(iii) The area is easily accessible and no costly road making would be necessary in its development. Kuwait Bay is a fair natural harbour and shipping costs should be low in comparison with those of Abadan.

(iv) The Territory of Kuwait is a British Protectorate and is therefore politically advantageous to a British Company.

CON. (i) Quality of the Oil. The probability of a high sulphur content, high viscosity and low percentage of more valuable fractions.

(ii) Lack of direct knowledge of the thickness or facies of the pre-Kuwait Series beds on account of the absence of exposures in the neighbourhood.

(iii) Improbability of well developed structures in the pre-Kuwait Series beds and lack of direct knowledge of extent and form of such structure as may exist in them.

2. On the evidence available it appears that:

(i) There is a reasonably good chance of finding an accumulation of oil at depth near one or other of the seepages and possibly elsewhere in the eastern part of the Territory of Kuwait.

(ii) on regional grounds, it is improbable that such oil would contain more than a low percentage of the lighter fractions and the odds are heavily against obtaining a field with oil of a quality comparable with that of Fields or Haft Kel.

In deciding whether the gamble is worth taking or not a most important factor to take into consideration is the market value of heavy crudes.

10. RECOMMENDATIONS

In our Preliminary Report we recommended

(1) detailed surface geological work to try to determine flat structures in the exposed Kuwait Series beds and

(2) Geophysical work to determine structure in the pre-Kuwait beds.

The former recommendation has been carried out with the results indicated above.

A consensus of opinion expressed about the second recommendation regards seismic methods as unlikely to give useful results on account of flatness of structures to be expected.

In the writer's opinion further surface geological work in Kuwait will not yield more conclusive evidence as to exposed structure nor data of vital importance on stratigraphical problems. For further information on the latter questions, a reconnaissance in the Province of Hasa is recommended but it is doubtful whether results obtained would lend much assistance in deciding whether or not to explore further the possibilities of Kuwait Territory. If it is the agreed opinion of geophysical experts that conditions are unfavourable to an application of any of their methods, then this decision must be made on the evidence now available.

If it be decided to take this gamble, one recommends that the next step in testing the area should be to drill a well capable of going to 4,000 or 5,000 feet and capable of coping with oil under pressure if found.

With only the slight data available, it would clearly be advisable to put such a test near one of the seepages and in view of the possibility of accumulation being controlled by a monoclinal structure, it should be located one or two miles to the east of the actual seep. Such a location should encounter any accumulation of useful size in either of the structural cases considered above.

The merits of the four chief seepage localities as sites for a first test well are as follows:—

Bahrah. Fluid oil here reaches the surface. This *may* indicate that a somewhat better quality oil is available here. The initial horizon is probably lower than at any of the other seepages and one or two hundred feet of dead drilling may here be avoided. On the other hand a considerable quantity of oil has been dissipated into the Kuwait beds and the original reservoir may be, to a large extent, drained. The locality is less easily accessible with heavy transport than either Madanayat or Burgan.

Madanayat. Only gas occurs here and the absence of surface oil or bitumen may indicate that there has been less loss by seepage into the Kuwait Series beds. The initial horizon may be somewhat higher than that of Bahrah but is probably lower than that of Burgan. The locality is easily accessible for heavy transport from Kuwait Harbour.

Burgan. The extensive bitumen deposit here probably indicates considerable waste from the original reservoir. The initial horizon is probably higher than that of either Bahrah or Madanayat. The locality is more accessible than Bahrah but less so than Madanayat.

Fowaris. Geological conditions are similar here to those of Madanayat except that the initial horizon is probably higher. The place is much less accessible than any of the other localities and lies in Neutral Territory, which could require special political provisions to be made.

Considering all factors the most suitable location for a first test well would be near Malah water (brackish) wells about two miles to the East of the Madanayat gas seepages.

It is perhaps unnecessary to add that one well would be insufficient to test fully the whole area but it is recommended that no further location be made until all possible information has been obtained from the first well.

11. ACKNOWLEDGEMENTS

Our warmest thanks are due to H.E. Shaikh Sir Ahmad al Jabir as Subah and to H.B.M. Political Agent in Kuwait, Lt.-Col H. R. P. Dickson for their continual assistance and cooperation during the field work in Kuwait Territory.

Masjid-i-Sulaiman,
May 12th, 1932.

Sgd/- P. T. COX

50.

720/7 *Abadan*
13th April 1932.

H.E. Sheikh Ahmed El Jabir El Sabah, C.I.E.
The Sheikh of Kuwait,
KUWAIT.

Your Excellency,

We much regret to have to inform Your Excellency that as the result of our present investigations in search of oil in Your Excellency's territory, our London Principals have decided that they are not justified in approaching Your Excellency at present with a view to acquiring a concession.

We are accordingly compelled to abandon work meantime and the necessary steps are being taken to evacuate personnel and material.

Our Principals desire us to express their sincere gratitude to Your Excellency for permitting us to undertake this exploration, and their most cordial appreciation of Your Excellency's protection and assistance at all times.

May we be permitted to take this opportunity also to associate ourselves with these sentiments and to say how disappointed we are that our work in Your Excellency's territory has to be abandoned for the present.

We have the honour to be,
Your Excellency's most obedient servants,
For ANGLO-PERSIAN OIL COMPANY, LTD.,
Sgd. E. H. O. Elkington.
General Manager.

---

51. *The EGS draft concession of 26th May 1932 was as follows:*

In the name of God The Merciful.

THIS AGREEMENT made the day of 19 ,
corresponding to day of 13 ,

At Kuwait between HIS EXCELLENCY SHEIKH SIR AHMED BIN JABIR AL SUBAH, K.C.I.E., C.S.I., SHEIKH OF KOWEIT in Arabia (hereinafter called THE SHEIKH which expression where the context so admits shall include HIS HEIRS, SUCCESSORS, ASSIGNS AND SUBJECTS) of the First Part, and true and Lawful Attorney of EASTERN & GENERAL SYNDICATE LIMITED whose Registered Office is at 19, St. Swithin's Lane, E.C.4. London, England (hereinafter called THE COMPANY which expression where the context so admits includes its ASSIGNS and SUCCESSORS) of the Other Part.

WHEREAS THE SHEIKH is desirous of developing the Oil and Petroleum Resources of His Territory, he has for that purpose, agreed to grant unto THE COMPANY the Concession hereinafter contained.

(1) In consideration of the rights, covenants and royalties hereinafter reserved and contained, THE SHEIKH, in exercise of His Powers, as Ruler and Sovereign of His Dominions for Himself, Heirs, Assigns, Successors and Subjects hereby grants unto THE COMPANY exclusively throughout the whole of the territories of THE SHEIKH, including the islands and territorial waters

subject to His control, (all hereinafter called 'KOWEIT TERRITORY'), the rights, privileges and interests enumerated in this Concession: and, THE SHEIKH hereby grants free access to all the agents, successors and servants of THE COMPANY necessarily employed by the latter to all parts of the KOWEIT TERRITORY under the control of THE SHEIKH, saving only sacred buildings, shrines, graveyards and the area within the present existing town wall of Koweit.

The term for which this Concession is granted unto THE COMPANY is of seventy (70) calendar years from the date of the execution of these presents, THE COMPANY yielding and paying therefore the fees, payments, royalties, privileges and rights to THE SHEIKH and subject to the provisions hereinafter enumerate.

If THE COMPANY, unless prevented by the Act of God or by war, fire, flood or lightning or some other thing beyond human control, shall not have commenced its exploratory work of the surface of the KOWEIT TERRITORY through examination and investigation by its geologists and engineers within a period of nine (9) calendar months from the date hereof, then the provisions of these presents shall lapse and this deed shall be null and void and neither party shall have any claim against the other in consequence thereof and no moneys already paid to THE SHEIKH shall be returnable.

(2) THE SHEIKH hereby grants to THE COMPANY, during the term of this Concession, the exclusive rights to explore the KOWEIT TERRITORY for petroleum, natural gas, asphalt, ozokerite, oil and its products, and grants to THE COMPANY the exclusive ownership of all said substances won from the KOWEIT TERRITORY, with the exclusive right to search for, own, exploit, develop, carry away, export and sell said substances so won from the KOWEIT TERRITORY, and for that purpose, and in connection therewith, exercise in, over and upon the said lands any or all of the following things:

(a) To drill, sink, make, erect, set, operate, construct and maintain oil and gas Wells, Water Wells, Pits, Buildings, Camps, Storage Facilities for Water and Gas and Oil, Water, Steam, Oil and Gas Lines, Roads, Waterways, Pipelines, Engines, Machinery, Furnaces, Brick-Kilns, Cement Ovens, Workmen's Cottages, Railways, Bridges, Tramways, Telephone and Telegraph Lines and Stations in connection therewith, and other Ways of Communication and Transportation, Canals, Wharves, Dams, Erections and other Works, and Dwelling Houses for THE COMPANY'S agents and workmen; and do generally whatever THE COMPANY may deem expedient for the proper exploitation of the KOWEIT TERRITORY provided that does not harm general interest.

(b) To exclusively erect Oil Refineries and Oil Tanks, wherever THE COMPANY shall deem suitable whether in proximity to discovered wells or otherwise, provided such action does not in any way harm the general interest.

(c) To have and use free for any purpose connected with the working of the said KOWEIT TERRITORY stone, cement, lime and other construction materials, and any water within the territory of THE SHEIKH and with the assistance of THE SHEIKH make and construct Water-Courses, Reservoirs and Ponds for collecting such water, provided no harm to the General public or to individuals is apparent.

THE COMPANY shall, in peace time throughout the period of Concession, accept and transmit, on its telegraph lines, THE SHEIKH'S GOVERNMENT'S telegrams whether in cypher or in clear and likewise allow him the use of its telephone lines, and also THE SHEIKH may use the railways, on special personal services throughout the period of the Concession during peace time, and have the full use thereof when His Country is at war.

THE SHEIKH hereby grants free of charge, rent or cost to THE COMPANY all surface rights, easements, and privileges over the KOWEIT TERRITORY which it may find necessary for any of the purposes herein mentioned. However, in the event that THE COMPANY might desire to occupy lands covered by village areas, gardens or private water wells within the KOWEIT TERRITORY, then THE COMPANY shall not be permitted to enter into possession thereof without first making arrangements to do so with THE SHEIKH and through THE SHEIKH with the owners.

(3) The port and buildings which THE COMPANY require shall be erected outside the existing town wall of Koweit. THE COMPANY shall have the right to construct and develop the harbours along the coast of KOWEIT TERRITORY and to erect and construct Wharves, Cranes, employ Dredgers, lay down Buoys and erect Lighthouses and do whatever may be necessary to make the harbours safe for the navigation of ships and barges, the loading and unloading of oil and its products, machinery and other goods belonging to or sent to THE COMPANY, THE SHEIKH granting to THE COMPANY the necessary and proper surface rights in connection with such harbours. The Customs Administration of the ports developed by THE COMPANY shall be under THE SHEIKH's local Customs Officials and THE COMPANY undertakes to erect a conveniently large building for a Custom House at each such point and a suitable residence for THE SHEIKH's Official Representative. Should it be necessary also to maintain guards for the protection of THE COMPANY's works (wells, etc.) inland or along the pipelines or other communications to the sea, THE COMPANY shall build suitable buildings for such guards at its own expense.

THE SHEIKH's Flag and no other shall be used within the KOWEIT TERRITORY.

(4) THE COMPANY shall be free to construct, operate and maintain one or more pipelines in the KOWEIT TERRITORY and across same for the purpose of carrying oil, gas and kindred substances, (whether produced from the KOWEIT TERRITORY or elsewhere), connecting the oil fields and works with each other and with the harbour or harbours which THE COMPANY

may establish or develop on the coast and shall have the right to establish and maintain one or more coaling or oil stations along the coast of Koweit, the use of such facilities being vested entirely in THE COMPANY. THE SHEIKH retains the right to grant permission to others besides THE COMPANY to import oil and coal and lay pipelines for these purposes.

(5) THE COMPANY shall be free and at liberty to export, sell and dispose to any place or people or country it may wish to and in any manner it may desire all the oil and its products handled through its pipelines and harbour facilities and THE SHEIKH and those acting under Him shall not interfere with the internal management of THE COMPANY, but THE SHEIKH shall have the right to keep a general eye over the doings of THE COMPANY. And He shall have the right to levy, and THE COMPANY undertakes to pay Him a custom duty of one per cent (1%) on all the oil and its products produced by it hereunder from the KOWEIT TERRITORY which THE COMPANY may export. In calculating such one per cent (1%) on the oil exported, the value of the oil at the wells producing same shall be used.

(6) THE COMPANY shall have the right to import oil, machinery, equipment, plant, timber, utensils, iron work, building materials and everything belonging to or consigned to it, including medicines and food supplies for the use of THE COMPANY and its employees but not for resale to others, free of customs or import duty and taxes or other charges, but it shall pay on all personal goods, clothing and general merchandise imported by THE COMPANY for the personal use of its employees the ordinary duty in vogue in the KOWEIT TERRITORY computed on the values shown in original invoices plus expenses.

(7) Saving as herein provided, THE COMPANY shall be exempt and free, during the period of this Concession, from all taxes, imposts and charges of any kind, tolls and land surface rent of whatever nature, it being understood that THE COMPANY has no right to lease any building to any but its employees and agents. Should ships other than those engaged and used solely for THE COMPANY's business hereunder, either bringing or taking away, make use of the harbours, THE SHEIKH has the right to collect the usual harbour dues and taxes from such ships, it being understood that the wharves and appurtenances erected by THE COMPANY during the period of the Concession hereby granted and can only, during such period, be used by ships on other than THE COMPANY's business with the written permission of THE COMPANY.

(8) The ownership of this Concession may be transferred to, or the rights sold to, another company and THE SHEIKH undertakes to sanction such transfer when so requested, PROVIDED always that the rights, privileges and interests accruing hereunder to THE SHEIKH shall not thereby be lessened, and PROVIDED FURTHER:

(a) That the assignee company shall be one organized and registered either in Great Britain or organized and registered in Canada and shall maintain an office in Great Britain which shall at all times be in charge of a British subject who shall be the recognized channel of communication between the assignee company and His British Majesty's Government in the United Kingdom of Great Britain and Northern Ireland.

(b) That of the five directors of the assignee company one director shall at all times be a British subject who shall be *persona grata* to His British Majesty's Government. His selection for appointment as director shall be made in consultation with His British Majesty's Government and his salary as director shall be provided by the assignee company. It is understood that a reasonable time shall be allowed for the replacement of the British director in the event of this post falling vacant.

(c) That the assignee company shall at all times maintain in the region of the Persian Gulf an official to be called the 'CHIEF LOCAL REPRESENTATIVE' of the assignee company whose appointment shall be approved by His British Majesty's Government in the United Kingdom of Great Britain and Northern Ireland, and who shall be the sole representative of the company empowered to deal direct with the local authorities and population in Koweit. All communications which that official may desire on behalf of the assignee company to address to THE SHEIKH of Koweit shall be through the British Political Agent in Koweit. For the first five years after the assignee company starts to operate in Koweit or for such lesser period as the assignee company may operate in that territory, their Chief Local Representative shall be Major Frank Holmes, provided the arrangement between THE COMPANY and Major Holmes continues to be mutually satisfactory to them during such five years, or such lesser period above mentioned, provided also that any sooner determination of the appointment of Major Holmes shall be subject to the consent of His British Majesty's Government, which shall not be unreasonably withheld. It is understood that a reasonable time shall be allowed for the replacement of the Chief Local Representative in the event of this post falling vacant.

(d) That as many of the employees of the assignee company in Koweit as is consistent with the efficient carrying on of the undertaking shall at all times be British subject or Subjects of THE SHEIKH of Koweit.

And if by any other ways or means the Concession is transferred or sold to a THIRD PARTY, this Concession will then become null and void, and THE COMPANY shall leave all the immovable property and wells intact and they will be the property of THE SHEIKH.

(9) THE COMPANY'S representatives in the KOWEIT TERRITORY shall be immune from local interference except with the leave of THE COMPANY, and (in matters concerning themselves but not where the Subjects of THE SHEIKH are concerned) shall be responsible for their conduct to THE COMPANY's Board of Directors.

(10) If after commencing its operations in the KOWEIT TERRITORY, THE COMPANY for

any reason other than THE ACT OF GOD, or War, Fire, Flood or Lightning or some other thing beyond Human Control, should discontinue the same for a continuous period of TWO CALENDAR YEARS (2 years) THE SHEIKH shall have the right to cancel THE AGREEMENT and no responsibility shall attach to either party.

(11) THE COMPANY or its employees shall not interfere in any manner or way with the Politics of THE SHEIKH's Dominions or with His Subjects.

(12) THE COMPANY shall employ only native labor (Arab) who should be under the supervision of THE COMPANY's European or other appointed officials, and THE SHEIKH agrees to assist with the help of His Amirs and other local Agents to procure and provide for THE COMPANY such native labor as THE COMPANY may require, and THE COMPANY on its part undertakes to make the fullest use of the local unskilled labor, in its judgment capable of performing the work, to the extent of the suitable supply for its requirements: but THE COMPANY has the right to import unskilled labor should the local supply prove insufficient or unsuitable to the extent of its requirements and THE COMPANY has the right at all times to import skilled workmen of every kind,—all subject to paragraph (d) of Article (8) above.

(13) THE COMPANY shall pay to the native workmen it employs a fair wage, such wage to be decided and stated by THE COMPANY's representative at the time the workman is engaged. THE COMPANY shall provide where possible medical attention and medicines free of charge, to its native workmen during the time they are in employ of THE COMPANY.

(14) THE SHEIKH shall always afford the officials and employees of THE COMPANY every facility, assistance, and protection in carrying out their plans and projects as far as lies in His Power, and shall allow them to excavate, dig, quarry or drill the Soil in the KOWEIT TERRITORY (saving only sacred buildings, shrines, graveyards and the area within the present existing town wall of Koweit) wherever they shall have reasonable prospects of discovering and winning petroleum or kindred products, and THE COMPANY by its officials shall be at liberty to abandon any excavation pit or well wherever and whenever they shall deem it expedient so to do, provided always that nothing in this article shall be presumed to give to THE COMPANY or its Assigns or Agents right of entry into, or on, to private properties without prior sanction of THE SHEIKH or His duly appointed representative.

(15) Within SIXTY DAYS (60 days) from the signature of This AGREEMENT, THE COMPANY in consideration of THE SHEIKH's granting this Concession and the assistance to be afforded to their employees, shall pay to THE SHEIKH the sum of RUPEES THIRTY THOUSAND (Rs. 30,000). But if the payment of this sum of RUPEES THIRTY THOUSAND (Rs. 30,000) is not made by THE COMPANY within 60 days specified, then this AGREEMENT will become null and void.

And after first payment of RUPEES THIRTY THOUSAND (Rs. 30,000) as specified above in this article on each anniversary of the date of the signature of This AGREEMENT, THE COMPANY shall pay to THE SHEIKH the sum of RUPEES TWENTY THOUSAND (Rs. 20,000).

The yearly payment of RUPEES TWENTY THOUSAND (Rs. 20,000) shall continue without fail whether THE COMPANY is working or not, until THE COMPANY should declare that oil has been found on the KOWEIT TERRITORY in commercially exploitable quantities, in which event it is agreed that this yearly payment shall cease after the expiry of the then current year for which same has been paid.

(16) Should THE COMPANY succeed in finding oil in commercially exploitable quantities, it agrees to pay to THE SHEIKH in lieu of the annual payment of RUPEES TWENTY THOUSAND (Rs. 20,000) provided for in Article (15) a royalty of RUPEES THREE AND ANNAS EIGHT ONLY (Rs. 3/8/-) per English ton of net crude oil got and saved (i.e. after deducting water and foreign substances, and oil required for the customary operations of THE COMPANY's installations in THE SHEIKH'S Territories).

(17) THE COMPANY hereby undertakes that the amount received by THE SHEIKH in respect of royalties shall not be less than RUPEES SEVENTY THOUSAND (Rs. 70,000) in any complete calendar year in which THE COMPANY continues work, such calendar year to begin at the end of the last day of the year for which the annual payment of RUPEES TWENTY THOUSAND (Rs. 20,000) has been paid, and it is only on the oil won after said day that THE SHEIKH'S royalty begins to accrue. In the event of THE SHEIKH, in consultation with the Political Resident in the Persian Gulf, disputing THE COMPANY's decision as to the commercial exploitation, THE COMPANY hereby undertakes its readiness to submit the matter to arbitration as provided in Article (21) below.

(18) In the event of THE COMPANY failing within SIX CALENDAR MONTHS (6 months) following the end of any calendar year to pay to THE SHEIKH the royalties due in respect of that calendar year or failing, save for causes beyond its control, to carry out its obligations under this AGREEMENT, THE SHEIKH shall have the power to terminate the Concession, in which case the provisions of Article (19), (b), shall apply.

(19) THE COMPANY for itself, Successors and Assigns hereby covenants with THE SHEIKH in a manner following:

(a) To pay the fees and payments required by this DEED at the time and in a manner appointed and also to observe the provisions herein contained.

(b) At the termination of the Concession whether by the expiration of its period of SEVENTY CALENDAR YEARS (70 years) stipulated, or before such expiration under Articles (10) or

(18) of This AGREEMENT, but after the lapse of THIRTY-FIVE (35) calendar years from the date hereof, to deliver to THE SHEIKH all buildings and erections of brick, stone or other materials whatsoever, the railways, telegraphs and telephones and other things standing and being on the KOWEIT TERRITORY, and all pits, wells, mines, waterways, pipelines, refineries, oil and water tanks, and all such other works and other things belonging to any of the mines and wells, fixed machinery, plants, railways and their cabins and wagons, telegraph and telephone lines and port appurtenances belonging to THE COMPANY, and to leave the ports and harbours as they are, and to relinquish all rights vested in it under Article (7) of This AGREEMENT, leaving also buoys and barges, in fact all fixed things belonging to it which are on the KOWEIT TERRITORY. Provided always that if this Concession shall terminate under Articles (10) or (18) of these presents within a period of THIRTY-FIVE (35) calendar years from the date hereof, THE COMPANY shall have the right to remove from said KOWEIT TERRITORY any or all its plant, machinery, tools, apparatus and other things belonging to it above mentioned.

(20) THE COMPANY shall do or cause to be done nothing in the KOWEIT TERRITORY which, unless expressly authorized by the provisions herein contained, shall be an infringement of or derogatory to the rights, privileges and prerogatives inherent in THE SHEIKH as Ruler of the KOWEIT TERRITORY, and in case any such infringement shall inadvertently have been committed by any of THE COMPANY'S officials, upon due proof of such infringement being received by THE COMPANY'S local representative, THE COMPANY shall forthwith make such amends, as may seem fair and reasonable and suitable, and in case of dispute, the Local Judge may be asked to arbitrate and in case of further disagreement, it may be referred direct to THE SHEIKH for judgment.

(21) If at any time during or after the currency of this AGREEMENT any doubt, difference or dispute shall arise between THE SHEIKH and THE COMPANY concerning the interpretation or execution hereof, or anything herein contained, or in connection herewith, or the rights and liabilities of either party hereunder, the same shall, failing any agreement to settle it in any other way, be referred to two arbitrators, each party choosing one of such arbitrators, and an Umpire who shall be the Political Resident in the Persian Gulf, or a person nominated by Him, before proceeding to arbitration. Each party shall nominate its arbitrator within THIRTY (30) DAYS after being requested in writing by the other party to do so. The decision of the arbitrators, or in the case of a difference of opinion between them, the decision of the Umpire, shall be final. The place of arbitration shall be such as may be agreed by the parties, and in default of agreement shall be Basra. If the question submitted to arbitration should involve the interpretation or execution hereof and the decision of the arbitrators is contrary to the contention of THE COMPANY, then THE COMPANY shall have a reasonable time thereafter within which to comply with such decision before it will be considered in default.

(22) THE COMPANY shall pay all monies that may become due to THE SHEIKH under this AGREEMENT into His account with the

Bank in

The bank receipt for such money shall be a full discharge for THE COMPANY in regard to due payment.

(23) THE COMPANY reserves the right to surrender back at any time to THE SHEIKH this CONCESSION on the KOWEIT TERRITORY and thereby terminate all its rights, privileges, responsibilities and obligations hereunder with respect thereto. In such event, the annual payments under Article (15) and minimum amount of royalty under Article (17) which THE COMPANY may then be paying to THE SHEIKH shall cease and terminate entirely. However, THE COMPANY shall make any payments then due hereunder for the then current year. Notice in writing shall be given to THE SHEIKH by THE COMPANY of its election to surrender as aforesaid and this CONCESSION and AGREEMENT shall be held as terminated as of the date of such notice. The provisions of Article (19), (b), shall apply in the event of such surrender.

(24) In the event of any discrepancy between the meanings of the English and Arabic versions hereof, the English version shall prevail.

(25) This AGREEMENT, which comprises the preamble and TWENTY-FOUR (24) Articles other than this, is made and signed by the parties hereto in Original, Duplicate and Triplicate, the Duplicate being retained by HIS EXCELLENCY THE SHEIKH and Both the Original and Triplicate by THE COMPANY.

This AGREEMENT extends over pages, all of which are signed by the parties at foot.

IN WITNESS WHEREOF the said parties have hereunto set their hands and sealed the day, month and year shown below their respective signatures, and GOD IS GRACIOUS.

---

52. *A. H. T. Chisholm (1902–), M.A. (Oxford, 1925), C.B.E. (Military, 1946); joined Anglo-Persian Oil Company 1927, serving 1928–32 in Persia (Masjid-i-Suleiman, Abadan, Teheran, Isfahan, Ahwaz); 1932–34 in Kuwait; 1935–36 in Persia (Abadan). Resigned from APOC 1936 because his request for transfer to its London head office staff was refused. Editor of* The Financial Times, *London, 1937–40. British Army (2nd Lieut. 1940–Colonel 1944) in U.K., Iran, Egypt, U.K., India, 1940–45. In 1945, at*

Anglo-Iranian Oil Co. (formerly APOC) invitation, joined their London head office staff, serving (as from 1946 as General Manager, Public Affairs and Information) until retirement 1962. Adviser to the British Petroleum Company Ltd. (formerly AIOC, APOC) 1962–72. Between 1946 and 1972 frequently visited Kuwait, being received on many occasions in private audience by the late Shaikh Ahmad up to 1949, by the late Shaikh Abdullah Salim up to 1964, and subsequently by Shaikh Sabah Salim.

---

53. *Cable received 6th June 1932 by Gulf (New York) from Gulf (London):*

June 6th cable received from Holmes by Syndicate dated June 6th stating May 27th he had long discussion with Sheikh who stated Political continually pressing him give concession to Anglo Persian Oil Co. and that A.P.O.CO. draft concession was not left with him but shown to him and taken away with statement it would be redrafted on more favorable terms; Anglo Persian Oil Co. representative and British Political Resident Persian Gulf calling see Sheikh in course of few days to present draft stop Sheikh voluntarily told Holmes 'No matter what terms A.P.O.CO. offer I will not discuss them nor will I upon any consideration grant A.P.O.CO. Koweit concession, I have promised Koweit concession to you and shall stand by my word' unquote Holmes states believes Sheikh will keep promise but fears political people working to cause delay fullstop
In letter dated May 26th Holmes wrote he had furnished B.P.A.K. copy new form draft concession stop

---

54. *The following State Department memorandum, after referring to Ambassador Mellon's position in London (see Note 47), refers to expected British Government policy towards Kuwait:*

Department of State,
Division of Near Eastern Affairs,
March 22nd, 1932

Dear Mr. Secretary:

I am informed that Mr. Mellon has communicated, through Mr. Finley, to the Department his hope that the negotiations now going on between the American Embassy in London and the British Government on behalf of the Gulf Oil Company, a Mellon interest, which is seeking a concession in the Shaikhdom of Kuwait in the Persian Gulf, will, if possible, be transferred to Washington and carried on exclusively between the Department of State and the British Embassy. I have discussed this matter with Mr. Castle and he concurs with me in believing that this suggestion is not feasible. To reverse the ordinary channels of communication and deal with this matter solely in Washington, thus ignoring our Embassy in London, would necessitate some explanation to the British Ambassador, and the true explanation would be somewhat embarrassing to give. I do not mean to say that it would be out of order for you, if you so desire, to mention the matter to the British Ambassador simultaneously with the sending of necessary instructions to London on this matter. However, it is my impression that your practice is to consult with foreign representatives in Washington as a rule only regarding matters of outstanding interest to this Government, and I do not believe that the Mellon concession can be so regarded.

I believe that the transfer of these negotiations entirely to Washington would be particularly undesirable for domestic reasons. It is obvious that if this fact became known either to the press or to our friends on the Hill it would be grossly misrepresented and misinterpreted. The inference might, I fear, be drawn that the Department of State has reversed an age-long procedure merely for the personal benefit of Mr. Mellon and in order to assure a more expert handling of the matter than might be possible through the ordinary channels of our Embassy staff at London. As you are of course aware, leaks of this kind have not been, unfortunately, a rarity in the Department recently.

Another reason for disapproving the procedure suggested, so I understand, by Mr. Mellon is that, in my opinion, the Gulf Oil Company has a very slim chance of acquiring an oil concession in the Shaikhdom of Kuwait. This Shaikhdom is in close proximity to the areas of South Persia now under exploitation by the Anglo Persian Oil Company, controlled by the British Government. The attached map will show you the precise location of Kuwait. If a foreign oil company should become established in this territory and find oil in any quantity it would obviously be in a position seriously to compete with the British-owned oil production in South Persia. It is also to be remembered that no concession has as yet been granted to any company in the Shaikhdom of Kuwait. The Gulf Oil Company is merely a rival concession-hunter with the British-controlled Anglo Persian Oil Company, and the Gulf Oil Company has an understanding with the Eastern and General Syndicate, a British Company, to take over any concession which may be acquired by that British company from the Shaikh of Kuwait. The Anglo Persian Oil Company, with the undoubted support of the British Government, has, on the other hand, been granted a right to explore for oil in the Shaikhdom and will, if it acquires a concession, not be over-riding, as far as I can see, any previously acquired vested interest of the Eastern and General Syndicate or of the Gulf Oil Company with which the Syndicate is associated in this concession scramble. In conclusion, it may be well to remember that the British have on a previous occasion (please see marked passage on page 34 of the attached pamphlet, entitled 'Mandate for Palestine') accused us of effecting the cancellation of oil concessions actually granted to British subjects in Haiti and Costa Rica. There can be no question in the present instance of the cancellation of any concession since none has been acquired, and it would be asking a great deal of the British Government to expect them to refrain from supporting the Government-controlled Anglo Persian Oil Company in establishing itself in British-controlled territory like the Shaikhdom of Kuwait in the close proximity of the Anglo-Persian Oil fields in South Persia.

For the above reasons, I think that we should let well enough alone in this matter and remain content with such efforts as we can legitimately make through London to assist the Gulf Oil Company in the present instance.

Wallace Murray.

---

55. *See page 21 and also Note 48.*

---

56. *The over-optimistic impression which Holmes continually conveyed to EGS and, from 1927 onwards, also to Gulf that Shaikh Ahmad would not give his concession to anyone but themselves was only partly due to his justifiable faith in the Shaikh's friendship and personal promises to himself. Though even as regards those promises, from 1927 and especially from 1931 until APOC and Gulf combined forces in late 1933, when communicating with his principals he disregarded, perhaps deliberately, the possibility that Shaikh Ahmad's mind was turning towards the desirability of such a combine (see page 14); an eventuality which, while fully honouring the Shaikh's promises to Holmes personally, would not similarly fulfil the expectations of the latter's principals.*

The main reason, however, underlying Holmes's over-optimistic assessment to his principals of their position on this and other occasions was the nature of the information, on which he had to rely, from both of his two intimate contacts with the Shaikh, Mullah Saleh and Muhammed Yatim. Both of them had such a strong personal interest (see Note 17) in his success in the negotiations that inevitably their reports to him of the Shaikh's attitude were consistently over-favourably coloured towards himself and correspondingly over-adverse to their APOC competitors.

---

57. *This action followed the 'Achnacarry Agreement', so called from the name of a country house in Scotland where the oil-companies' representatives met in 1928 for discussions of the world oil situation.*

---

58. *In 1945 Holmes was asked at a London meeting of oil-companies' representatives how he had been so certain, for ten years before oil was struck in Bahrain, that the unanimous verdict of the world's leading petroleum geologists that oil in commercial quantities would not be found on the Arab side of the Gulf was wrong. He tapped his finger on his nose and said* 'this *was my geologist*'.

The fact is that the surface oil-seepages which he saw in 1922 in Saudi Arabia (he told the writer of this Record in 1932 that he first heard of them when he was in Abyssinia in 1918 from an Arab trader) and subsequently in 1923 in Bahrain and Kuwait, were sufficient evidence for him, after his extensive experience as a mining engineer of the vagaries of theoretical advice however expert, to back his practical judgement against even the best geologists.

This resolute confidence in his own opinion, triumphantly justified in this instance, was less successful for Holmes on a later occasion in connection with Kuwait. He prophesied to the writer of this Record in 1934 during their joint negotiations in Kuwait, that future oil production would be enormous in Saudi Arabia, Qatar and the Neutral Zone, but quite moderate in Kuwait and Bahrain. He was to be proved right as regards Saudi Arabia and Bahrain, but wrong about Qatar, the Neutral Zone and especially Kuwait. His opinion about the latter must have been a main factor in deciding EGS to commute for a comparatively small payment their entitlement to an overriding royalty on future Kuwait production when terminating in 1933 their previous contract with Gulf Oil Corporation (see Note 97).

---

59. *The surprise caused throughout the world petroleum industry by the discovery in June 1932 of oil in commercial quantities in Bahrain, conclusively disproving the previous prevailing geological theory that Arabia was oil-dry, was responsible for a story widely circulated since that one of the industry's leading geologists was on record only a few months previously that 'he would drink all the oil ever to come out of Arabia'.*

This story, and its strikingly incorrect forecast, is not accurate although founded on fact, and as it has been wrongly attributed to various prominent oilmen, the truth about it is worth recording. This is that the late Dr George Lees, then Chief Geologist of the Anglo-Persian Oil Company, after summing up the evidence for and against finding oil in commercial quantities in Bahrain as an even chance either way included himself among the doubters by adding that he 'would drink any commercial oil found in Bahrain'.

In addition to its effect on the Kuwait concession negotiations, as described in this Record, the discovery of oil in Bahrain in June 1932 resulted in Standard Oil of California and IPC sending negotiators to Saudi Arabia (where Ibn Saud granted the concession to the former in May 1933) and APOC sending negotiators, including the writer of this Record in 1933, to Qatar (on behalf of IPC as Qatar was in its 'Red Line Agreement' area; see Note 34) where they were granted the concession by Shaikh Abdullah al Thani in May 1935.

---

60. *In 1932 Mr N. A. (later Sir Neville) Gass (d. 1965) was Deputy General Manager of APOC in Iran and Iraq. In 1938 he was transferred to its head office in London, becoming a Managing Director in 1940 and Chairman of the Company from 1957 to 1960.*

---

61. *Exchange of letters between APOC Abadan and Shaikh Ahmad, June 1932:*

No. 72–C/7
H.E. Shaikh Sir Ahmad Al Jabir
As-Subah, K.C.I.E., C.S.I.,
Ruler of Koweit, *Abadan*
KOWEIT. 25th June, 1932.

Your Excellency,

We have the honour to refer to our letter of the 15th June last and to inform Your Excellency that we have now received a telegram from our Chairman, Sir John Cadman, in which he directs us to express to Your Excellency the very deep sense of gratification it has given to him personally to receive Your Excellency's message.

In deference to Your Excellency's wishes, Sir John Cadman has given immediate instructions for an Agreement embodying a comprehensive concession to explore, prospect and mine for oil within Your Excellency's territory to be prepared and submitted to Your Excellency for consideration. The preparation and translation of this document will necessarily take a short time but we desire to assure Your Excellency that it will be undertaken without delay and submitted as soon as possible.

Sir John Cadman has made, in his telegram, a particular reference to the grateful recollections he retains of the hospitality and courtesy he received from the Rulers and people of Koweit during his visit to Koweit 20 years ago and he expresses the earnest hope that the contact so happily inaugurated then, and continued since may develop shortly into a still closer association between Your Excellency's State and this Company.

The cordial reception accorded to the undersigned by Your Excellency prompts us to add with what pleasure we shall look forward to continuing the discussions on lines which Your Excellency desired.

We have the honour to be,
Your Excellency's most obedient servants,
For ANGLO-PERSIAN OIL COMPANY LTD.,
N. A. Gass
DY. GENERAL MANAGER.

TRANSLATION
No. 479

The General Manager, *Kuwait,*
Messrs. A.P.O.C. Ltd., Dated 20/6/32
Abadan.

After compliments.

I have received with much pleasure your kind and courteous letter of 15th June 1932 under No. C.7/72 in which you had thanked me for the reception afforded to Mr. N. A. Gass during his visit to Kuwait. I have done nothing beyond my duty, and his appreciation of my assistance to him is out of his generosity and courtesy. His communicating my views regarding oil fields in my territory and your referring to Company's principals and your proceeding to London noted. I wish you continuous success and a very pleasant voyage.

Yours truly,
Ahmed Aljaber Alsabah.

62.

72–C/7. 13th August, 1932.

His Excellency
Shaikh Sir Ahmad Al Jabir
As-Subah, K.C.I.E., C.S.I.,
Ruler of Koweit,
KOWEIT.

Your Excellency,

May I be permitted to introduce by this letter Mr. A. H. T. Chisholm, who will present for Your Excellency's consideration the draft concession in connection with the exploration and development by my Company of the oil resources within Your Excellency's State.

Mr. Chisholm is empowered to discuss the terms of this draft concession with you and will take up residence in Koweit and will remain at Your Excellency's disposal for the purpose.

I hope to visit Koweit on the 18th instant and trust it will be convenient to Your Excellency to accord me an interview during my visit.

With an assurance of my very high esteem.

I have the honour to be,
Your Excellency's obedient servant,
For ANGLO-PERSIAN OIL COY. LTD.,
N. A. Gass
DY. GENERAL MANAGER

---

63. *The APOC draft concession document presented to Shaikh Ahmad by Mr Chisholm on 16th August 1932 was as follows:*

This is an AGREEMENT made at Kuwait on the 1932 between His Excellency Shaikh Sir Ahmad al-Jabir as-Subah, Knight Commander of the Most Eminent Order of the Indian Empire and Companion of the Most Exalted Order of the Star of India, the SHAIKH OF KUWAIT (hereinafter called 'the Shaikh') on his own behalf and in the name of and on behalf of his heirs and successors in whom is or shall be vested the responsibility for the control and government of the Principality of Kuwait and the ANGLO-PERSIAN OIL COMPANY LIMITED (hereinafter called 'the Company') relating to the search for production and handling of petroleum within the territories of Kuwait.

Article 1. The Shaikh hereby grants to the Company the exclusive right to explore throughout the whole of his territories as shown on the map annexed hereto (excepting only in the town of Kuwait, and in graveyards), and search for petroleum and cognate products (hereinafter collectively referred to as 'petroleum') for a maximum period of two years, that is until 1934 at the latest, on the conditions set forth below.

Article 2. In consideration of the exclusive right above granted and of the assistance and protection which the Shaikh shall afford to the employees of the Company the Company shall pay to the Shaikh Rs. 50,000 immediately on the signature of this agreement and shall pay Rs. 20,000 on the 1933 unless the drilling period mentioned in Article 3 shall commence on or before 1933 in which event this payment of Rs. 20,000 shall be cancelled.

Article 3. From the date of the expiration of the exploration period mentioned in Article 1, that is, from the 1934 at the latest, the Company shall have, for a maximum period of two years, the exclusive right to import machinery and workmen for searching for petroleum and shall drill in a workmanlike manner to a total depth overall of at least 4,000 feet at such and so many places as the Company decide, and at the commencement of this drilling period and on the same date in the year immediately succeeding, that is, on the 1934 and on 1935 at the latest, the Company shall pay to the Shaikh Rs. 25,000 in consideration of the exclusive rights above granted and of the assistance and protection which the Shaikh shall afford to the employees of the Company.

Article 4. As from the date of expiration of the drilling period mentioned in Article 3, that is from the 1936 at the latest, the Shaikh undertakes to grant for 70 years to the Company on the request of the Company the exclusive right to search for and produce petroleum (excepting only in the town of Kuwait and in graveyards) and the right to refine petroleum within and to export petroleum from the territories of Kuwait on the conditions set forth below.

Article 5. On the date when the Shaikh grants to the Company the exclusive right mentioned in Article 4, that is on the 1936 at the latest, and on each subsequent anniversary thereof during the currency of this Agreement and pending the declaration set forth below the Company shall pay to the Shaikh Rs. 25,000 or royalty as set forth below, whichever shall be the greater sum. The Company shall pay to the Shaikh a royalty of Rs. 2/10/- for every 1 ton of petroleum which the Company shall produce over and above the petroleum required for its operations hereunder, and so soon as the Company shall declare that petroleum has been found in commercial quantities and quality the annual payment by the Company to the Shaikh shall be Rs. 50,000 plus the total of the royalty accrued with a minimum payment of Rs. 65,000 in the first year of Rs. 80,000 in the second year and Rs. 100,000 yearly thereafter, save only in the event of the interruption of the Company's operations hereunder by reason of events outside the control of the Company, in which event payment shall continue to be made at the annual rate of Rs. 25,000.

Article 6. In connection with its operations hereunder the Company shall have the right freely to construct in the territories of Kuwait and to operate power stations, refineries, pipelines and storage tanks, telegraph, telephone and wireless installations, roads, railways, tramways, buildings, wharves and jetties with such lighting as may be requisite and any other works found necessary, and for such purposes to use freely (but not for export) any stone, sand, lime, gypsum, clay or water which may be available and may be required for its operations hereunder. The Company shall at its discretion but in consultation with the Shaikh select the position of any such works. The Company may likewise freely utilise all such means as may be necessary for the effective conduct of its operations hereunder.

Article 7. In consideration of the payments by the Company herein prescribed and to assist in the development of the natural resources of Kuwait the Company shall have the right for the purposes of its operations under this Agreement freely to import or export without any taxes duties or payments to the Shaikh all machinery materials equipment or goods which may be required.

Article 8. The Company shall have the right to import water free and to use local supplies as far as available for its operations and employees and the right to purchase at current market rates fuel and food and other supplies of every kind for its operations hereunder and the Company shall employ subjects of the Shaikh as far as possible for all work for which they are suited under the supervision of the Company's skilled employees.

Article 9. The Shaikh grants to the Company free of cost the free use and occupation of all land belonging to the Shaikh which the Company may need for the purposes of its operations hereunder; it is furthermore agreed between the Shaikh and the Company that the Company may buy or lease for such purposes by agreement with the proprietors any lands houses or buildings on conditions to be arranged with such proprietors but at rates not in excess of those ordinarily current in their respective localities.

On the termination of this Agreement all lands granted by the Shaikh and any lands or buildings which the Company may have bought and any houses or buildings constructed by the Company within the territories of the Shaikh shall be handed over to the Shaikh free of cost.

Article 10. All lands granted by the Shaikh to the Company and all lands houses and buildings acquired by the Company, all petroleum produced as well as any profits therefrom and all machinery materials equipment and goods imported or subsequently exported shall be free of all taxes and duties or payments to the Shaikh during the period of this Agreement.

Article 11. Throughout the period of this Agreement the Shaikh shall give to the Company and all its employees all protection in his power from theft, highway robbery, assault, wilful damage and destruction, but the Company in consultation with the Shaikh shall appoint and itself pay permanent trustworthy guards under a reliable man of good standing and these guards shall protect the property of the Company and its employees at all times.

If in spite of the efforts of the said guards thefts should occur the Shaikh undertakes to take all reasonable measures to recover the property stolen and to compensate the Company for any damage sustained in his territory. In the event of non-recovery of the stolen property, except for reasons beyond the control of the Shaikh, the value thereof shall be deducted by the Company from any payments due from time to time by the Company to the Shaikh.

In the case of serious offences the said guards are subject to dismissal or stoppage of pay by the Company. In the event of any heavier punishment being needed, the offender shall be handed over to the representative of the Shaikh, and the Shaikh undertakes that punishment shall be inflicted and that he will use his utmost endeavours to uphold the authority of the Company.

Article 12. If the Company shall fail to carry out its obligations under this Agreement or shall fail within six months after any anniversary of this Agreement to pay to the Shaikh the annual payments due or the royalties which may have become due the Shaikh acting on the advice of the Political Resident shall have the power to terminate this Agreement and all the property of the Company then in Kuwait territory for the purposes of this Agreement shall then but on no other grounds become the property of the Shaikh without payment. In the event of the Company failing to make the declaration provided in Article 5 on or before the   1941 the Shaikh shall have the power to call upon the Company either to make the declaration forthwith or to surrender all rights under this Agreement which shall then terminate.

Article 13. With the approval of the Shaikh acting on the advice of the Political Resident the Company may transfer the obligations and benefits of this Agreement to any British Company.

Article 14. The Company shall have the right at any time after the completion of drilling mentioned in Article 3 to give the Shaikh one year's notice in advance to terminate this Agreement and the Company shall on expiry of such notice have no further liabilities hereunder after making due payment of all monies due to the Shaikh up to the date of termination.

Article 15. In the event of there arising between the Shaikh and the Company any dispute or difference in respect of the interpretation of this Agreement or the rights or responsibilities of the Shaikh or of the Company therefrom resulting, such dispute or difference shall be submitted to two arbitrators one of whom shall be named by the Shaikh and one by the Company and to an umpire who shall be appointed by the arbitrators before they proceed to arbitrate.

The decision of the arbitrators, or in the event of disagreement between the arbitrators the decision of the umpire, shall be final.

Article 16. This Agreement is written in English and translated into Arabic. If there should at any time be disagreement as to the meaning or interpretation of any clause in this Agreement the English text shall be regarded as binding.

In witness whereof the parties to this Agreement have set their hands the day and year first above written.

ON BEHALF OF THE SHAIKH OF KUWAIT	ON BEHALF OF THE ANGLO-PERSIAN OIL COMPANY LIMITED
IN THE PRESENCE OF	IN THE PRESENCE OF

---

64. *Sir Hugh Biscoe, Political Resident in the Gulf since 1929, had died suddenly on 18th July 1932. His successor, Lt-Col. (later Sir) Trenchard Fowle, lacking his predecessor's close acquaintance with the complexities of the concession negotiations, needed to familiarise himself with them before formulating his own views and requesting London's instructions.*

---

65. *Exchange of correspondence between APOC New York and Gulf Oil Corporation, July–October 1932:*

17 Battery Place
New York
October 4, 1932

Sir John Cadman
Anglo-Persian Oil Company, Ltd.
Britannic House, Finsbury Circus
London, E.C.2

KUWAIT

My dear Sir John:

I enclose herewith for your information and records, copies of letters which I have exchanged with Mr. Wallace of the Gulf Exploration Company.

I suggest these should be handed to Lefroy for his files.

Yours sincerely,
B. R. JACKSON
Representative in U.S.A.
of the Anglo-Persian O.C. Co. Ltd.

Mr. W. T. Wallace,
The Venezuela Gulf Oil Company
17 Battery Place,
NEW YORK

26th July, 1932

Dear Mr. Wallace,

KUWEIT

With reference to our recent conversation in New York, I have now consulted my colleagues and am writing to give you our views on this matter.

To explain our attitude clearly I propose to give you a very brief account of what has transpired since we became interested in this area approximately twelve years ago.

At that time we first realised that a survey of Kuweit might supply the solution to certain geological problems confronting us in Southern Persia. Accordingly an attempt was made to secure governmental support in approaching the Shaikh, but our efforts were not then—nor have they been at any time since—successful.

To an American this may well seem incredible; but I can assure you that this is typical of the attitude of our Government Departments, who, in our view, are consistently over-scrupulous to avoid giving basis for any suggestion that preferential treatment was being extended to this Company. Indeed I might well quote the recent withdrawal of the 'nationality clause' in support of this fact.

Some drifting then took place, and we became preoccupied with more immediate problems elsewhere—influenced no doubt by the fact that no other interests were then active in the area.

In these circumstances we could well afford to wait until economic and general considerations justified investigation. The next event of any importance was the appearance on the scene of the E. & G.S., and here I must frankly admit that seen from an outside point of view our inactivity might lend itself to criticism. At the same time, it should be remembered that acceptance of the royalty and other conditions which were associated with both Bahrein and Kuweit would have seriously prejudiced certain major negotiations, not only in Persia, but also and more particularly in Iraq—so that to some extent our inactivity was directly beneficial to the IPC negotiations.

Whilst, however, giving due weight to these considerations, it must be admitted—on looking back—that we stood aloof too long on a matter which is really of considerable importance to us inasmuch as our special position in Persia and the Middle East give us—we feel—some substantial reason to regard the whole of the Gulf as within our sphere of influence. We have now assumed an activity too long delayed, and I cannot feel that the E. & G.S. should be animated by anything other than gratitude towards us for leaving them so long without competition.

In all my discussions with my colleagues, and in any examination of the correspondence, I can find no evidence of anything having been said or done which has directly or indirectly affected the attitude of the Foreign or Colonial Offices towards the E. & G.S.

I have written frankly, as you desired, and I think that while you may not sympathise with our attitude you will appreciate its logic; indeed any other policy, if oil were found, would lay us open to grave criticism from our shareholders.

Should there be any point on which you would like to communicate with me please address me here, as I do not expect to return to New York before September.

With kind regards,

Yours sincerely,
B. R. Jackson

GULF EXPLORATION COMPANY
17 Battery Place
NEW YORK
September 8, 1932

Mr. B. R. Jackson,
Anglo-Persian Oil Company,
17 Battery Place,
New York, New York.

Dear Mr. Jackson,

Your interesting letter of July 26th was awaiting me on my return to the office after a somewhat protracted necessary absence.

Regarding the attitude of the British Government your information and mine seem quite at variance. Some of the information I have is, of course, indirect; but it all confirms the facts known to me, which in themselves permit of no other conclusion than that your Company has received support and highly preferential treatment from the Government. Indeed, such an inference would be inescapable merely from the circumstances surrounding the presentation on behalf of your Company of the several draft concessions to the Sheikh, one before the Syndicate presented its draft and others afterward; and the procedure of the Colonial Office in that connection.

The Government has accorded such preferential treatment both against American interests, in striking contrast to an open door policy and to the treatment expected and received by British subjects in this country, and against another British interest which by its diligence and by reason of the substantial investment it has made in this business in good faith would seem to deserve something better from its own Government.

This preferential treatment has perhaps been more, rather than less, in evidence since the withdrawal of the nationality clause than before.

The situation is to say the least an anomalous one, whether it be looked at as one in which a Government shows decided partiality as between two groups of its own citizens, or as one in which the Government, being financially interested in one of the two groups and constituting itself the judge, uses its position to force a result or render a decision in favor of the group in which it is itself interested. Even the events of the past year have not convinced me, however, that the British Government, which while obstructing its efforts has still encouraged the Syndicate to hope for fair treatment, will so stultify itself. I should rather expect that in such a situation the Government will be inclined, as you suggest, to act finally with particularly strict impartiality.

Such strict impartiality would seem to require that the Syndicate (and thereby, of course, the American interests involved) should be put back into the same position as if the nationality clause, which the Government now says is not necessary, had never been insisted upon. That clause, as you perhaps know, was first insisted upon by the British Government about four years ago, just after the offer of the Bahrein concession (a matter which was, of course, confidential as between the groups of the T.P.C.) was made to the Turkish Petroleum Company. Since then, and to this good day, the Syndicate's hands have been effectively tied.

Under the circumstances, the Syndicate would seem entitled to different consideration. Over six years ago it offered your Company an opportunity, which your Company declined, to interest itself in the Koweit territory. It offered a similar opportunity to other British groups before finally bringing the matter to us in 1927. From the beginning it has carried on its negotiations with the

Sheikh with the full knowledge and official approval of the British Government. When it entered into negotiations with us in 1927, and thereafter until about a year ago, it had no reason for anticipating any such developments as those with which it had found itself confronted.

While I have found the delays and other developments in this matter sorely trying to one's faith at times, I still cannot believe that the British Government will in the final analysis ignore every rule of fair play, and by sheer force and awkwardness push the Syndicate out of the position it should rightfully occupy, or put pressure upon the Sheikh to act contrary to his own wishes.

Nor do I believe that your Company, after reflection, would wish to take advantage of any such situation.

I am still hopeful, therefore, that when the acts of the British officials and representatives of your Company are reviewed and the position and policy of the Government and your Company are finally declared by the more responsible officials of each the Syndicate and we shall receive the consideration to which I feel we are entitled both from the British Government and your Company.

Very truly yours,
(Signed) Wm. T. Wallace.

October 3, 1932

Mr. William T. Wallace
Gulf Exploration Company
17 Battery Place
New York City, New York
Dear Mr. Wallace:

Your letter of September 8th relating to Kuwait reached me on my return on the 1st instant.

If I may emulate your candour, the terms of your letter hardly suggest a disposition to discuss the question at issue dispassionately, indeed, there is such complete conflict between us, not only of opinion, but also as to facts, that I do not feel any useful purpose will be served by my replying in detail to your charges.

Your chief quarrel would appear to be with the British Government whose defence is fortunately no concern of mine.

In your concluding paragraph you express the hope that when the policy of my Company is reviewed by its 'more responsible officials', both the Syndicate and your own Company will receive the consideration to which you feel they are entitled. I cannot conceive of any grounds upon which either the Syndicate or your Company can claim special consideration from the Anglo-Persian Oil Company and you are under a serious misapprehension if you believe that the policy of my Company in Kuwait has not long since been carefully considered by my colleagues on the board.

I must re-iterate that my Company very definitely regard Kuwait as within their natural and particular sphere of influence and will not passively look on while any other companies, with no reasonable interest whatsoever in that area, are endeavoring to secure concessions there.

On the reflection which you commend to me, I do not believe that your own Company, would in similar circumstances, view the matter in any other way.

Yours truly,
B. R. Jackson

---

66. *On 27th November 1932 Reza Shah of Persia (d. 1944) declared his unilateral annulment of APOC's oil concession in Persia (see Note 2) which, originally granted in 1901 and modified in 1920, had been inconclusively discussed with a view to radical revisions for several months previously between his Ministers and the Company. The dispute was referred in December by the Persian and British Governments to the Council of the League of Nations in Geneva where its discussion in late January 1933 led to negotiations being resumed between the Persian Government and the APOC. These negotiations were eventually finalised on 29th April 1933 in Tehran where a new Concession Agreement was signed by the Persian Government and APOC after several weeks of hard bargaining, in which both the Shah and the Company's Chairman (Sir John, later Lord, Cadman) were personally involved. (See Note 77).*

67.

Office of the Political Resident
in the Persian Gulf,
Camp Kuwait.
Dated the 14th January 1933.

A. H. T. Chisholm, Esquire,
Anglo Persian Oil Company Ltd,
Kuwait.
Dear Sir,

I beg to inform you that I have given the Comments of His Majesty's Government on your proposal for an Oil Concession in Kuwait, to His Excellency the Shaikh of Kuwait, and that your Company may now commence negotiations with His Excellency.

A similar letter will be given to Major Frank Holmes of the Eastern and General Syndicate, Limited, on his arrival here, but as the latter is at present in London, I am telegraphing to His Majesty's Government to inform the Eastern and General Syndicate, Limited in the above sense.

Yours faithfully,
T. C. Fowle (Lieut-Col.),
Political Resident in the Persian Gulf

---

68. *Mr Chisholm's report from Kuwait to Abadan dated 9th January 1933 included the following:*

On 6th inst. I had a long interview with the Sheikh, to summarise briefly,

(a) I gave the Sheikh the latest news as to the Anglo-Persian dispute.

(b) I mentioned that while in London I had been able to discuss various outstanding Kuwait problems with influential personages with whom I had come into contact over Anglo-Persian affairs; and that I hoped that this might aid his own discussions.

(c) I stated that as a result of my conversations with Sir John Cadman, our revised draft concession which we had in readiness for the reopening of negotiations would contain, *inter alia*, certain increased payments.

The Sheikh was very cordial and interested, and the interview appeared to have a good effect.

---

69. *The text of APOC's revised concession draft of January 1933 was as follows:*

This is an AGREEMENT made at Kuwait on the (hereinafter called the 'Agreement date') in the year 1933 between His Excellency Shaikh Sir Ahmad al-Jabir as-Subah, Knight Commander of the Most Eminent Order of the Indian Empire and Companion of the Most Exalted Order of the Star of India, the SHAIKH OF KUWAIT (hereinafter called 'the Shaikh') on his own behalf and in the name of and on behalf of his heirs and successors in whom is or shall be vested the responsibility for the control and government of the Principality of Kuwait and the ANGLO-PERSIAN OIL COMPANY LIMITED (hereinafter called 'the Company') relating to the search for production and handling of petroleum within the territories of Kuwait.

Article 1. The Shaikh hereby grants to the Company the exclusive rights throughout the whole of his territories as shown generally on the map annexed hereto, excepting only the area within the present town wall of Kuwait as well as mosques sacred buildings and graveyards but including all islands appertaining to Kuwait, (hereinafter called 'the defined area') to explore and search for petroleum and cognate substances (hereinafter collectively referred to as 'petroleum') for a maximum period of two years, that is until the Agreement date in 1935 at the latest, on the conditions set forth below.

Article 2. Until the Agreement date in 1937 the Company shall have the exclusive right to drill within the defined area and within a period of two years (hereinafter called the 'drilling years') terminating prior to that date the Company shall so drill in a workmanlike manner to a total depth overall of at least 4,000 feet at such and so many places as the Company may decide. The Company shall duly close any unproductive holes drilled by it and subsequently abandoned.

Article 3. As from the Agreement date in 1937 or from the Agreement date next following the termination of the drilling years the Shaikh undertakes to grant for 70 years to the Company on the request of the Company the exclusive rights to search for and produce petroleum within the defined area and the right to refine petroleum within and to export petroleum or its products from the territories of Kuwait on the conditions set forth below.

Article 4 (A). In consideration of the rights granted by the Shaikh to the Company hereunder and of the assistance and protection which the Shaikh shall afford to the Company and its employees and property the Company shall pay to the Shaikh on the Agreement date in each year the following sums:—

On signature of the Agreement	Rs. 65,000
On each subsequent Agreement date:—	
(a) prior to the drilling years	Rs. 20,000
(b) during the drilling years	Rs. 25,000
(c) after the grant by the Shaikh of the rights specified in Article 3 above but prior to the declaration mentioned in (d) below.	Rs. 25,000 or Royalty of Rs. 2/10/- for every Ton of petroleum obtained and saved from the ground over and above the petroleum required for the Company's operations hereunder—whichever shall be the greater sum.

(d) after the Company shall have declared that petroleum has been found in commercial quantities and quality. — Rs. 60,000 in addition to Royalty of Rs. 2/10/- as above defined subject to a minimum in total of
(i) Rs. 75,000 on the first Agreement date after the declaration has been made
(ii) Rs. 90,000 on the second such Agreement date and
(iii) Rs. 120,000 on the third and each subsequent such Agreement date

save only in the case of interruption of the Company's operations hereunder by reason of events outside the control of the Company in which case the payment prescribed under (d) shall be made at the annual rate of Rs. 25,000.

(e) A Royalty of two Annas per thousand cubic feet of all natural gas sold.

(B) The Company shall measure by approved methods all petroleum produced under Article 3 above and shall keep full and correct accounts of all petroleum so produced and of the petroleum consumed in the Company's operations hereunder.

Article 5. In connection with its operations hereunder the Company shall have the right in the territories of Kuwait without hindrance to construct and to operate power stations, refineries, pipelines and storage tanks, telegraph telephone and wireless installations, roads, railways, tramways, buildings, wharves and jetties, with such lighting as may be requisite and any other works found necessary, and for such purposes to use free of all taxes and payments to the Shaikh (but not for export) any stone, sand, lime, gypsum, clay or water which may be available and may be required for its operations hereunder, provided always that the water supply of the local inhabitants and nomad populations who may be dependent on the same is not endangered. The Company shall at its discretion but in consultation with the Shaikh select the position of any such works. The Company may likewise utilise without hindrance all such means as may be necessary for the effective conduct of its operations hereunder.

Article 6. In consideration of the payments by the Company prescribed in Article 4 hereof and to assist in the development of the natural resources of Kuwait the Company shall have the right for the purposes of its operations hereunder freely to import or export without any taxes duties or payments to the Shaikh all machinery materials equipment or goods which may be required.

Article 7. The Company shall have the right to import water free for its operations and employees and the right to purchase at current market rates fuel and food and other supplies of every kind in connection with its operations hereunder and the Company shall employ subjects of the Shaikh as far as possible for all work for which they are suited under the supervision of the Company's skilled employees.

Article 8. The Shaikh grants to the Company free of cost the free use and occupation of all land belonging to the Shaikh which the Company may need for the purposes of its operations hereunder; it is furthermore agreed between the Shaikh and the Company that the Company may buy or lease for such purposes any lands houses or buildings on conditions to be arranged with the proprietors thereof but at rates not in excess of those ordinarily current in their respective localities.

On the termination of this Agreement all lands granted by the Shaikh and any lands or buildings which the Company may have bought and any houses or buildings constructed by the Company within the territories of Kuwait shall be handed over to the Shaikh free of cost.

Article 9. All lands granted by the Shaikh to the Company and all lands houses and buildings acquired by the Company, all petroleum produced as well as any profits therefrom shall be free of all taxes and duties or payments (excepting always the payments provided in Article 4 hereof) to the Shaikh during the period of this Agreement.

Article 10. Throughout the period of this Agreement the Shaikh shall give to the Company and all its employees all protection in his power from theft, highway robbery, assault, wilful damage and destruction, but the Company in consultation with the Shaikh shall appoint and itself pay trustworthy guards under a reliable man of good standing and these guards shall protect the property of the Company and its employees at all times.

If in spite of the efforts of the said guards thefts should occur the Shaikh undertakes to take all reasonable measures to compensate the Company for any damage sustained in his territory and for the recovery of the property stolen.

In the case of serious offences the said guards are subject to dismissal or stoppage of pay by the Company. In the event of any heavier punishment being needed, the offender shall be handed over to the representatives of the Shaikh, and the Shaikh undertakes that punishment shall be inflicted and that he will use his utmost endeavours to uphold the authority of the Company.

Article 11. If the Company shall fail to carry out its obligations hereunder or shall fail within six months after any Agreement date to pay to the Shaikh the payment due under Article 4 hereof the Shaikh acting on the advice of the Political Resident shall have the power to terminate this Agreement and all the property of the Company then in Kuwait territory for the purposes of this Agreement shall then but on no other grounds save as provided in Article 8 above become the property of the Shaikh without payment. In the event of the Company failing to make the declaration provided in Article 4 (A) (d) hereof on or before the Agreement date in 1942 the Shaikh shall have the power to call upon the Company either to make the declaration forthwith or to surrender all rights under this Agreement which shall then terminate.

Article 12. With the approval of the Shaikh acting on the advice of the Political Resident the Company may transfer the obligations and benefits of this Agreement to any British Company.

Article 13. The Company shall have the right at any time after the termination of the drilling years to give the Shaikh one year's notice in advance to terminate this Agreement and the Company shall on expiry of such notice have no further liabilities hereunder after making due payment of all monies due to the Shaikh up to the date of termination.

Article 14. Nothing in this Agreement shall be read as restricting in any way the right of the Shaikh to grant to other parties concessions for substances other than petroleum provided that the operations and rights of the Company hereunder are not thereby injuriously affected.

Article 15. No failure or omission on the part of the Company to carry out or perform any of the conditions of this Agreement shall give the Shaikh any claim against the Company or be deemed a breach of this Agreement in so far as the same arises from *force majeure*, and if through *force majeure* the fulfilment by the Company of any of the conditions of this Agreement be delayed the period of such delay, together with such period as may be necessary for the restoration of any damage done during such delay shall be added to the periods fixed by this Agreement.

Article 16. In the event of there arising between the Shaikh and the Company any dispute or difference in respect of the interpretation of this Agreement or the rights or responsibilities of the Shaikh or of the Company therefrom resulting, such dispute or difference shall be submitted to two arbitrators one of whom shall be named by the Shaikh and one by the Company and to an umpire who shall be appointed by the arbitrators before they proceed to arbitrate.

The decision of the arbitrators, or in the event of disagreement between the arbitrators the decision of the umpire, shall be final.

Article 17. This Agreement is written in English and translated into Arabic. If there should be at any time disagreement as to the meaning or interpretation of any clause in this Agreement the English text shall be regarded as binding.

In witness whereof the parties to this Agreement have set their hands the day and year first above written.

SHAIKH OF KUWAIT	ON BEHALF OF THE ANGLO-PERSIAN OIL COMPANY LIMITED
IN THE PRESENCE OF	IN THE PRESENCE OF

---

70. *For Mr Gabriel (d. 1950) and the remarkable role he was to play in the negotiations from April 1934 onwards, see page 55 et seq.*

---

71. *For Mr Yusuf al Ghanim see Note 27.*

---

72. *Memorandum drawn up by His Majesty's Government for submission to His Excellency the Shaikh of Kuwait showing a comparison between draft concessions submitted by the Eastern and General Syndicate, and the Anglo-Persian Oil Company.*

Note. The comparison is made between the *third* draft submitted by the Eastern and General Syndicate in June 1932, and the *first* draft (revised form) submitted by the Anglo Persian Oil Company in August 1932.

Eastern and General Syndicate	Anglo Persian Oil Company
FIRSTLY *Period.* 70 years (5 years of which are stated to be for exploratory work and selection of areas).	*Period.* Exploration 2 years. Test drilling 2 years. Lease 70 years.

Comment.

Exploration and Prospecting.

Whichever Company is granted the Concession, will find it necessary to carry out exploratory work as the geology of the country is much obscured by superficial deposits. The Eastern and General Syndicate refer rather vaguely to 'Exploratory work' by geologists and engineers, which they undertake to begin within 9 months of the date of Concession, failing which the Sheikh may cancel the Concession. Nothing is said as to drilling and the Syndicate require a period of 5 years for their exploratory work.

The Anglo-Persian Oil Company ask for a maximum period of two years for exploratory work, which will be reduced if they undertake drilling within that period, to be followed by a maximum period of two years, during which they undertake to drill at least 4,000 feet.

NOTE 72

Period of Lease.

As regards the period of lease both companies ask for approximately the same terms. In the case of a country such as Kuwait where the existence of oil has not been proved, it is considered desirable that the period of Lease should be a long one. The work of proving and developing an oilfield involved very heavy capital expenditure and companies are not likely to be attracted unless there is a guarantee of their being able to obtain a long period in which they may have an opportunity of securing a return on the money expended. There are precedents for such a period as is now asked for in Iraq, where the period granted by the Iraq Government is 75 years, in Persia where the period is 60 years, and in Bahrein where a similar period is provided for. In his letter of 25th August 1932 (36/1932) the Political Resident mentions that he understands that the Shaikh himself asked for a period of 70 years.

SECONDLY

Eastern & General Syndicate	Anglo-Persian Oil Co.
Area. The right to explore the whole territory of Kuwait and select an area for ultimate development not exceeding 1,640 square miles divided into 6 or more blocks.	*Area.* The draft agreement covers the whole of Kuwait territory for the whole period of the concession.

Comment.

The Anglo-Persian Oil Company's draft concession provides for the grant of the whole of the area of Kuwait for the period of the concession. On the other hand, the Eastern and General Syndicate provide for the selection within 5 years of an area amounting in the aggregate to 1,640 square miles, this area may be divided up into *six or more* blocks, and by such an arrangement they presumably hope to secure all promising oil bearing territory which is worth having. The value of their proposal to limit their activities to 1,640 square miles is, therefore, more apparent than real.

There would appear to be definite advantages in having the whole of this rather limited territory developed by one company of good standing. Competitive development of an oilfield by two or more companies seldom produce the best results, as it tends to lead to unnecessary loss of gas pressure and premature exhaustion of the field, as has been the case in many fields in the United States. The operations of the Anglo-Persian Oil Company in Persia are generally accepted as being in accordance with the best modern practice of unit development.

It may be mentioned that in the case of Iraq there was in the original concession a provision for the selection areas. So much difficulty arose in regard to this matter that the Iraq Government agreed in the revised Concession granted last year, to accept the Iraq Petroleum Company's proposal for the grant to them of exclusive rights over a large area amounting to about 36,000 square miles, or nearly twice the area of Kuwait.

Incidentally the Eastern and General Syndicate refer to the area to be reserved to them as 'conceded Territory'. There are obvious objections to the use of such a term and this should be replaced by 'Lease Area'.

THIRDLY

Eastern and General Syndicate	Anglo-Persian Oil Company
Working Obligations. The Concession to become void unless the Company has commenced exploratory work within a period of 9 months from the date of the Concession. If, after commencing operations following the exploratory period of 5 years the Company discontinue the same for a continuous period of two years, the Sheikh shall have the right to cancel the agreement.	*Working Obligations.* After a maximum period of two years for exploration the Company undertake within a period of 2 years to drill in a workmanlike manner a total depth overall at least 4,000 feet. If the Company fails to carry out its obligations, or to make the necessary payments, the Sheikh has the power to terminate the agreement.

Comment.

Nothing is said in the Eastern and General Syndicate's draft concession as to drilling obligations, and as the draft now stands they would have carried out these obligations if they only did surface and shallow geological drilling. It may be suggested, that they will have to do such work as will enable them to select their areas, but as the comment on this question indicates they can select any number of blocks in choosing their 1,640 square miles.

In Bahrein the Eastern and General Syndicate obtained a Concession in 1925. This provided for an exploration period of two years and a prospecting period of two years, subject to extensions. In 1927 the Syndicate assigned their rights to American interests, and they have not yet taken a

Mining Lease. They have recently applied for a further extension of the prospecting period and have recently begun work on a second well.

On the other hand, to undertake to drill at least 4,000 ft. Overall in a period which under the terms of the draft cannot exceed 4 years, and may be less, the Anglo-Persian Oil Company does give an indication of serious endeavour to prove oil.

Further the provision in the Eastern and General Syndicate's draft that they may discontinue operations for continuous period up to 2 years before they run the risk of losing their concession is unusual, and hardly indicates an intention to carry on exploration as speedily as possible. A provision of this kind is included in some oil concessions, but six months is a much more usual period, although a period up to 12 months is sometimes provided for.

FOURTH Eastern and General Syndicate.	Anglo-Persian Oil Coy.
Payments. *On signature of the Agreement Rs. 30,000.* Annually thereafter Rs. 20,000, until the Company declares that oil has been found in commercially exploitable quantities when tonnage royalty begins to operate. Royalty Rs. 3–8 annas per ton of oil won, with minimum of Rs. 70,000 in any complete calendar year in which the Company continues work. The question of commercial exploration to be referred to arbitration in the event of dispute.	*Payments.* *On signature of the Agreement Rs. 50,000.* and a year later Rs. 20,000. On commencement of drilling Rs. 25,000 per annum and a similar sum annually thereafter until the Company declares that petroleum has been found in commercial quantities when payments will be as follows:— A Royalty of Rs. 2–10 per ton of petroleum subject to the following payments. During the first year Rs. 50,000 plus any royalty accrued with minimum payment of Rs. 65,000; during the 2nd year a minimum of Rs. 80,000 and annually thereafter a minimum of Rs. 100,000 unless operations are interrupted through causes outside the control of Company, when payment will continue at annual rate of Rs. 25,000. If the Company fail to make a declaration as to commercial production by a date in 1941 to be specified, the Sheikh to have power to call upon the Company either to make the declaration forthwith or to lose the concession.

Comment.

It will be seen that the Anglo-Persian Oil Company offer a higher amount on signature of the Concession and a higher annual amount when they commence drilling (which must happen within 2 years of signature of the concession). Except for the first year after the declaration that oil has been produced in commercial quantities, their minimum royalty is also higher than that offered by the Eastern and General Syndicate. It will also be observed that the Anglo-Persian Oil Company must make a declaration by a date to be specified in 1941, that oil has been found in commercial quantities or surrender the concession. In the Eastern and General Syndicate's draft this question has to be settled by arbitration. Bearing in mind the absence of any specific drilling obligations during the period of 5 years provided by the Eastern and General Syndicate's draft for the exploration period, it seems unlikely that there would be much prospect of taking the Syndicate to arbitration at an earlier date than that by which the Anglo-Persian Oil Company must make their declaration.

After the declaration that commercial production has been obtained the Syndicate offer a tonnage royalty of Rs. 3. annas 8 per ton and the Anglo-Persian Oil Company a royalty of Rs. 2.10-annas. To decide which of these is better in the interest of the Sheikh depends on the production which is likely to be secured. And here the amount of minimum royalty *is important. Generally speaking* the Coy. with the higher minimum royalty will have the greater inducement to produce oil. It may also be said that too high a tonnage royalty may retard production, and here the provision in the Eastern and General Syndicate's draft, allowing a cessation of operations should not be lost sight of. The rate of royalty offered by the Eastern and General Syndicate is the same as that in India, but there is a local market to absorb all the production. The rate offered by the Anglo-Persian Oil Company is approximately the same as that levied by the Iraq Government.

In Venezuela the rate of royalty is understood not to exceed 2/- a ton in most instances, while in Trinidad there are many leases with a royalty of from 1/6 to 2/- although in recent years a royalty of 3/- has been imposed. In the case of both these countries, over 90 per cent of the oil has to be exported.

FIFTH Eastern and General Syndicate	Anglo-Persian Oil Company.
Customs Duty. The Company undertakes to pay the Sheikh a Customs Duty of 1% on all oil produced from Kuwait territory calculation on the value of the oil at the wells.	*Customs Duty.* No similar provision.

Comment.

It is not usual to impose a Customs Duty in addition to royalty. Such charge might with some justification be levied in a country where the State does not own the mineral rights and desires to secure some financial consideration for the loss of wasting asset such as Petroleum. But where the State is the owner of the minerals it is regarded as preferable that the State should take the whole of its financial return in the shape of royalty. Apart from the principle involved there is bound to be some difficulty in assessing the value of the oil for purposes of the duty, whether at the well head, or port of shipment. There are some cases in which royalty is charged on the basis of a percentage of the value of the oil won, and endless difficulty has arisen in finding a satisfactory method of determining the value. In Trinidad, where the Government owns the greater part of the minerals and the question has been thoroughly considered, the idea of an export tax has been definitely abandoned as unsuitable.

GENERAL OBSERVATIONS.

The foregoing notes deal with the more important provisions of the draft concessions and will it is thought enable the Sheikh to reach a decision as to which application is likely to be more acceptable to him.

Besides the points dealt with there are certain provisions which ought to be in any concession which may be granted of which the following are the more important. Some of these may be covered in part by the existing drafts:

1. During the exploratory period, the work of exploration should be commenced within a specified period, and carried on continuously under the supervision of a fully qualified petroleum geologist. Correct geological plans and records should be kept, which should be open to inspection by the Sheikh or his representative at all reasonable times.

2. The Company should give an undertaking at all times during the continuance of the Concession to carry on all operations for winning and working crude oil and natural gas in a skilful and workmanlike manner and in accordance with modern approved methods; to maintain all productive borings in good repair, working order and condition and to take all practical measures to prevent the injurious access of water to the oil bearing formations.

3. The Company should furnish an annual report of their operations, together with a plan showing the location of all wells drilled. They should also keep accurate records of the drilling 'logs' of all wells showing all casing inserted and any water horizons passed through.

4. All abandoned boreholes should be plugged so as to prevent ingress of water, and excavations fenced off if the Company is required to do so.

5. There should be a provision for the measurement of any oil won, for purposes of ascertainment of royalty. A draft of a suitable clause is appended.

6. It may be desirable to provide for the revision of the rate of royalty at intervals of, say, 10 years on the application of either party, such revision to be based on the general trend of world prices for petroleum products. Minimum and maximum rate of royalty should be laid down in the first instance.

7. It is usual to provide for a royalty on any natural gas sold. The equivalent of 2d. (? 2 annas) per thousand cubic feet is suggested as a suitable rate. It might also be provided that in the event of casing-head gasoline being produced the Sheikh may fix a royalty at a rate not exceeding the equivalent of ½ a gallon.

8. Minerals other than petroleum and cognate products, should be reserved to the Sheikh, who should retain the right to lease them to other parties, provided that the operations of the Company are not thereby obstructed.

9. The Company should undertake to offer no unnecessary or reasonably avoidable obstruction or interruption to the development and working of any minerals (other than petroleum or cognate products) including water and to afford to the Sheikh and to the holders of prospecting licences or mining leases in respect of other minerals reasonable means of access to such minerals for the purpose of working and carrying away the same.

10. Provision should be made for local office and resident local representative.

11. On the termination of the Concession the Company should undertake to deliver up the Sheikh without payment of compensation in good order and repair and fit for further working all productive wells or borings together with all casings and fixtures below ground levels, as well as all buildings, wharves, roads, pipelines and railway lines.

Other points will need consideration but as some arise on one draft concession and some on the other it will save time if these are deferred until a decision has been taken by the Sheikh as to which application he desires to accept.

---

73. *In a report from Kuwait to Abadan dated 13th February 1933 Mr Chisholm included the following:*

Major Holmes, by announcing to the Sheikh (who has not yet heard it officially) that the Colonial Office will now cease to deal with Kuwait affairs, which action he says follows his representations in London (as did the withdrawal of the 'Nationality Clause' last year), has scored a diplomatic success with the Sheikh.

In this connection, see also Note 7.

---

74. *On 28th February 1933 the State Department telegraphed to the American Embassy in London as follows:*

Please advise the Foreign Office that the American interests concerned in the Kuwait Petroleum Concession have received information which indicates that the British Government is now supporting the efforts of the Anglo-Persian Oil Company to obtain an exclusive and monopolistic concession for the entire territory of Kuwait. Add that the Department is reluctant to credit this report and request a statement as to its accuracy.

On the day after this telegram was sent APOC's Chairman, Sir John Cadman, was continuing in London (see Note 77) his discussions of the previous November in Gulf's Pittsburgh head office of possible APOC/Gulf collaboration in Kuwait with Mr Andrew Mellon (see Note 47), who was to return to the U.S.A. on 23rd March 1933 on vacating his Ambassadorship in London.

---

75. *Mr E. H. O. Elkington (d. 1963) was General Manager of APOC in Persia and Iraq from 1927 to 1939 when he was transferred to the Company's head office in London, becoming a Managing Director of the Company from 1948 until his retirement in 1955.*

76.

His Excellency
Shaikh Sir Ahmad Al Jabir
As-Subah, K.C.I.E., C.S.I.,
Ruler of Kuwait & Dependencies,
KUWAIT. 15.2.1933

Your Excellency,

I write to thank your Excellency most sincerely for the kindness and cordiality which Your Excellency extended to me during my visit to Kuwait. I need hardly say how impressed I was by Your Excellency's friendship and the pleasant aspect of your territory, and it is my fervent hope that we shall strengthen our association in the near future and have the pleasure and honour to work on Your Excellency's behalf and continue to enjoy Your Excellency's friendship and protection for many years to come.

With renewed assurances of my high esteem

I beg to remain,
Yours sincerely,
E. H. O. Elkington
General Manager
Anglo-Persian Oil Company Limited.

General Manager,
Messrs Anglo Persian Oil Company Limited, No. 2/827/R.
Abadan. 18–2–1933

In acknowledging receipt of your welcome letter dated 15th instant, contents of which have been duly noted by us with pleasure, we would state that we have, as a matter of fact, done nothing deserving your thanks and gratefulness—and this is of course the outcome of your courtesy.

Our cherished desire is to see you in the best of health and happiness.

In concluding our letter we pray for your continuous progress and prosperity.

Yours sincerely,
Sd/- Ahmad al Jabir

---

77. *Sir John (later Lord) Cadman (1877–1941) who had visited Kuwait in 1912–13 (as a member of the Slade Commission, see Note 4) was Chairman of APOC from 1926 until his death in 1941.*

Between July and October 1932 there had been correspondence (see Note 65) between APOC's representative in New York Mr B. R. Jackson (d. 1957: Chairman of APOC 1956–57) and Mr W. T. Wallace, Vice-President of Gulf Exploration Company (Note 81) in which Gulf asserted, and APOC denied, that H.M.G. was unduly favouring APOC's resumed (and by Gulf unexpected) negotiations in Kuwait. In October 1932 when Colonel Drake (Chairman of Gulf Oil Corporation) was in London, he discussed with Cadman their respective attitudes to each other's activities in Kuwait. This discussion was resumed by Cadman, during a visit to the U.S.A. from November to January 1933 at Houston and Pittsburgh with Mr Leovy (Vice-Chairman and President of Gulf; Note 81), when he suggested that their companies might collaborate in Kuwait; and during the next three months Mr Jackson had further discussions of such a possibility, without reaching agreement, with Gulf in New York and Pittsburgh (see below State Department memorandum dated 16th March 1933).

On 1st March 1933, Cadman, the day before he left for Persia (visiting Kuwait en route *on 25th March, see page 32) had again advocated some form of APOC/Gulf collaboration over Kuwait (see below his letter of 1st March 1933 to Sir J. B. Lloyd, a Director of APOC) when talking to Mr Andrew Mellon (see Note 47) who was then shortly to return to the U.S.A. (where Gulf Oil Corporation was one of his main business interests) on vacating his appointment on 20th March as American Ambassador in London. Mr Jackson subsequently reported on 30th March to APOC in London from New York that oil circles there considered ' "that Andy Mellon had returned determined to keep his hands on Kuwait" '; a week previously Mr Jackson had reported, following a discussion with Gulf in Pittsburgh, that Leovy, on being told of Cadman's conversation with Mellon, had then said ' "that, although he would be interested to hear Mellon's views, he and Mellon had worked together for a number of years and Mellon would never interfere in a matter which he, Leovy, had been handling; and therefore I could take it that Gulf's policy would not be changed by anything Mellon had to report".'*

Following Cadman's return from Persia to London in May 1933, by which time (see page 33 et seq.) his large offer to Shaikh Ahmad for the Kuwait Concession on 25th March had been countered by a larger one from Gulf and the Shaikh had suspended further negotiations as from 14th May, discussions of possible Gulf/APOC joint action in Kuwait was resumed with Cadman by Mr Wallace (see below his cable of 18th May 1933) who was then in London. After Wallace had returned to New York on 24th May 1933 following the APOC/Gulf 'standstill' agreement of 23rd May (see page 37) the discussions, which were to culminate in the APOC/Gulf agreements of 14th December 1933 (Note 96) were continued in London from June onwards, Gulf's principal representative being Mr Stevens (Note 89) and APOC's Mr (later Sir William, and Lord Strathalmond) Fraser (Note 92).

Department of State
DIVISION OF NEAR EASTERN AFFAIRS

March 16, 1933.

Mr. Stevens of the Gulf Oil Company called today regarding the status of the Kuwait concession. Mr. Stevens said that he had hoped to have some definite news to give the Department from Mr. Jansen, the British representative of the Eastern and General Syndicate, who had proceeded to the Near East early in March. Mr. Jansen stopped at Kuwait for a short time in the early part of March, before going on to Bahrein, and in a conversation the Shaikh of Kuwait told him that upon his (Jansen's) return from Bahrein on March 9th it would be possible to discuss the question of the concession. Apparently Mr. Jansen had been delayed in Bahrein, however, and did not return to Kuwait until March 14th. Consequently, there had not yet been time to hear anything from him. Mr. Stevens added that his company was nevertheless encouraged with the general situation. He said that he would like to give us certain background which had developed in the past few months.

Last autumn Colonel Drake, Chairman of the Board of the Gulf Oil Company, was in London and had an opportunity to meet Sir John Cadman at a luncheon. Sir John brought up the question of the Kuwait concession and Colonel Drake said that he was surprised that the APOC was hampering the endeavors of the Gulf Company to enter that area, inasmuch as the APOC had been offered the territory and had declined it. Sir John said that Colonel Drake must be mistaken, since his company had never refused to go into Kuwait and since Colonel Drake was not familiar with all the details of the negotiations he was unable to controvert Sir John's statement. Early this year, however, Sir John came to the United States for a meeting of the American Petroleum Association at Houston and, while there, discussed the question further with Mr. Leovy, Vice

Chairman of the Gulf Oil Company. Mr. Leovy had been furnished with photostat copies of the letters written by the APOC to the Eastern and General Syndicate definitely declining to go into the Kuwait concession. Sir John was apparently surprised to see these letters and had to admit to Mr. Leovy that the APOC was not really justified in objecting to the Gulf Company's present endeavors to obtain a concession.

In subsequent conversations with Mr. Leovy, Sir John stated that his company considered Kuwait to be within its sphere of influence and that in spite of its early decision not to attempt to obtain a concession, it was now determined to obtain such a concession and would use every means in its power to do so. At the same time Sir John suggested the possibility of a joint development of the field by the APOC and the Gulf Oil. Mr. Leovy inquired upon what basis Sir John proposed such a development. Sir John answered that he had in mind a fifty–fifty share. When Mr. Leovy said that it was impossible to divide the responsibility on a fifty–fifty basis and that one or the other should be in control, Sir John said that of course the APOC would appoint a chairman of the Board of Directors and would furnish the field staff for the operations. Mr. Leovy said he did not think his company would be interested in such an arrangement.

Within recent weeks, Mr. Jackson, New York representative of the APOC, had been in to see the Gulf Oil Company on several occasions and has continued the discussion of the possibility of a joint operation in Kuwait. Mr. Jackson has suggested the possibility of having the Gulf Company appoint the chairman of the Board for one year and the APOC appointing their nominee the next year. In this way each company would have control of the combined operations in alternative years. While the Gulf Company has not turned this proposal down definitely, it has intimated that it is not particularly interested at the present time.

Mr. Stevens interprets these suggestions of the APOC as indicating that that company is no longer certain of its position in Kuwait and that, feeling somewhat weakened, it is preparing a basis upon which it can participate, at least to some extent, in the development of the Kuwait resources. Mr. Wallace, Vice President of the Gulf Company, has informed Mr. Jackson in no uncertain terms that the Gulf Company also was going to use its utmost endeavors to obtain the concession in question and that if the APOC did eventually obtain the concession it would have to pay a high price for it.

Asked whether the British Government presumably would wish to pass further upon any concession that might be signed, Mr. Stevens said that it was their feeling that since the British Government had now prepared its formal comments upon the two draft concessions, the Shaikh was free to sign whichever one he might choose and the only remaining formality for the British Government was to pass upon the wording of the article under which British interests in Kuwait would be guaranteed. Mr. Stevens added that this company had given instructions to Mr. Jansen to endeavor to have the Shaikh sign the proposed concession in its present form, leaving to future negotiations any minor changes that might be necessary. It was expected that something would be heard from Kuwait within the next few days and the Gulf Oil Company would be glad to keep the Department informed of further developments.

32, West Hill,
Highgate, N.6.
March 1st 1933.

Sir John Lloyd
My dear J.B.,

I am jotting down very hurriedly a short note to you following on my conversation with the American Ambassador, Mr. Mellon tonight. I found he was fully informed and all out for lending a hand to the Edmund Davies Group in their effort to negotiate the concession in Kuwait. I told him that any policy that put the Anglo-Persian Company and the Gulf Company in competition would lead to a form of bidding which would be costly to whichever concern secured the concession, and as it was the desire of the Anglo-Persian Oil Company to make the most friendly overtures to the Gulf Company I wondered whether his good offices could be used to stimulate the bringing together of the two companies for the purpose, and exploit this region on a fifty–fifty partnership basis, leaving the Anglo-Persian Oil Company to work the concession at such a rate as was compatible with world market conditions. I told Mr. Mellon that I thought it would be quite possible to work out some solution on these lines, and I also told him that I did not like the procedure now being adopted, whereby the Gulf Company was bidding for the concession through a third party, and I thought that we were both capable of handling this matter without outside assistance. He promised to discuss the matter on his arrival in New York, and in consequence, I think it would be as well if you concur, to send a telegram to Jackson on the following lines:—

> 'We are continuing to bid for concession but for your guidance I have had interview with Mellon here and asked him to get Gulf to consider proposition on following lines. They should withdraw leaving us free to negotiate on basis of fifty fifty partnership.
>
> APOC to work concession and at such rate compatible with world market conditions each group taking and selling its own share within "as is" position or on such other basis within "as is" may from time to time be mutually agreed. When you meet Leovy try to develop arrangements on these lines.'

Another telegram to Elkington on the following lines:—

'Please continue actively to urge our claims for Kuwait and adopt such lines as you think best. For your information I am again endeavouring to come to terms with Gulf Company but it is very desirable that we should put all pressure at Kuwait to make Gulf Company realise we are in dead earnest to secure concession and shall go to any lengths to do so.'

Yours ever,
John Cadman

Sir John Lloyd

Cable received by Gulf in Pittsburgh from Mr Wallace in London, 18th May 1933:

16 May 18th telegram from Holmes indicates [see Note 84] Sheikh again on rampage and we suspect for the sole purpose again bringing about a raise in the bid stop Feel that in the absence of agreement with APOC process will continue stop Might approach APOC have meeting with them Monday see if some plan can be worked out whereby each of us might obtain satisfactory concession on onehalf the territory leaving entirely in abeyance for present any plan pooling our interests stop In the absence of such approach and possible settlement controversy between us believe we otherwise blocked stop In event no such move is made plan leaving olympic May 24th.

---

78. *The object of Holmes's visit to Jeddah in March–April 1933 was twofold. He hoped to influence Ibn Saud on the strength of their 1922–23 acquaintance (see Note 15) both to favour Standard Oil of California (who since 1928 had been operating his original EGS concession of 1925 in Bahrein) in their current competition with IPC for the Saudi Arabia concession, and also to include in its area (as SOCAL eventually succeeded in doing in their concession agreement of 29th May 1935) Ibn Saud's interest in the Kuwait–Saudi Arabia Neutral Zone. Holmes's hopes were disappointed and his stay in Jeddah was cut short when he found on arrival there that, far from having Ibn Saud's goodwill, the King was resentful that Holmes's three-year option-contracts both for El Hasa of 1923 and for the Neutral Zone of 1924 had lapsed with various payments due to him unpaid.*

---

79. *APOC Minutes of Sir John Cadman's interview with Shaikh Ahmad on 25th March 1933:*

KUWAIT

On 25th March 1933 the Chairman accompanied by Messrs. Elkington, Gass and Allen visited Kuwait.

An interview took place with the Shaikh at 11 a.m. the following being present:—

The Chairman,
Messrs. Elkington,
Gass &
Chisholm
Interpreters:—Haji Williamson
&
Mr. Helmy.

The interview lasted about 2¼ hours and the following is a brief summary of the salient points which were covered in the conversation.

1. *Exploitation of Kuwait*

The Chairman emphasised that it was the Company's intention to exploit Kuwait energetically and quickly. He had heard that an impression had been conveyed to the Shaikh that the APOC wanted the concession with the sole object of preventing others from exploiting it and with no intention of doing so themselves. He could entirely disabuse the Shaikh's mind from any such impression.

2. *H.M.G.'s Interest*

It was again emphasised to the Shaikh that we were an all British Company, the largest British Oil Producing Company in the world and that H.M.G. were our partners in that they possessed a controlling financial interest in the Company. Their object was to secure supplies of oil for the British Navy. They never interfered with the commercial management of the Company nor did they ever bring pressure to bear on the Company to obtain Concessions in territories in which the Company were not themselves interested on commercial grounds or grounds of general policy; nor would they alternatively attempt to influence an owner of property to concede the oil rights to the APOC as this would lay them open to the criticism of taking advantage of a privileged position. They had necessarily on grounds of higher policy to adopt a strictly impartial attitude and this was the reason that they had indicated to the Shaikh that he had a free hand to offer his oil rights to whomsoever he wished.

3. *Marketing Arrangements*

The Chairman gave a brief explanation of the quota arrangements throughout the world for marketing oil, stressing the necessity for such arrangements owing to over-production of oil. He pointed out that the APOC were a party to these arrangements with the most powerful oil Companies in the world and were in the best position to dispose of any production that might be obtained from Kuwait territory to the best advantage of Kuwait. He touched on the difficulties which the Bahrein Oil Company were already experiencing in this respect. He warned the Shaikh that if Kuwait oil was in the hands of others, the APOC and their partners to these marketing arrangements would be compelled to take steps to hinder the marketing of Kuwait oil as this new source of supply would inevitably be to the detriment of their own markets.

4. *The Eastern General Syndicate*

The Chairman explained to the Shaikh that this Syndicate was an intermediary whose sole object was to obtain the Concession in order to sell it to an Oil Company who were in a position to provide the funds and to operate the Concession; that they had in fact approached the APOC on the grounds, as they alleged, that they were aware that the Shaikh was averse to the APOC and that they therefore would obtain the Concession on behalf of the APOC. The Chairman had turned this offer down because he knew the allegations of the EGS were untrue and he made it clear to their representatives that he preferred to deal direct with the Shaikh and intended that the Shaikh should himself receive the full payments up to the limit he was prepared to go which would not be possible when dealing through an intermediary to whom a commission would have to be paid. He emphasised therefore that the EGS were merely brokers and that it was unfair for the Shaikh to place him, the Chairman of a Company which was prepared and possessed the finance to exploit the territory themselves, in a position of bidding against a broker.

5. *Ibn Saud's Territory*

The Chairman touched on possible developments in Ibn Saud's territory pointing out that here again his representative was on the spot negotiating for a Concession and that the activities of the EGS were again merely those of brokers who would in turn have to find a group that was prepared to finance and operate the Concession from whom they would receive an overriding commission.

6. *American Participation*

The Chairman mentioned that he was in a far better position to invoke American participation than ever the EGS could be. He was on terms of intimate friendship with many of the leaders of the American Oil Industry and had had reason to discuss Kuwait affairs with some of them during his recent visit to the States. He questioned however if it was in the Shaikh's interests to seek an International partnership to work his Concession and for his part he would most strongly advise the Shaikh against it. He was able to draw comparisons between the ease of development of the Anglo-Persian Concession with the I.P.C. Concession and pointed out the constant delays and difficulties on account of the number of interests which had to be consulted, which were encountered in any movement which took place and before any decision could be reached in connection with IPC affairs.

In the course of this conversation opportunity was constantly given to the Shaikh to express his views on the points which were raised.

To a direct enquiry as to whether the Shaikh would prefer to have an All British Company operating his territory the Shaikh replied that it was a matter of indifference to him what nationalities were involved so long as the payments stipulated in the Agreement were made.

To an enquiry again whether he was not influenced to some extent by the fact that our proximity to his operations would enable us to exploit his territory forthwith he replied again that he did not regard the point of importance in that an Agreement would be signed, he would receive his payments and if the Concessionaire failed to observe any of his obligations he would be in a position to cancel the Agreement and reopen negotiations with others.

Asked if the fact that an intermediary must obviously receive his commission thus reducing the total amount which the actual operating Company would be in a position to pay direct to the Shaikh was not a less favourable proposition than a Company dealing direct with him and therefore able to pay the total amount to the Shaikh, he replied that he was again indifferent to this issue as the Concession would go to the highest bidder and from his knowledge of the alternative terms which were offered he was able to satisfy himself that he was making the best deal.

The Shaikh then explained the procedure he proposed to follow. He had recently received an offer from the EGS and he was in a position to tell us that it was better than ours. He enquired if we were prepared to raise our offer; if we were and it was better than the revised offer of the EGS he would then give the EGS another opportunity to revise their offer. If their subsequent revision was less than ours they would be told so and given a last opportunity. If the final terms offered were not so good as ours the matter would be closed, but if on the contrary they were above ours we would be given another opportunity to revise ours and so on until he was able to tell one or the other party definitely that the Concession would be granted to them. At that stage he would discuss the actual conditions of the document and he stated, so far as ours was concerned, that he would certainly ask for some modifications to our present document if we were the party which was ultimately successful. The Shaikh gave his assurance that under no circumstances would the offer of one party be disclosed to the other.

The Chairman then made an offer of Rupees two lacs for the initial payment and Rupees two lacs for the final guaranteed minimum after the 'declaration' mentioned in the Agreement with an undertaking that the other annual payments would be scaled up in much the same proportion.

He told the Shaikh that he was prepared to double these two figures if the Shaikh was prepared to sign the Agreement forthwith but that he was unable to leave his higher offer open.

The Shaikh regretted his inability to accept the higher offer of double the amount specified as he had given his promise to the EGS that he would not close with the Company without giving them an opportunity to revise their offer.

The initial payment and guaranteed minimum each of Rupees two lacs therefore stands as a verbal guarantee from the Chairman to the Shaikh which will not be confirmed in writing unless he insists upon it owing to the risk of a leakage of information from a written offer.

The question of the Royalty which stands in our Agreement at Rupees 60,000 plus Rs. 2/10/- per ton was not raised by either the Shaikh or the Chairman and it is hoped that no occasion will arise to discuss any revision in the tonnage figure.

Abadan, 26/3/33

---

80. *In this proposed division north–south of Kuwait territory the southern area (adjoining the Neutral Zone and Saudi Arabia) was to be offered to American interests only, reflecting Holmes's hope that the Saudi and Neutral Zone concessions also would be held by American interests (see Note 78) connected with EGS.*

---

81. *Mr W. T. Wallace was Vice-President in New York of Gulf Exploration Company, the Gulf Oil Corporation subsidiary company dealing with its Kuwait interests, which had been acquired from EGS in November 1927 through another subsidiary Eastern Gulf Oil Company of Delaware. It was through Wallace that Holmes and T. E. Ward had first interested Gulf (see page 13) in Kuwait when Holmes visited New York in October 1926. From then until the end of the concession negotiations in December 1934 Wallace, together with Mr F. A. Leovy, Vice-Chairman and President of Gulf Oil Corporation (and of Eastern Gulf Oil Company) in Pittsburgh, directed Gulf's Kuwait interests and negotiations.*

---

82. *See Note 66.*

---

83. *The total territory of Kuwait was approximately 6,000 square miles until 1970 when, by the addition of the north half of the Divided Zone (see Note 20) it increased to become 7,000.*

In 1933 APOC's proposed concession agreement covered the whole territory for the whole concession period of seventy years.

Holmes's proposed concession agreement on behalf of EGS/Gulf had hitherto provided for exploration rights over the whole territory for not more than five years followed by selection of an area, for development during the rest of the concession period of seventy years, not exceeding 1,640 square miles divided into six or more blocks.

---

84. *Exchange of letters between A. H. T. Chisholm and Shaikh Ahmad, May 1933:*

Abadan Refinery,
Persian Gulf.
26th May, 1933.

His Excellency Sir Ahmad Al Jaber,
Assubah, K.C.I.E., C.S.I.,
Ruler of Kuwait.

Your Excellency,

At Mr. Elkington's interview with Your Excellency on 14th inst. Your Excellency requested him to convey a message from yourself to Sir John Cadman.

Your Excellency's message was sent to London to await Sir John Cadman's arrival there, and we have now received the following telegram dated 24th inst. from Sir John Cadman, which we have the honour to communicate as follows:—

'I was delighted to receive Your Excellency's message of friendship and congratulations which recalls your very friendly and hospitable reception of me at Koweit 2 months ago. I hope our friendship will eventually be strengthened by bonds of mutual interests in the development of Your Excellency's Territories. Accept my assurances, esteem and my sincere wishes for Your Excellency's good health and prosperity.'

We beg to remain,
Your Excellency's obedient servants,
A. H. T. Chisholm
For Anglo-Persian Oil Company, Ltd.

Translation

No. 1060/4

28th May, 1933.

Dear Mr. Chisholm,

I hope you are well and enjoying sound health.

I have pleasure in acknowledging receipt of your friendly writing dated 26th instant. I am exceedingly delighted to go through the beautiful contents of the wire despatched by the friend of all, Sir John Cadman, for which I sincerely thank him for his appreciation and good hopes. We trust, God willing, that this friendship will be a strong and permanent one, everlasting with amity and sincerity more than you anticipate.

In conclusion I wish you good health and success.

Sincerely yours,
Sd/- Ahmed El Jabir

---

85. *The text of Elkington's letter to Shaikh Ahmad acknowledging the Shaikh's decision to suspend negotiations:*

Anglo Persian Oil Co. Ltd.,
Kuwait.
14th May, 1933.

His Excellency Sir Ahmed Aljaber Alsabah,
K.C.I.E., C.S.I.,
Ruler of Kuwait.
Your Excellency,

We beg to acknowledge with thanks Your Excellency's letter No. R/938 dated the 29th April, 1933 addressed to our Representative Mr. Chisholm, in reply to which, we now have the honour to confirm our conversation with Your Excellency this morning regarding the decision of Your Excellency to suspend all negotiations concerning an Oil Concession in Kuwait Territory for the time being and until further notification is received from Your Excellency.

We note that Your Excellency has already communicated the same decision to Major Holmes, the Representative of the Eastern and General Syndicate, also.

We are most grateful for the expression of Your Excellency's views with which we were honoured at our interview with Your Excellency to-day and we will now await the further notification referred to above in accordance with Your Excellency's wishes.

We are addressing copies of this letter to the Political Agent, Kuwait and to Sir John Cadman as agreed with Your Excellency.

We beg to remain,
Your Excellency's Obedient Servants.
Sgd. E. H. O. Elkington
General Manager
For Anglo Persian Oil. Co. Ltd.

---

86. *On 17th May 1933 Major Holmes cabled EGS in London from Kuwait as follows:*

I arrived here noon today fullstop I requested interview with the Sheikh of Kuwait and the Sheikh of Kuwait sent back word to me that he had no business to talk of to me for the present fullstop Secretary of the Council of State informed me first that the Anglo-Persian Oil Co., Ltd. Agents had told fullstop that the Eastern and General Syndicate Ltd. London was trying to dispose of its interest to the Anglo-Persian Oil Co. Ltd. and that Mr. W. T. Wallace had also approached the Anglo-Persian Oil Co. Ltd. with a view to combining forces in obtaining and working the Kuwait Oil Concession jointly Secondly that the Sheikh of Kuwait was angry with and distrusted Frank Holmes because of these secret deals fullstop The Secretary of the Council of State asked me if I had heard of such rumours fullstop I answered certainly I have every month for the past 3 years fullstop Lt. Col. Fowle, Political Resident in the Persian Gulf has been here and threatened that the British Government would cease support the Sheikh of Kuwait if he did not give the Kuwait Oil Concession to the Anglo-Persian Oil Co. Ltd. fullstop Lt. Col. Fowle, Political Resident in the Persian Gulf reached Bushire 16th May fullstop Please keep away from the Anglo-Persian Oil Co. Ltd. in London for the present and give me few days in which to try and appease the Sheikh of Kuwait fullstop It is unfortunate that such a position has arisen and it may be difficult and take some time to adjust but it is not hopeless if you cease to complicate matters by even seeing let alone discussing terms for disposing of Kuwait over which we hold no title fullstop I shall telegraph you in a few days—Frank Holmes

---

87. *While this applied equally to both APOC and Gulf, it did not apply similarly equally to their Kuwait representatives Chisholm and Holmes. The former, and all his staff, were long-term employees of APOC identifying its interests with their own; the latter, and all his staff, were only temporarily employed by Gulf and had their own interests and position to consider as well as their employer's. This disparity can account for the way in which Shaikh Ahmad was apparently kept informed by Mohammed Yatim and Holmes about the impending Gulf/APOC combine, while Chisholm and his staff continued to ignore or discourage local speculation on such a possibility.*

---

88. *In addition to the size of its initial payments the Saudi Arabia concession agreement of 29th May 1933 had two other aspects which were to affect the terms negotiated in 1934 for the Kuwait Concession, namely:*

(1) the initial payments, and also the royalty of four shillings per ton payable on crude oil production, were stipulated by Ibn Saud to be paid in gold or its equivalent as a safeguard against sterling devaluation (see Note 103).

(2) the royalty was to be paid on crude oil 'won and saved' (see page 58 et seq.).

---

89. *Mr Guy Stevens (d. 1945), Assistant to the Vice-President of the Gulf Exploration Company in New York (Mr Wallace; see Note 81) was Gulf's senior representative in London in connection with the Kuwait negotiations from June 1933 onwards; on the incorporation of Kuwait Oil Company in February 1934 he became its senior Gulf director.*

---

90. *The text of Holmes's letter to APOC was as follows:*

Balfour House,
119 to 125 Finsbury Pavement,
London, E.C.2.
20th November, 1933.

B. R. Jackson, Esq.,
The Anglo Persian Oil Co, Ltd,
Britannica House,
Finsbury Circus, E.C.2.
Dear Mr. Jackson,

I send you herewith a copy of a telegram that was received this morning from my representative in Kuwait, Mr. Mohammed Yateem. This telegram is in reply to one I sent on the 14th November requesting Mohammed Yateem to find out from His Excellency The Shaikh of Kuwait, whether he had any preference as to what the Oil Company to operate his territory should be called and whether he minded who the English groups were that would join with the Americans in forming the Company.

I told Mohammed Yateem to mention that the Company with whom the Americans would most probably work would be the Anglo Persian Oil Company Ltd.

You will see from the reply to this telegram that there is no difficulty about the name of the Company, nor is there any difficulty regarding the Americans and the Anglo Persian Oil Company forming a Company, subject, of course, to the safeguards regarding the interests, both politically and commercially, of the Shaikh and his people being protected, mentioned in the cable.

There are one or two points in the Agreement that I would like to discuss with you, so I propose to ring you up tomorrow for an appointment at a time suitable to you.

Yours sincerely,
Frank Holmes

Enc.

---

91. *The minutes of the meeting on 21st November 1933 between Mr Fraser and Major Holmes (with Mr Jackson and Mr Chisholm also present) were as follows:*

On seeing Major Holmes' letter to Mr. Jackson of 20th November (q.v.) covering a copy of a telegram from Kuwait, Mr. Fraser requested Major Holmes to call on 21st November.
Kuwait: Mr. Fraser stated that Major Holmes' telegram from Kuwait was interesting, but he

wished to make it clear that, in view of the position of our negotiations with the Gulf, it was advisable to communicate as little as possible with Kuwait for the present. Mr. Fraser suggested that Major Holmes' communication to the Shaikh of Kuwait of matters relating to our proposed combine with the Gulf was contrary to our interpretation of the standstill agreement with the Gulf; though he recognised that Holmes had doubtless acted in good faith. Nevertheless he requested Major Holmes to give him an assurance that he would not communicate with the Shaikh regarding the Kuwait oil concession without first consulting him.

Major Holmes replied that he entirely appreciated Mr. Fraser's point of view. He had only telegraphed to the Shaikh because the latter had telegraphed to him through Yatim asking what progress was being made by him (Holmes) in the matter of his concession negotiations, to which it was difficult not to reply. He gave his assurance however that he would not communicate with the Shaikh further on the subject either directly or indirectly, without informing Mr. Fraser as to what was being done.

Neutral Zone: Major Holmes said that the S.O. of California (with whom he and his Syndicate were having no further dealings, considering that S.O.C. had not treated them fairly in a matter connected with the Hasa concession) would be anxious to get a concession over the Kuwait–Hasa Neutral Zone, which area he understood had been excluded from their Hasa concession. He believed Mr. Philby might endeavour to assist them, by getting an option on the territory from Ibn Saud and then approaching Shaikh Ahmad for a similar option. He believed that, if IPC wished to obtain a concession for the Neutral Zone, his syndicate (the EGS), by reason of their connection with Ibn Saud and Shaikh Ahmad, would be able to get it for them. He also referred to a Mr. Ydlibi, with whom EGS had recently had dealings concerning the Neutral Zone, and promised to send Mr. Fraser more information on this subject (see later letter received from Major Holmes). Mr. Fraser, in thanking Major Holmes for his interesting information, said that he would bear it in mind for future reference.
Bahrein: Major Holmes stated that he had just heard that the Bahrein Petroleum Company had obtained an extension of their prospecting licence up to December 1934. In answer to a question of Mr. Fraser's, he said that it looked as if 100,000 acres should fully cover the only oil structure yet indicated in Bahrein, so that any remaining territory would be a highly speculative proposition.
Qatar: Major Holmes said that although his syndicate was not in the market for and was not interested in a Qatar Concession, he believed S.O.C. would spare no pains to obtain it.

---

92. *Mr William Fraser, later Sir William and from 1955 Lord Strathalmond (1888–1970) was Deputy-Chairman of APOC from 1928 to 1941 and Chairman from 1941 to 1956. On the incorporation of Kuwait Oil Company in February 1934 he became its senior APOC director and first Chairman (the chairmanship alternated yearly between APOC and Gulf).*

---

93. *Holmes's self-confident and assertive character made him an exceptionally able and forceful lone negotiator, but an egocentric and difficult member of a negotiating team. With increasing age the effects of the internal malady from which he eventually died (see Note 15) could sometimes intensify those defects, outstanding examples being his conversation with the Political Agent in Kuwait on 22nd May 1933 (see page 36 et seq.) and his behaviour on hearing about Traders Ltd. in October 1934 (see page 65 et seq.).*

A letter to Mr Fraser (see Note 92) in London from Mr Chisholm in Kuwait dated 22nd February 1934, five days after his first arrival there with Holmes as KOC's joint negotiators, concludes as follows:

I am relieved that this difficult first week has gone off as well as it has. Holmes has all the defects of his robust qualities and is consequently a difficult colleague; maybe he would say the same of me. He attacked me violently this morning, accusing me of every sort of treachery and pro-APOC self-seeking and threatening to throw up his brief forthwith if I did not toe his line. Fortunately in Kuwait the surroundings keep me perpetually in the worst of tempers and on tip-toe for squalls, so I joined vigorous battle and we eventually went off arm-in-arm for our interview with the Shaikh. Probably we shall be all the better for this blood-letting.

As the last sentence indicates, Holmes's colleagues and associates could, and often did, condone his occasionally eccentric or brusque behaviour in view of his record of achievement and determination to succeed.

---

94. *The following exchange of correspondence between the India Office, APOC and Gulf (Guy Stevens) in January 1933 considered the question of Major Holmes's suitability as a negotiator:*

Personal and Confidential *India Office,*
Whitehall, S.W.1.
10th January, 1933.

W. Fraser, Esq., C.B.E.,
Deputy Chairman,
Anglo-Persian Oil Coy. Ltd.,
Britannic House,
Finsbury Circus,
E.C.2.

Dear Mr. Fraser,

There is one point in connexion with our talk on Monday not mentioned in my other letter to you today.

Nothing was said in the conversation about the doubts felt as to the suitability of Major Holmes for the joint negotiations with the Sheikh or about our hope that in practice Mr. Chisholm would do most of the work. The reason why I did not say anything as to this, is that since we have decided to take Major Holmes along with us so long as he does not give trouble, it seemed unnecessary to raise any question of his suitability with Mr. Stevens, and we are satisfied with your assurance that you will be able to secure Major Holmes' replacement if he shows signs of giving trouble.

Yours sincerely,
J. C. Walton

J. C. Walton, Esq., C.B., M.C.,
India Office,
Whitehall, S.W.1. 11th January, 1934.

Dear Mr. Walton,

In acknowledging your letter of the 10th January may I, in the first place, say how much I appreciate the readiness shown by you and your colleagues in the India Office to understand the points of view which we have urged from our side in respect not only to the Kuwait, but also to the Qatar Concession. I feel sure that we shall soon emerge from our discussions with a clear entente regarding a satisfactory form of procedure and the general substance of the Agreements which we are trying to secure in the Gulf.

With respect to the question raised regarding Major Holmes, you will remember that during the discussion at which Mr. Stevens was present he himself volunteered the statement that, in the event of any trouble arising due to Major Holmes, he would take steps to withdraw him. However, neither he nor I think that this is likely to occur and you may rest assured that I shall watch this particular aspect of the negotiations with the greatest of care, in order to ensure that difficulties do not arise. Before Major Holmes goes out I shall have a personal talk with him, in which I propose to stress strongly the necessity for him doing everything to assist, and nothing to compromise, the successful progress of the negotiations.

Yours sincerely,
W. Fraser

Guy Stevens, Esq., *Britannic House,*
Savoy Hotel, *London E.C.2.*
Strand, W.C.2. 11th January, 1934.

Dear Mr. Stevens,

A propos of our discussion this morning regarding future procedure *vis-à-vis* the India Office, I send you herewith a copy of a letter which has been sent to Mr. Walton to-day, which is self-explanatory.

Yours sincerely,
(signed) A. C. Hearn

---

95. *Gulf's negotiating instructions to Holmes, similar to those given by APOC to Chisholm, were as follows:*

London
February 4, 1934.

Memorandum to Major Frank Holmes

It is expected that you will be leaving by aeroplane, Wednesday, February 7th. There are, however, one or two matters still unsettled the further discussion and settlement of which may possibly necessitate a postponement of your departure for a few days.

There will be a meeting of the Board of Directors of the Kuwait Oil Company Limited to-morrow (Monday) afternoon at 3 o'clock. It will probably be determined at that time whether it will be possible for you to leave on Wednesday. If you are to leave, Mr. Fraser and I will endeavour to see you and go over matters fully with you immediately after the Directors' meeting. If that should not be possible we will, assuming that you go on Wednesday, arrange definitely to see you on Tuesday afternoon.

There are, however, some things which will undoubtedly be gone over at our Conference with Mr. Fraser which I should like to emphasize and some matters which will probably not be touched upon in our conference with Mr. Fraser which I should like to mention, and the importance of which I should like to impress upon you.

We shall, I think, be able to hand you before you leave a Draft of proposed concession in substantially the final form so far as present views here are concerned. The amounts of the payments and the rate of royalty will not be specified in this Draft; and it is understood that in having the Draft translated and printed in Cairo, the amounts to be paid and the rate of royalty will likewise be left in blank to be filled in by the negotiators at the appropriate time. These amounts and the rate of royalty are to be expressed in terms of Rupees and are the same figures which we have heretofore discussed and will be supplied to you in a separate memorandum.

A joint power of attorney in favour of yourself and some other negotiator to be selected by our associates, and giving you and him jointly ample authority for your purposes, will be signed immediately after the meeting of Directors to-morrow afternoon. This power of attorney may be handed to you or it may be sent to the other negotiator; but in any event a copy can be furnished to you, which if you desire, you can have translated in Cairo.

Up to the moment, I have not been advised who the other negotiator will be. Whoever he is he will be advised by our associates here of your movements of which, for this purpose, you should keep Major Davis currently and fully advised. If the other negotiator happens to be Mr. Chisholm you should in addition to any advice which may be sent to him by our associates here, communicate with him and advise him of your movements in accordance with your understanding with him.

Mr. Janson has agreed that, if desired, the Syndicate will advance funds for your expenses for which either Gulf or the Kuwait Oil Company will reimburse the Syndicate; and as the funds of the Kuwait Oil Company may not be deposited and available before Tuesday, it might be well for you to arrange for the expenses for your trip with the Syndicate.

It is my understanding that you desire your salary to continue to be deposited to your account in New York as heretofore.

Your mission is to be a very important one and your selection as one of the negotiators implies a high degree of confidence in you and places upon you a large responsibility.

Our relations with our associates in this venture are, of course, set out in a formal agreement and in the formal documents creating the Kuwait Oil Company, but the real foundation of our association and the success of the Kuwait Oil Company must rest upon mutual confidence and goodwill and the fullest co-operation at every step and in the widest sense. It happens that one of the early occasions calling for the observance of the spirit of co-operation underlying this venture will be the conduct of the negotiations in Kuwait in which you are to take part; and the very fact that you have been selected as a part of this important mission indicates our confidence that you will fully meet the requirements of the situation.

You should approach the matter without any mental reservations whatever as to the good faith of your associate negotiator or of our associates here. There may be, and very likely will be, differences of opinion between you and your associate negotiator and between our associates and ourselves; but all those our associates and we are determined shall be resolved between us in the spirit upon which our relations have been founded. As Mr. Fraser has already told you a code address for the Kuwait Oil Company will be registered here. Cables sent to that address by the negotiators will be delivered to Mr. Fraser who, for the time being, will be the spokesman at this end for the Kuwait Oil Company. Matters that require consideration by the two associated Companies will be taken up with us by Mr. Fraser and decisions or instructions communicated to the negotiators by Mr. Fraser are to be considered as the decisions and instructions of the Kuwait Oil Company. As Mr. Fraser stated to you, however, if upon any point respecting which the negotiators communicate with Mr. Fraser they are not in agreement between themselves, each negotiator should feel free, in the cable to Mr. Fraser, to express his disagreement with his associate negotiator and set forth his own views. There will therefore be no occasion for any argument or dispute between the two negotiators which can by any chance tend to create any feeling between them or interfere with their complete and frank co-operation. Any point, whether of greater or less importance, will be settled by the associated Companies themselves or the Directors of the Kuwait Oil Company acting for them and the conclusions duly communicated to the negotiators.

Mention has been made in some of our talks with Mr. Fraser and in your last letter to Mr. Fraser of other matters beside the negotiation of the Concession for the Kuwait Oil Company. I wish to impress upon you again and to remind you of your undertaking that until the Kuwait Concession matter is successfully concluded to the satisfaction of the Kuwait Oil Company, this matter is to have your exclusive attention. Any discussion by you of other matters with the Sheikh or any other person and even any suggestion on your part that you should be concerned in any other matter will, under the circumstances which I have undertaken to explain to you on several occasions, be most prejudicial. We expect that you will be very meticulous about this and have nothing to do, directly or remotely, with any other matter for any reason or upon any occasion whatever until the negotiations for the express purpose of which you are going to Kuwait shall have been satisfactorily concluded.

You and the other negotiator should currently keep the Political Agent at Kuwait fully advised of your conversations with the Sheikh and the course of your negotiations. Careful attention should be given to this aspect of your work and the negotiators should do everything in their power to establish relations with the Political Agent on the basis of the utmost cordiality and frankness. We have reason to believe that if this course is followed by the negotiators they will have the assistance and support of the Political Agent.

You should never, under any circumstances, have any appointment or discussions with the Sheikh or Political Agent directly or indirectly unless accompanied by your associate negotiator or with his previous knowledge and full approval.

As Mr. Fraser explained to you at one of our conferences with him in November last, it is expected that the other negotiator will individually communicate with his superiors if he should desire to do so; and you will be free to communicate in like manner directly with Major Davis. Before you leave, Major Davis will give you his cable address and arrange with you about a code; and you should, in addition to the communications which the two negotiators jointly send Mr. Fraser, keep Major Davis closely in touch with developments by letter or cable as circumstances may require.

You will see from the foregoing, among other things, that it is contemplated that the negotiators themselves shall be relieved completely of the necessity of engaging between themselves in any argument or dispute which might tend to interfere with their cordial co-operation. The two associated Companies, or the Directors of the Kuwait Oil Company, will be prepared to settle all such matters and they should be duly referred to them for settlement.

Likewise, of course, while power of attorney will authorise the negotiators jointly to sign a Concession, no proposal differing from anything contained in the Draft Concession supplied you for negotiation purposes is to be submitted by the negotiators without previous express approval from here and no Concession is to be signed until after its terms are fully known and signature expressly authorised.

If the suggestions which will be made at our conference with Mr. Fraser and those set out in this Memorandum are carefully followed your negotiations will, I believe, be pleasant and the results successful. Personally, I entertain no doubt that you will with consideration and diplomacy and patience make your mission a success and fully justify the confidence reposed in you; and I wish you a very pleasant journey and success to your efforts.

Guy Stevens

---

96. *The two agreements signed on 14th December 1933 were as follows:*

Memorandum of an Agreement made the 14th day of December Nineteen Hundred Thirty-Three (December 14, 1933)

by and between

Anglo-Persian Oil Company, Limited, a Company registered under the Companies (Consolidation) Act (1908) of Great Britain and having its registered office at Britannic House, Finsbury Circus, London, (hereinafter called 'Anglo-Persian'), its permitted successors and assigns,

and

Gulf Exploration Company, a Company organized under the laws of the State of Delaware, U.S.A., a wholly owned subsidiary Company of Gulf Oil Corporation of Pennsylvania, and having an office and place of business at Gulf Building, Pittsburgh, Pennsylvania (hereinafter called 'Gulf'), its permitted successors and assigns:

RECITALS

On or about November 30th, 1927, Eastern Gulf Oil Company, a wholly owned subsidiary Company of Gulf Oil Corporation of Pennsylvania, U.S.A., entered into an Agreement with Eastern and General Syndicate, Limited, of London, under which Eastern Gulf Oil Company secured an option upon certain oil concessions then held or to be negotiated for by Eastern and General Syndicate, Limited.

Among the concessions included in such option agreement was one which Eastern and General Syndicate, Limited, expected to obtain covering the whole or a substantial part of the territory of Kuwait.

In the event of Eastern and General Syndicate, Limited, securing a Kuwait concession satisfactory to Eastern Gulf Oil Company, and the option being exercised and such concession being transferred to Eastern Gulf Oil Company or its nominee, Eastern and General Syndicate, Limited was to receive the sum of Fifty Thousand Dollars (from which was to be deducted a portion of certain expenses incurred in securing the concession) and an over-riding royalty of one shilling per ton on oil produced under such concession in excess of an average of seven hundred fifty tons daily.

Pursuant to its undertaking in the option agreement, Eastern and General Syndicate, Limited, has endeavoured to secure from the Shaikh of Kuwait a concession that would be satisfactory and could be transferred to Eastern Gulf Oil Company or its nominee, and has gone to considerable expense in connection with such endeavours.

Anglo Persian has also been interested in obtaining an oil concession in Kuwait, and has conducted negotiations to that end with the Shaikh of Kuwait.

Anglo-Persian and Gulf have reached the conclusion that it would be to their mutual advantage to join in a common endeavour to secure an oil concession or concessions in Kuwait and arrange for operations in Kuwait in their joint interest; and have, accordingly, made the following

AGREEMENT

In consideration of the assurances, promises and undertakings of the other as herein expressed, and for other good and valuable considerations, Anglo-Persian and Gulf each agrees with the other as follows:—

1. *Titles of Paragraphs:—* The titles used at the beginning of paragraphs herein are for convenience only, and are not be be considered in interpreting the meaning of the text.

2. *Meaning of Terms:—* As hereafter used in this Agreement, each of the following terms shall unless the context indicates otherwise have the meaning stated in this Paragraph, namely:—

THE COMPANY shall mean the company to be organized pursuant to the provisions of Paragraph 9 hereof;

THE CONCESSION shall mean the concession or concessions obtained or that may be obtained as contemplated by Paragraph 3 hereof for the exploration and exploitation of the oil lands or resources of Kuwait;

KUWAIT shall mean the territory of that name on the western side of the Persian Gulf (including mainland, islands and territorial waters), as shown on the attached map;

SHAIKH shall mean the Shaikh or other the Ruler for the time being of Kuwait.

3. *Negotiations for Concession:—*

(a) After the signing of this Agreement, neither party shall carry on any negotiations or make any efforts, directly or indirectly or otherwise and/or either alone or jointly with others to secure a concession or any interest in any concession covering the whole or any part of Kuwait except as may be mutually agreed by the parties or as may be determined by the Company; and any Kuwait oil concession or interest in any Kuwait oil concession heretofore or hereafter obtained by or on behalf of either party shall be considered as held in trust for the benefit of both parties and be transferred forthwith to the Company.

(b) Each party undertakes to employ, in such manner and to such extent as may be mutually agreed or as may be determined by the Company, the agencies and facilities at its disposal to secure from the Shaikh a concession in terms satisfactory to the parties hereto; and, in the absence of agreement between the parties on other terms, each party consents to the negotiation and acceptance of a concession on the terms or terms not substantially more onerous to the concessionaire than those of the draft concession attached hereto marked Appendix A.

(c) In the event that the first concession obtained does not cover the entire territory of Kuwait, efforts will be continued as may be mutually agreed or as may be determined by the Company to secure another concession or concessions which shall eventually cover all of Kuwait, and this Agreement shall apply to all concessions or interests in concessions obtained by or on behalf of either party or held by the Company covering any part of Kuwait.

4. *Expenses and Commitments:—*

(a) Each party hereto shall bear all expenses incurred by it prior to the date of this Agreement in connection with efforts to obtain a Kuwait discussion.

(b) From the date of this Agreement, all expenses incurred by either party in connection with efforts mutually agreed upon or determined by the Company to secure any Concessions or in carrying out the provisions and purposes of this Agreement shall be for equal joint account of the parties hereto or for account of the Company.

(c) With respect to the commitments or potential commitments of Eastern Gulf Oil Company to the Eastern and General Syndicate, Limited, the latter has executed an Agreement (a copy of which is attached hereto marked Appendix B) granting an option under which such commitments may be discharged and satisfied for the sum of thirty-six thousand pounds sterling; and it is agreed that within the period of ninety days stipulated in such option, Gulf shall cause such option to be exercised by Eastern Gulf Oil Company and each of the parties hereto will thereupon reimburse Eastern Gulf Oil Company to the extent of one half of the said sum of thirty-six thousand pounds sterling.

5. *Financing of the Company:—*

In the absence of agreement between the parties or decision of the directors of the Company to meet the financial requirements of the Company by other methods, the parties hereto will supply to the Company from time to time in equal shares and against capital stock of the Company all funds needed by the Company to carry on its operations as required by the Concession and this Agreement.

6. *Operation:—*

(a) Each party agrees that the directors of the Company nominated by it shall give their timely approval to such expenditures, operations and measures as shall be requisite to enable the Company to determine the location and extent of the oil resources of Kuwait, and if oil be present in commercial quantities to put the Company in position efficiently to produce, store, transport and handle such oil and generally to assure due compliance on the part of the Company with the terms of the concession and of this Agreement.

(b) Either of the parties shall have the right to require the Company at any time and from time to time after oil in commercially workable quantities has been found or determined to exist to produce such quantity of crude oil as may be desired by the party making the request if or to the extent that the quantity demanded is consistent with the scientific development of the concessionary area.

(c) All production from Kuwait produced at the request of both Gulf and Anglo-Persian shall be allocated by the Company 50–50 to Gulf and Anglo-Persian at cost, and each party shall pay the Company at cost for its share of such oil.

All production from Kuwait produced at the sole request of Gulf or Anglo-Persian shall be allocated in full to the party making the request and such party shall pay the Company at cost for all such oil.

(d) 'Cost', as referred to in the preceding subparagraphs (c) shall be defined as the actual out-of-pocket expenses incurred by the Company, including exploration, drilling, royalties, duties, taxes and all other expenses whatsoever applicable to such oil, depreciation, amortization and interest on capital at reasonable rates to be agreed upon.

7. *Marketing:—*

(a) Anglo-Persian is desirous of being assured that Kuwait oil, if discovered in commercial quantities, will not be used to upset or injure its own trade or marketing position directly or indirectly at any time or place. Gulf considers this an entirely legitimate and reasonable desire on the part of Anglo-Persian and for its part gives such assurance. Anglo-Persian recognises, however, that Gulf will wish to have outlets for Kuwait oil if and when produced; and therefore has no desire that Gulf should assume any restrictions with respect to the marketing of such oil and products therefrom which would in any way interfere with Gulf's freedom to obtain such outlets consistently with the observance of the above assurance.

(b) Reciprocally, Anglo-Persian gives assurance that Kuwait oil will not be used to upset or injure Gulf's trade or marketing position directly or indirectly at any time or place, while remaining free to obtain consistently with the observance of the above assurance outlets for Kuwait oil if and when produced.

(c) Both parties recognise the difficulty of foreseeing at the present time the conditions that may obtain in the future and of prescribing in detail the application of the principles set forth and agreed to above in this paragraph 7, and undertake to confer from time to time as either party may desire and mutually settle in accordance with such principles any question that may arise between them regarding the marketing of Kuwait oil and products therefrom.

(d) Since under its arrangement with the Burmah Oil Company the Anglo-Persian's marketing position in India is of an in and out nature dependent on the relation from time to time between Burmah Oil Company's and/or the Burmah-Shell Company's outlet there and the volume of indigenous production, Anglo-Persian is hereby recognized as having, by virtue of its said arrangement with Burmah Oil Company and/or Burmah-Shell Company, a 'trade or marketing position', in India, within the meaning of the preceding sub-paragraphs even though in pursuance of such arrangement it may not at any given time actually be supplying oil or the full range of its products to that market.

8. *Supply of Persian or Iraq Oil to Gulf*

The parties have in mind that it might from time to time suit both parties for Anglo-Persian to supply Gulf's requirements from Persia and/or Iraq in lieu of Gulf requiring the Company to produce oil or additional oil in Kuwait.

Provided Anglo-Persian is in position conveniently to furnish such alternative supply, of which Anglo-Persian shall be sole judge, it will supply Gulf from such other sources with any quantity of crude thus required by Gulf provided the quantity demanded does not exceed the quantity which in the absence of such alternative supply Gulf might have required the Company to produce in Kuwait—at a price and on conditions to be discussed and settled by mutual agreement from time to time as may be necessary—such price F.O.B., however, not to be more than the cost to Gulf of having a similar quantity produced in and put F.O.B. Kuwait.

9. *Operating Company:—*

(a) The parties will cause to be organised in such manner and in such jurisdiction as they may agree to be most advantageous from taxation and other standpoints a Company for the purpose of obtaining and holding the concession and carrying on all operations thereunder. Organisation of such Company will be proceeded with promptly in order that, if deemed advisable, negotiations for the concession may be conducted in its name. The initial capital of the Company shall be £50,000 or an approximate equivalent in the currency of the country where the Company is organised; and the parties hereto shall be bound to subscribe for the same equally. All increases in the capital of the Company shall be subscribed to equally by the parties.

(b) The Directors of the Company shall at all times be of an even number, one half of whom shall be nominated by each of the parties. Any one or more of the Directors nominated by either party attending any meeting of Directors shall have the right to exercise the entire voting power of such party. Directors having the right to exercise a majority of the entire voting power of all directors shall be required to constitute a quorum.

(c) The Directors may elect a Chairman, who, however, shall not be empowered as such to cast, in addition to the vote to which he may be entitled as a director, a deciding vote. Each person, if any, so elected as Chairman shall hold office until the next succeeding annual meeting of the Company, and the person, if any so elected as Chairman shall be nominated for the alternate periods by the parties hereto respectively.

(d) The Directors shall elect or appoint the officers, managers and other personnel of the Company; define the authority and fix the term of office of each; and exercise exclusive control over expenditures and all operations and affairs of the Company; and they shall take such steps as may be required to authorise and cause the Company and its officers and agents to carry out the terms of the present agreement.

(e) The Memorandum, Articles of Association, Charter Bye-Laws or other instruments defining the powers and rules of the Company shall be mutually agreed upon.

(f) Neither party shall assign or transfer its interest under this Agreement or dispose of its shares in the Company either in whole or in part except to, or with the written approval of, the other party, unless (A) such assignment, transfer and/or disposal be incident to a merger of interests or properties of such party with interests or properties of another responsible oil company or companies and the transferee takes over the whole of the undertaking of such party under this Agreement and enters into an agreement taking of such party under this Agreement and enters into an agreement with the other party hereto to become bound by all terms of this Agreement; or (B) the other party be unwilling to pay the price of consideration obtainable therefor from a responsible third party after being given notice in writing of the price or consideration at which the first party is willing to dispose of said shares and/or interest, which notice shall also include the names and addresses of any or all the third party potential purchasers whom it may be intended to approach, and a reasonable period of time (not being less than 30 days) within which to determine whether or not to purchase the said shares and/or interest. Should either party receiving such an offer fail to exercise the same, then the party serving the same shall be entitled for a period of sixty days thereafter to dispose of the shares and/or interest referred to in the notice to any one or more of the responsible third parties named at the price named in the notice but not at any other price or to any third party not so named without first offering them in like manner to the other party at such other price, and always provided any such third party purchaser or purchasers enters into an undertaking to become bound by all the terms of this Agreement. All rights of the transferor party under this Agreement shall inure to the transferee of such party's interest and/or shares in the Company which has entered into an agreement with the other party hereto to be bound by the terms of this Agreement.

10. Force Majeure

No failure or omission on the part of either party hereto to carry out or perform any of the stipulations, covenants or conditions of this Agreement shall give the other party any claim or be deemed a breach of this Agreement insofar as the same arises from *Force Majeure*.

11. *Arbitration:—*

(a) If at any time during the currency of this Agreement any difference or dispute shall arise between the parties hereto concerning the interpretation or execution hereof, or anything herein contained, or in connection herewith, or the rights or liabilities of either party hereunder, or if the parties are unable to agree upon any matter which is by this Agreement to be agreed between them, the same shall, failing any agreement to settle it in any other way, be referred to two arbitrators, one of whom shall be chosen by each party and a referee, who shall be chosen by the arbitrators before proceeding to arbitration.

(b) Each party shall nominate its own arbitrator within 60 days after the delivery at its registered office of a request so to do by the other party, failing which its arbitrator may at the request of the other party, be designated by the President for the time being of the Court of International Justice at the Hague. In the event of the arbitrators failing to agree upon the referee within 60 days after being chosen or designated, the President for the time being of the Court of International Justice at the Hague may appoint a referee at the request of the arbitrators or either of them.

(c) The decision of the arbitrators, or in case of a difference of opinion between them, the decision of the referee, shall be final and binding upon both parties.

(d) The place of arbitration shall be such as may be agreed by the parties and in default of agreement shall be London.

12. *Term of This Agreement*

The term of this agreement shall be the period of the concession and any extension or renewal thereof.

IN WITNESS WHEREOF the parties hereto have caused these presents to be duly executed the day and year first above written.

ANGLO-PERSIAN OIL COMPANY, LIMITED

ATTEST: (SEAL)
(Sgd.) JNO. CLARK
Secretary

BY: (Sgd.) W. FRASER
Director.

GULF EXPLORATION COMPANY,

ATTEST: (SEAL)
(Sgd.) W. J. GUTHRIE

BY: (Sgd.) F. A. LEOVY
President.

MEMORANDUM of an Agreement made the fourteenth day of December, 1933 by and between Anglo-Persian Oil Company, Limited, a company registered under the Companies (Consolidation) Act (1908) and having its registered office at Britannic House, Finsbury Circus, London, (hereinafter called 'ANGLO-PERSIAN') of the one part and Gulf Oil Corporation of Pennsylvania, a company organised under the Laws of Pennsylvania and having an office and place of business at Gulf Building, Pittsburgh, (hereinafter called 'GULF') of the other part.

Whereas Anglo-Persian is entering into an Agreement with Gulf Exploration Company a wholly owned subsidiary company of Gulf in the form annexed hereto.

NOW in consideration of Anglo-Persian entering into an Agreement in the form annexed with Gulf Exploration Company, of Gulf procuring Gulf Exploration Company to enter into the said Agreement with Anglo-Persian and of the undertakings of the respective parties herein, the parties hereto mutually agree as follows:—

1. Gulf agrees with Anglo-Persian that Gulf will procure its Associated Companies for the time being and each of them fully to observe and be bound by and will itself fully observe and be bound by all the provisions of Paragraphs 3 and 7 of the annexed Agreement between Anglo-Persian and Gulf Exploration Company as amplified by this present Agreement as if Gulf and each of its Associated Companies were parties to that Agreement jointly with Gulf Exploration Company, and Gulf will procure Gulf Exploration Company to carry out all the terms and provisions of the said Agreement, and Gulf agrees that Clause 7 of the said Agreement shall be read as applying not only to the marketing position of Anglo-Persian but also to that of all and each of Anglo-Persian's associated Companies.
2. Anglo-Persian agrees with Gulf that Anglo-Persian will procure its associated Companies for the time being and each of them fully to observe and be bound by all the provisions of Paragraphs 3 and 7 of its Agreement with Gulf Exploration Company in the form annexed hereto as amplified by this present Agreement as if they were respectively parties to that Agreement jointly with Anglo-Persian, and Anglo-Persian agrees that Clause 7 of the said Agreement shall be read as applying not only to the marketing position of Gulf Exploration Company but also to that of Gulf and all and each of Gulf's Associated Companies.
3. For the purposes of this Agreement, 'ASSOCIATED COMPANY' shall in the case of Gulf mean and include any subsidiary Company or Associated Company of Gulf which is in any manner howsoever directly or indirectly under the control of Gulf.
4. For the purposes of this Agreement, 'ASSOCIATED COMPANY' shall in the case of Anglo-Persian mean and include any subsidiary Company or Associated Company of Anglo-Persian which is in any manner howsoever directly or indirectly under the control of Anglo-Persian.

IN WITNESS WHEREOF the parties hereto have caused these presents to be duly executed the day and year first above written.

ATTEST: (SEAL)
(Sgd.) JNO. CLARK,
Secretary

ANGLO-PERSIAN OIL COMPANY, LIMITED
BY: (Sgd.) W. FRASER
Director

ATTEST: (SEAL)
(Sgd.) W. J. GUTHRIE

GULF OIL CORPORATION OF PENNSYLVANIA
BY: (Sgd.) F. A. LEOVY
Vice Chairman of the Board

---

97. *The terms of Gulf's agreement of 30th November 1927 with EGS acquiring the latter's interests in negotiating a Kuwait Concession (see page 14) had included a provision that, if and when Gulf obtained the concession, an overriding royalty would be payable to EGS of one shilling per ton on oil produced under that concession in excess of 750 tons per day.*

On 6th November 1933, Gulf, as a necessary preliminary to their Agreement of 14th December with APOC whereby they would become sole and equal partners in negotiating a Kuwait Concession through their Kuwait Oil Company, concluded an agreement with EGS to terminate their 30th November 1927 agreement and to pay them, on or before 31st January 1934, £36,000 in full quittance of all obligations to date under that agreement including the overriding royalty.

The smallness of this payment in respect to royalty rights which would, if retained, have become of enormous value (e.g. some £7,000,000 p.a. at Kuwait's present annual production) subsequently caused comment, especially when Kuwait's huge oil reserves were proved and annual production rapidly increased from 1946 onwards. Apart from the general uncertainty in 1933 as to Kuwait's future oil potential, Holmes's own pessimism on the subject (see Note 58) was certainly a main factor.

---

98. *The text of the agreement between the British Government and the Kuwait Oil Company was as follows:*

THIS AGREEMENT dated the fifth day of March One thousand nine hundred and thirty four is made BETWEEN HIS MAJESTY'S GOVERNMENT IN THE UNITED KINGDOM (hereinafter called 'His Majesty's Government') of the one part and the KUWAIT OIL COMPANY LIMITED (hereinafter called 'the Company' which expression shall where the context so admits be deemed to include its successors and/or assignees) of the other part

WHEREAS in the event of the Kuwait Oil Company obtaining a concession from the Sheikh of Koweit (hereinafter called 'the Sheikh') certain responsibilities will devolve on His Majesty's Government, the Company has agreed with His Majesty's Government as follows:—

1. THE Kuwait Oil Company, any transferee Company and any subsidiary Company that may be created shall be and remain a British Company registered in the British Empire.

2. NOTWITHSTANDING anything contained in the Agreement between the Company and the Sheikh the obligations and benefits of that agreement shall not be transferred to any other company without the prior consent in writing of His Majesty's Government, and shall not be transferred to any company in which more than fifty per centum of the capital and voting power is directly or indirectly controlled by persons other than British subjects.

3. THE employees of the Company in Koweit shall at all times so far as is consistent with the efficient carrying on of the undertaking be British subjects or subjects of the Sheikh. With the consent of His Majesty's Government, which consent shall not be unreasonably withheld, persons of other nationalities may be employed if in the opinion of the Company they are required for the efficient carrying on of the undertaking.

NOTWITHSTANDING anything contained in the Agreement between the Company and the Sheikh, the importation of foreign native labour shall be subject to the approval of the Political Resident in the Persian Gulf.

4. ONE of the superior local employees of the Company shall be designed chief local representative of the Company in Koweit. The approval of His Majesty's Government shall be required for the person so designated. He will be ordinarily resident at Koweit and will be responsible for the Company's local relations with the Koweit authorities. These local relations shall always be conducted through the Political Agent at Koweit, except as regards routine commercial business, which may be transacted through the official representative (if any) whom the Sheikh may appoint under the agreement between the Company and the Sheikh.

5. SUBJECT to the terms of the concession the Company undertakes at all times to pay due deference to the wishes of the Sheikh and to the advice of the Political Agent and the Political Resident in the Persian Gulf.

6. THE right given to the Company by its agreement with the Sheikh to utilise means of transportation by air shall be subject to any general regulations for civil aircraft made by the Shaikh on the advice of His Majesty's Government.

7. IN the application of the right given to the Sheikh by his Agreement with the Company to make full use free of charge of the Company's wireless and telegraph installations and railways for governmental purposes in times of national emergency, the Sheikh acting on the advice of His Majesty's Government shall be the sole judge whether a 'national emergency' has arisen.

8. NOTWITHSTANDING anything contained in the agreement between the Company and the Sheikh the Company shall not have the right to use or occupy, and shall not include in the areas selected for the purposes of its operations, any sites which may have been selected by or on behalf of the Sheikh or His Majesty's Government for defence purposes, for aerodromes, aeroplane or seaplane bases or for wireless and telegraph installations or in connection with the development of harbours, provided that with the consent of His Majesty's Government which shall not be unreasonably withheld the Company shall have the right to use for the purposes of its operations such harbours as may be developed by the Sheikh or His Majesty's Government if there is not reasonable harbour accommodation available elsewhere. Harbours developed by the Company shall be under its complete and exclusive control.

9. TELEGRAPH, wireless and telephone installations, if any, maintained by the Company shall be for use only in its business and as provided in the concession, and shall be so constructed and operated that their operations shall not interfere with the operations of such wireless, telegraph or telephone installations as may be established by the Sheikh or His Majesty's Government.

10. THE Company declare that it is their intention, should they consider that commercial conditions justify it, to erect a refinery at Koweit. If and when the Company is satisfied that commercial production is assured, the Company agrees to examine with His Majesty's Government the question of establishing a refinery in Koweit of suitable type and capacity.

11. IN the event of a state of national emergency or war (of the existence of either of which His Majesty's Government shall be the sole judge) His Majesty's Government shall have the right of pre-emption of all the oil produced in Koweit in accordance with the terms of the schedule hereto.

12. IN the event of notice of termination of the Agreement between the Company and the Sheikh being given under the terms of that Agreement on the ground that the Company has failed to observe any of the terms of this Agreement between the Company and His Majesty's Governments, the arbitration provisions of the said Agreement between the Company and the Sheikh shall apply if the Company considers that notice of termination on such grounds under that Agreement is not justified.

IN WITNESS whereof Sir Louis James Kershaw, K.C.S.I., C.I.E. on behalf of His Majesty's Government has hereunto set his hand and seal and the Company has hereunto caused its Common Seal to be affixed the day and year first above written.

THE SCHEDULE above referred to

Pre-emption Clause.

In the event of a state of national emergency or war (of the existence of which His Majesty's Government shall be the sole judge)

(1) His Majesty's Government shall have the right of pre-emption of all crude oil gotten under the Concession granted by the Sheikh to the Company and of all the products thereof and shall have the right to require the Company to the extent of any refining capacity it may have in Koweit to produce oil fuel that shall comply with the Admiralty specifications at the time provided that Koweit Oil be of a suitable kind and quality for this purpose.

(2) The Company shall use its utmost endeavours to increase so far as reasonably possible with existing facilities the supply of oil and/or products thereof for the Government to the extent required by the Government.

(3) The Company shall with every reasonable expedition and so as to avoid demurrage on the vessel or vessels engaged to convey the same, do its utmost to deliver all oil or products of oil purchased by the Government under their said right of pre-emption in the quantities at the time and in the manner required by the Government at a convenient place of shipment or at a place of storage in Koweit to be determined by His Majesty's Government. In the event of a vessel employed to carry any such oil or products thereof on behalf of His Majesty being detained on demurrage at the port of loading the Company shall pay the amount due for demurrage according to the terms of the charter party and/or the rates of loading previously agreed with the Company unless the delay is due to causes beyond the control of the Company. Any dispute which may arise as to whether the delay is due to causes beyond the control of the Company shall be settled by agreement between His Majesty's Government and the Company, and, in default, of such agreement, the question shall be referred to two arbitrators, one to be chosen by His Majesty's Government (or the Political Resident) and the other by the Company, with power to appoint an umpire in case of disagreement, such arbitration to be held in England and to be deemed a reference to arbitration under the provisions of the Arbitration Act of 1889 (52 and 53 Vict. C.49) of the Imperial Parliament, or of any statutory modification or re-enactment thereof for the time being in force.

(4) The price to be paid for all oil or products of the refining or treatment of oil taken in pre-emption by His Majesty's Government shall be either (a) as specified in a Separate agreement or (b) if no such agreement shall have been entered into, a fair price for the time being at the point of delivery as the same shall be settled by agreement between His Majesty's Government and the Company, or in default of such agreement by arbitration in the manner provided by the last preceding sub-clause.

To assist in arriving at a fair price at the point of delivery the Company shall furnish for the confidential information of His Majesty's Government, if so required, particulars of the quantities, descriptions and prices of Koweit oil or products sold to other customers and of charters or contracts entered into for carriage and shall exhibit to His Majesty's Government original or authenticated copies of contracts or charter parties entered into for the sale and/or carriage of such oil or products.

(5) His Majesty's Government shall be at liberty to take control of the works, plant and premises of the Company in Koweit, and in such event the Company shall conform to and obey all directions issued by or on behalf of His Majesty's Government. Compensation shall be paid to the Company for any loss or damage that may be proved to have been sustained by the Company by reason of the exercise by His Majesty's Government of the powers conferred by this sub-clause. Any such compensation shall be settled by agreement between His Majesty's Government and the Company or, in default of agreement, by arbitration in the manner provided by sub-clause 3.

SIGNED SEALED AND DELIVERED BY the said Sir Louis James Kershaw on behalf of His Majesty's Government in the presence of John Charles Walton India Office Civil Servant	L. J. Kershaw
THE COMMON SEAL of the Kuwait Oil Company Limited was hereunto affixed in the presence of	A. C. Hearn Guy Stevens H. T. Kemp

---

99. *The draft agreement (dated 11th January 1934) taken to Kuwait by Chisholm and Holmes was as follows:*

IN THE NAME OF GOD THE MERCIFUL

This is an AGREEMENT made at Kuwait on the in the year 135 corresponding to day of 135 between His Excellency Shaikh Sir Ahmad al-Jabir as-Subah, Knight Commander of the Most Eminent Order of the Indian Empire and Companion of the Most Exalted Order of the Star of India, the SHAIKH OF KUWAIT in the exercise of his powers as Ruler of Kuwait on his own behalf and in the name of and on behalf of his heirs and successors in whom is or shall be vested for the time being the responsibility for the control and government of the State of Kuwait (hereinafter called 'the Government') and the KUWAIT OIL COMPANY LIMITED a Company registered in Great Britain under the Companies Act, 1929, its successors and assigns (hereinafter called 'the Company').

Article 1.

The Government hereby grants to the Company the exclusive right to explore search drill for produce and win natural gas asphalt ozokerite petroleum and their products and cognate substances (hereinafter referred to as 'petroleum') within the State of Kuwait including all islands and territorial waters appertaining to Kuwait as shown generally on the map annexed hereto, the exclusive ownership of all petroleum produced and won by the Company within the State of Kuwait the right to refine transport sell for use within the State of Kuwait or for export and export or otherwise deal with or dispose of any and all such petroleum and the right to do all things necessary for the purposes of those operations. The Company undertakes however that it will not carry on any of its operations within areas occupied by or devoted to the purposes of mosques sacred buildings or graveyards or carry on any of its operations except the sale of petroleum housing of staff and employees and administrative work within the present town wall of Kuwait.

The period of this Agreement shall be 75 years from the date of signature.

Article 2.

(A) Within nine months from the date of signature of this Agreement the Company shall commence geological exploration.

(B) The Company shall drill for petroleum to the following total aggregate depths and within the following periods of time at such and so many places as the Company may decide:—

(i) 4,000 feet prior to the 4th anniversary of the date of signature of this Agreement.
(ii) 12,000 feet prior to the 10th anniversary of the date of signature of this Agreement.
(iii) 30,000 feet prior to the 20th anniversary of the date of signature of this Agreement.

(C) The Company shall conduct its operations in a workmanlike manner and by appropriate scientific methods and shall take all reasonable measures to prevent the ingress of water to any petroleum-bearing strata and shall duly close any unproductive holes drilled by it and subsequently abandoned.

Article 3.

In consideration of the rights granted by the Government to the Company by this Agreement and of the assistance and protection which the Government hereby undertakes to afford to the Company and its operations employees and property the Company shall pay to the Government the following sums:—

(a) Within thirty (30) days after the signature of this Agreement Rupees .

(b) On each anniversary of the date of signature until after the declaration mentioned in (c) below shall have been made by the Company:— Rupees or Royalty of Rupees Annas for every English ton of 2,240 lbs. of Kuwait petroleum exported from or sold for consumption in Kuwait during the year ended 3 months prior to the anniversary of the date of signature—whichever shall be the greater sum.

(c) On each anniversary of the date of signature after that immediately following a declaration by the Company that petroleum has been found in commercial quantities and quality Royalty of Rupees Annas as above defined or Rupees whichever shall be the greater sum.

Article 4.

On each anniversary of the date of signature of this Agreement the Company shall deliver to the Government a return of petroleum if any on which royalty is payable for the year ended three months prior to such anniversary and a statement of the amount of royalty if any due to the Government for such year. The Government shall have the right to check such returns and statements which shall be treated as confidential by the Government.

Article 5.

(A) For the purposes of its operations hereunder the Company shall have the right without hindrance to construct and to operate power stations, refineries, pipelines and storage tanks, facilities for water supply including boring for water, telegraph telephone and wireless installations, roads, railways, tramways, buildings, ports, harbours, harbour works, wharves and jetties, oil and coaling stations, with such lighting as may be requisite and any other facilities or works which the Company may consider necessary, and for such purposes to use free of all payments to the Government any stone, sand, gravel, gypsum, clay or water which may be available and may be required for its operations hereunder, provided always that the water supply of the local inhabitants and nomad population who may be dependent on the same is not endangered. The Company shall at its discretion but in consultation with the Government select the position of any such works. The Company may likewise utilise without hindrance all such means of transportation by land air and water communication or operation as may be necessary for the effective conduct of its operations hereunder.

(B) The Company shall under normal conditions accept and transmit free of charge on its wireless and telegraph installations such Government messages as will not interfere with the Company's business, and in times of national emergency the Government shall have full use free of charge of the Company's wireless and telegraph installations and railways for governmental purposes.

(C) The Government's ships shall have the right to use harbours utilised or constructed by the Company, provided that such use in no way hampers the Company or interferes in any way with the safety of its operations of which the Company shall be sole judge. Any wharves or appurtenances constructed by the Company shall be for its exclusive use.

Article 6.

(A) The Company shall maintain in the region of the Persian Gulf a Chief Local Representative to represent it in matters relating to this Agreement with the Government.

(B) The Government shall have the right to appoint in consultation with the Company a suitable person to act as its official representative. The salary of the representative shall not exceed Rupees   a month and shall be paid by the Company.

Article 7.

The Company shall have the right to import water, petroleum, fuel, machinery, motor cars and lorries, equipment, plant, timber, utensils, iron work, building materials, food supplies, medicines, medical supplies, office equipment and household furniture, and all other materials, equipment and goods of whatsoever nature required by the Company and its employees for the purposes of its operations hereunder but not for resale to others, and to export its petroleum and articles previously imported by the Company free of customs or import or export duty and taxes or other charges, but it shall pay on all personal goods, clothing and general merchandise imported by the Company for the personal use of its employees or for resale to them, and ordinary duty in force for the time being in the State of Kuwait. Saving as in Article 3 and in this Article provided, the Company, its operations, income, profits and property including petroleum shall be exempt and free during the period of this Agreement from all present or future harbour duties, import duties, export duties, taxes, imposts and charges of any kind whether state or local, tolls, and land surface rent of whatever nature; and in consideration thereof the Company shall pay to the Government on each anniversary of the date of signature of this Agreement   annas per ton of petroleum on which royalty is payable.

Article 8.

(A) The Company shall have the right to purchase at current market rates fuel, water, food, building and constructional materials and other supplies of every kind in connection with its operations hereunder.

(B) The Company shall employ subjects of the Government as far as possible for all work for which they are suited under the supervision of the Company's skilled employees, but if the local supply of labour should in the judgment of the Company be inadequate or unsuitable the Company shall have the right to import labour. The Company shall also have the right to import skilled and technical employees. The Company shall pay to the workmen it employs a fair wage, such wage to be decided and stated by the Company at the time the workmen are engaged.

(C) The Company shall provide free of charge medical service for its employees, and the Ruler of Kuwait and his family shall have the right to such medical service and necessary medical supplies free of charge.

Article 9.

The Government grants to the Company free of cost the unrestricted use and occupation of and surface rights over all uncultivated land belonging to the Government which the Company may need for the purposes of its operations and in particular the Company shall have the right to select in consultation with the Government an area or areas of land chosen by the Company outside the present town wall of Kuwait with exclusive surface rights upon which to erect oil refineries, storage, terminal and shipping facilities and any other works required for the Company's operations; and the Company may buy or lease for such purposes any lands, houses or buildings on conditions to be arranged with the proprietors thereof but at rates not in excess of those ordinarily current in their respective localities.

Article 10.

The Government shall give to the Company and its employees and property all protection in its power from theft, highway robbery, assault, wilful damage and destruction, and the Company may appoint and itself pay trustworthy guards to assist in protecting the property of the Company and its employees. The Company may erect at its own expense suitable buildings for the accommodation of such guards at such places as the Company shall decide.

Article 11.

(A) Before the expiration of the period specified in Article 1 hereof this Agreement shall come to an end either by surrender as provided in paragraph (B) of this Article or in Article 12 or in one of the three following cases only:

(a) If the Company shall fail to fulfil its obligations under Article 2 hereof in respect of geological exploration or drilling.

(b) If the Company shall fail within six months after any anniversary of the date of signature of this Agreement to make to the Government any payments agreed to be due under Article 3.

(c) If any sum awarded to the Government by the arbitrators shall not have been paid as provided in Article 18.

In any one of the above mentioned cases and in no other the Government shall be entitled to terminate this Agreement and all the property of the Company within the State of Kuwait shall become the property of the Government.

(B) In the event of the Company failing to make the declaration provided in Article 3 within 12 years of the date of signature of this Agreement the Company shall at its option either pay to the Government the minimum annual payment provided in Article 3(C) or surrender all rights under this Agreement.

Article 12.

(A) The Company shall have the right at any time after it has drilled the 4,000 feet provided in Article 2(B) (i) or after the expiry of 2 years from the date of signature of this Agreement whichever shall be the later date to give the Government one year's notice in advance to terminate this Agreement and the Company shall on expiry of such notice have no further liabilities except to make payment of all monies which may be due to the Government up to the date of termination.

(B) Should this Agreement be terminated by the Company under this Article 12, then:—

(a) If such termination occurs within 35 years from the date of signature of this Agreement all lands granted by the Government and any lands or buildings which the Company may have bought and any houses or buildings constructed by and other immovable property of the Company within the State of Kuwait shall be handed over to the Government free of cost, but

(b) if such termination occurs after 35 years from the date of signature of this Agreement all the property of the Company in the State of Kuwait shall be handed over to the Government free of cost in such working condition as the property then is.

Article 13.

On the expiry of this Agreement at the end of the period of 75 years provided in Article 1 or of any extension or renewal of that period all the property of the Company in the State of Kuwait shall be handed over to the Government free of cost in such working condition as the property then is.

Article 14.

The Government hereby agrees that the Company may transfer the obligations and benefits of this Agreement to any Company registered within the British Empire.

Article 15.

(A) Nothing in this Agreement shall be read as restricting in any way the right of the Government to grant to other parties concessions for substances other than petroleum provided that the operations and rights of the Company hereunder are not thereby injuriously affected.

(B) The Company shall use the Government's flag within the State of Kuwait.

Article 16.

Failure on the part of the Company to fulfil any of the conditions of this Agreement shall not give the Government any claim against the Company or be deemed a breach of this Agreement in so far as such failure arises from *force majeure*, and if through *force majeure* the fulfilment by the Company of any of the conditions of this Agreement be delayed the period of such delay shall be added to the periods fixed by this Agreement.

Article 17.

The Government shall not by general or special legislation or by administrative measures or by any other Act whatever annul this Agreement, except as provided in Article 11, or alter the terms of this Agreement.

Article 18.

(A) If at any time during the currency of this Agreement any difference or dispute shall arise between the parties hereto concerning the interpretation or execution hereof, or anything herein contained, or in connection herewith, or the rights or liabilities of either party hereunder, the same shall, failing any agreement to settle it in any other way, be referred to two arbitrators, one of whom shall be chosen by each party, and a referee, who shall be chosen by the arbitrators before proceeding to arbitration.

(B) Each party shall nominate its own arbitrator within 60 days, after the delivery of a request so to do by the other party, failing which its arbitrator may at the request of the other party be designated by the British Political Resident in the Persian Gulf. In the event of the arbitrators failing to agree upon the referee within 60 days after being chosen or designated, the British Political Resident in the Persian Gulf may appoint a referee at the request of the arbitrators or either of them.

(C) The decision of the arbitrators, or in case of a difference of opinion between them the decision of the referee, shall be final and binding upon both parties.

(D) In giving a decision the arbitrators or the referee shall specify an adequate period of delay during which the party against whom the decision is given shall conform to the decision and that party shall be in default only if that party has failed to conform to the decision prior to the expiry of that period and not otherwise.

(E) The place of arbitration shall be such as may be agreed by the parties and in default of agreement shall be London.

Article 19.

The Company shall make all payments that become due to the Government under this Agreement into the Government's account at the Bank in and the Bank's receipt shall be a full discharge for the Company in respect to the payment of the sum stated in the Bank's receipt. The Government may from time to time designate in writing another Bank or Banks for the purpose of this Article.

Article 20.

This Agreement is written in English and translated into Arabic. If there should at any time be disagreement as to the meaning or interpretation of any clause in this Agreement the English text shall prevail.

In witness whereof the parties to this Agreement have set their hands the day and year first above written

SHAIKH OF KUWAIT	ON BEHALF OF THE KUWAIT OIL
IN THE PRESENCE OF	COMPANY LIMITED IN THE PRESENCE OF

---

100. *Joint power of attorney granted to Chisholm and Holmes by Kuwait Oil Company, 5th February 1934:*

KNOW ALL MEN BY THESE PRESENTS THAT THE KUWAIT OIL COMPANY LIMITED a Company registered in England under the Companies Act 1929 and having its Registered Office at 71 Queen Street in the City of London England (hereinafter called 'the Company') HEREBY APPOINT Archibald Hugh Tennent Chisholm of Abadan in the Empire of Persia and FRANK HOLMES of Great Baddow, Chelmsford, in the County of Essex, England, Major (retired) JOINTLY but not SEVERALLY to be ATTORNEYS of the Company in the name and on behalf of the Company to negotiate for and obtain a grant or concession from the Ruling Sheikh of Kuwait of the exclusive right to explore, search, drill for, produce and win, and of the right to refine, transport and export, sell and otherwise deal in mineral oil natural gas asphalt ozokerite petroleum and their products and cognate substances in and from the State of Kuwait on and subject to such terms and conditions as the said Attorneys may approve and to execute any deed instrument or other document that may be necessary or expedient for carrying the above purpose or any of them into effect.

The Company hereby ratify and agree to confirm all acts done or caused to be done by the said Attorneys in exercise of the powers hereby conferred.

IN WITNESS whereof the Company have caused their Common Seal to be hereunto affixed this fifth day of February One thousand nine hundred and thirty four

THE COMMON SEAL of Kuwait Oil Company Limited was hereto affixed by order of the Board in the presence of:—

W. Fraser Director
J. Cookson Acting Secretary

---

101. *Holmes's and Chisholm's letter to Shaikh Ahmad after their meeting with him on 22nd February 1934:*

Kuwait Oil Company, Limited
Kuwait,
Persian Gulf.
22nd February, 1934.

To:—
His Excellency Shaikh Sir Ahmad al-Jabir as-Subah,
K.C.I.E., C.S.I., Ruler of Kuwait.
KUWAIT.

Your Excellency,

After compliments. In accordance with Your Excellency's request at our interview this morning, we have pleasure in sending you herewith on behalf of the Kuwait Oil Company Limited two copies of a Concession Agreement.

We also enclose one copy of our joint Power of Attorney for the Kuwait Oil Company Limited.

In accordance with the usual procedure we are sending one copy of the Concession Agreement to the Political Agent.

We remain,
Your Most Obedient servants,
(Signed) FRANK HOLMES
A. H. T. CHISHOLM
For Kuwait Oil Co. Ltd.

---

102. *The monetary maxima which the KOC negotiators were initially authorised to offer were, as described later in this Record, to be considerably increased during the course of the negotiations.*

A noteworthy fact, and one of which the negotiators themselves were never aware, is that even the final terms eventually accepted by Shaikh Ahmad, and incorporated in the

Concession Agreement of 23rd December 1934, were less than the top limit to which APOC and Gulf had agreed, at the beginning of the negotiations, to go if necessary. This is shown by the following memorandum tabled at a Board Meeting of the Kuwait Oil Company on 21st March 1934:

Information for Board

	Negotiators' authorised maxima Rs.	Figures submitted to Sheikh Rs.	Maxima agreed between APOC & Gulf Rs.	Sheikh's demands (20.3.34) Rs.
(a) Initial payment	200,000	200,000	666,666	500,000
(b) Annually	65,000	50,000	100,000	100,000
before declaration or royalty per ton	2·10 rupees	2·10 rupees	2·10 rupees	3·4 rupees
(c) After	200,000	150,000	266,666	250,000
declaration or royalty	2·10	2·10	2·10	3·4
(d) Tax exemption	2 annas	2 annas	11 annas (approx.)	4 annas

For easier reference, the terms incorporated in the Concession Agreement of 23rd December *1934* were:—

(a) Rs. 475,000
(b) Rs. 95,000 or royalty of 3 rupees per ton
(c) Rs. 250,000 or royalty of 3 rupees
(d) 4 annas

---

103. *The royalty per ton in both the APOC Concession Agreement of 29th April 1933 with the Persian Government (see Note 66) and the contemporary IPC Concession Agreement in Iraq (and also in Standard Oil Company of California's Saudi Arabian Concession Agreement of 29th May 1933, see Note 88), was four shillings, the sterling equivalent of the Rs. 2.10 proposed to Shaikh Ahmad by the negotiators. But in all those concession agreements provision was made that, in the event of any depreciation in the gold value of sterling, the royalty amount would be correspondingly increased. The negotiators felt that the demand for a higher royalty by Shaikh Ahmad, who they knew to be well informed regarding the terms of the Persian, Iraqi, and Saudi Arabian concessions, might be because no such provision was included in their own concession proposals.*

---

104. *See table opposite.*

---

105. *List of general alterations received by the negotiators from Shaikh Ahmad on 23rd March 1934.*

After signature of the agreement the company shall deliver to us each year 5,000 gallons of petrol, and when the company declares the existence of oil shall increase the quantity to 10,000 English gallons.

Clause

3. The possible assistance and protection.

C. That oil has been found in commercial quantity is enough.

4. The word petrol to be cut out and 'oil and its extracts' to be stated.

5. The Company may use whatever it may need for its operations of sand, stones, gypsum, clay and what resembles them.

C. The company shall not be allowed to use harbours, etc. that are along the coast and their appurtenances that are at present being utilised as harbour or wharf for the ships belonging to the Kuwaitis.

6B. We shall appoint a man through whom the company shall engage labourers and his salary shall be paid by the company and shall not be less than Rs. 800/-.

7A. All what the company import such as machinery, drilling material, equipment, building material and everything concerning its operations is exempted from current custom duty. As regards other goods which has no concern with operations such as food supplies, clothings, and furniture, custom duty shall be paid. The Company shall have no right to import any firearms etc. for use without our permission.

7B. Officials to be appointed under the authority of the local custom official in all the harbours utilised by the company and the company shall have to pay the salaries of such officials and build places for their accommodation.

Table showing the Negotiating Terms as at 27th March 1934

	Negotiators' authorised maxima Rs.	Figures submitted to Shaikh Rs.	Maxima agreed between APOC & Gulf Rs.	Shaikh's demands (20.3.34) Rs.	Shaikh's demands (21.3.34) Rs.	Negotiators' estimates of figures acceptable to Shaikh	
						Major Holmes 26.3.34 Rs.	Mr Chisholm 26.3.34 Rs.
Initial payment	200,000	200,000	666,666	500,000	475,000	450,000	425,000
Annually before declaration	65,000	50,000	100,000	100,000	100,000	90,000	80,000
or							
Royalty per ton	2·10	2·10	2·10	3·4	3·0	3·2*	2·14**
Annually after declaration	200,000	150,000	266,666	250,000	250,000	250,000	250,000
or							
Royalty	2·10	2·10	2·10	3·4	3·0	3·2**	2·14**
Tax exemption	2 annas	2 annas	11 annas (approx)	4 annas	4 annas	—	—
Annual free petrol allowance							
Before declaration	—	—	—	—	*5,000 glns.	—	—
After declaration	—	—	—	—	10,000 glns.	—	—

* Retail price of petrol in Kuwait is approx. Rupees 1·0 per gallon.
** Including Taxation Exemption.

C. When the company sells any of the material and equipment which has been imported for its operations it shall have to pay custom duty on what it sells in auction or privately.

8B. All unskilled labourers shall be from the subjects of Kuwait and all such labourers should be engaged through the man whom we appoint.

C. If the need presses for more labourers the company shall have to consult us and after our agreement they will have to be imported from the neighbouring Arab countries provided they obey the local laws. As for European employees attached to the company if any act of misconduct on the part of any of them shall cause a breach of the peace such as a quarrel or dispute the company shall be informed at the time and the company shall have to give him the shortest possible notice and send him away from the country.

9. The lands, houses and buildings which the company abandons should be returned to us without cost.

10. The possible assistance and protection.

B. The head guard to be appointed by us likewise the guards and the company shall have to pay their salaries and to build places for their accommodation in the places it fixes for them.

12. If the expiry occurs after 35 years from the date of signature then all the movable and immovable properties of the company to be handed over to us without cost.

15. If we grant a concession to any other company for other substances and the company claims that this is harming its interests we shall have to consider the matter and consult experts in minerals and mines and if the company finds any mines such as gold, silver, copper, lead and such like the company shall have to cease work in it and inform us at the time.

16. If any failure arises from *force majeure*. Should be made clear what is beyond human nature, for example war, drowning, flood, lightning, or fire.

18. If any disagreement occurs between the two parties and is difficult to solve directly, they shall consult about it with the representative of the British Government in Kuwait or any of the neighbouring British representatives.

19. Payment of money to any bank we select.

---

106. *The negotiators' letter to Shaikh Ahmad detailing the two additional clauses was as follows:*

His Excellency Shaikh Sir Ahmad Al-Jabir
As-Subah KCIE. CSI.
Ruler of Kuwait,
KUWAIT.

10th April, 1934.

Your Excellency,

After compliments.

Our London Office has informed us that the Political Resident has informed Your Excellency of an agreement which has been made between His Britannic Majesty's Government and the Kuwait Oil Company in London concerning political matters.

In accordance with the terms of that agreement our London Office wishes us to make the following two additions to our draft concession document.

(1) At the end of article 11 A(c) add the following words: (d) 'If the Company shall fail to observe any of the terms of the agreement between the Company and His Majesty's Government signed in London on March 5th 1934, and, if the matter is referred to arbitration under article 18, fail to remedy such failure within the reasonable time which shall be fixed by the arbitrators for so doing.'

(2) Renumber the existing article 20 as Article 21 and insert the following new article:

'Article 20: It is hereby declared that should any of the terms of this agreement be inconsistent or in conflict with the terms of the agreement between the company and His Majesty's Government signed in London on March 5th 1934, this agreement shall, to the extent of the inconsistency or conflict, be subordinate to and controlled by the terms of that agreement between the Company and His Majesty's Government.'

We have also received replies to our recent telegrams to London, so if your Excellency can receive us to morrow we can discuss them and the above points with Your Excellency then.

We have the honour to be
Your Excellency's obedient servants
Sgd. A. H. T. Chisholm. Frank Holmes.
For KUWAIT OIL COMPANY LIMITED

---

107. *For full details of Traders Ltd. see Note 139.*

---

108. *See Note 134.*

---

109. *Lord Lloyd (1879–1941), Governor of Bombay Province of British India from 1918 to 1923 and British High Commissioner in Egypt from 1925 to 1929, had visited Basrah and elsewhere in the Near and Middle East as Mr George Lloyd, a Conservative Member of Parliament with large business interests, on several occasions before and during the 1914–18 War. From October 1931, when the (predominantly Conservative) National Government succeeded the previous Labour administration in Britain, Lord Lloyd was prominent among the large group of Conservative Members of the House of Commons and the House of Lords who favoured strong right-wing 'Imperialist' policies in foreign affairs. He became Colonial Secretary in Mr Winston Churchill's first government in 1940.*

---

110. *See Note 139.*

---

111. *British Oil Development Ltd. (B.O.D) was registered in London in 1928 with mainly British and Italian shareholding. It was reorganised with increased capital after obtaining from the Iraq Government in May 1932 an oil concession west of the Tigris river; Mosul Oilfields, Ltd., with British, Italian, German, Dutch, French and Iraqi shareholding, was registered in December 1932 to acquire the share capital of B.O.D., while leaving the latter to operate the concession. Drilling operations there proceeded until 1936 when discussions began with Iraq Petroleum Co. Ltd. (see Note 29) which culminated in the latter company taking over B.O.D.'sc oncession in 1941, in which year both B.O.D. and Mosul Oilfields Ltd. were dissolved.*

---

112. *See Note 105.*

---

113. *See Note 127.*

---

114. *The American Legation in Baghdad wrote on 25th June 1934 to the State Department in Washington as follows:*

I have the honour to report that, as anticipated in the final paragraph, page 16, of my despatch No. 302 dated May 3, 1934, Major Frank Holmes has passed through Baghdad en route from Kuwait to London.

From my conversation with Major Holmes during his brief stay here, it appears that the negotiations for the oil concession in Kuwait are now at a standstill. He was frank in saying, but confidentially, that the failure to conclude the negotiations is due entirely to British Government interference; that the Shaikh of Kuwait has been, and still is, ready to come to a definite agreement with the new composite company and to conclude the negotiations on terms already acceptable to him and to the company; but that when the signatures are about to be affixed, the British Government always endeavors to inject some non-essential amendment requiring further consideration and thus drags out the negotiations interminably.

Mr. Chisholme, the representative of the Anglo-Persian Oil Company who first opposed and who is now cooperating with Major Holmes since the formation of the new joint company, also passed through Baghdad a few days ago and gave me his version of the cause of the delay. He states that the difficulty lies with his own company (the APOC) and with the American group represented by Major Holmes; and that these groups will not authorize certain concessions to the Shaikh essential to the conclusion of the negotiations. He said that in all probability he and Major Holmes, after discussing the matter in London with the Anglo-Persian Oil Company, would proceed to New York to consult with the American group.

A third version has been given to me by Colonel Dickson, the British Political Officer at Kuwait, who, with his family, stayed with me during their twenty-four hour stop in Baghdad en route to England. Colonel Dickson says that the delay is due to the Shaikhs demand for more remunerative royalties. The Shaikh maintains that to accept terms less favorable than those secured by the Shaikh of Bahrain and Ibn Saud from the Standard Oil Company of California would make him the laughing-stock of Arab countries and would injure his prestige.

Holmes, who, I know, has the confidence of the Shaikh, insisted, however, that the Shaikh is ready to sign and that he is just as annoyed by the British obstructiveness and interference as he (Holmes) is. Holmes, who is an Australian, had previously told me that he had been forced some time ago to secure the support of the Australian High Commissioner in London in order to prevent the British Government from blocking his negotiations in Kuwait. During his recent visit here he intimated that he would have to thrash out in London for a second time his own position vis-à-vis the British Government.

The versions of Colonel Dickson and Mr. Chisholme seem to coincide in essential particulars. That of Major Holmes, however, leads one to the conclusion that the British Government is possibly endeavoring by obstructionist methods to delay negotiations and so discourage Holmes (in reality the American group he represents) as to cause him to withdraw, thus leaving the concession open to a purely British enterprise—The Anglo-Persian Oil Company.

In the meantime, negotiations are at a standstill; and will not be reopened, at the Kuwait end at least, until next October or November.

Respectfully yours,
P. Knabenshue

Copy to American Embassy, London

---

115. *The texts of Holmes's and Chisholm's cable of 18th July 1934 and letter of 20th July to Shaikh Ahmad from London were as follows:*

We send our respectful greetings and best compliments to Your Excellency and the State of Kuwait. It is now one month since we left Your Excellency's presence but we always remember your kindness and friendship which we look forward to enjoying again when we return to Kuwait. In a letter we are writing all our news and now send our best personal wishes and salutations.

Holmes, Chisholm,
Kuoco, London.

H.E. Shaikh Sir Ahmad al Jaber as Subah,
K.C.I.E., C.S.I.,
Shaikh of Kuwait. 20th July, 1934.
Your Excellency,

We sincerely hope that you are in the best of health and that the affairs of Kuwait go forward with all prosperity and good fortune.

We ourselves arrived here after a comfortable journey and without delay began discussions with your friends of the Kuwait Oil Company in London. The English and American Directors of the Company participated in the discussions which were continued until one week ago. We feel that these discussions have been very helpful. We are preparing things so that we can be ready to return to Kuwait when we receive your command to do so. When that time comes, we sincerely hope and believe that an agreement will be reached between you and our Company. We most earnestly wish for that result.

Mrs. Holmes joins with us in sending most cordial greetings to you. She and Major Holmes are staying in England where the weather this summer is unusually good. Mr. Chisholm is shortly going to France for two or three weeks, returning to London early in August.

If you have any other commands for us besides those which you gave us when we left Kuwait, we shall be very glad to receive them.

In the meanwhile we pray for your good health and prosperity, and look forward to meeting you again.

Your sincere friends,
A. H. T. Chisholm Frank Holmes

---

116. *See page* 54.

---

117. *Chisholm and Holmes left London equipped with the following letter and documents from Kuwait Oil Company:*

Major Frank Holmes,
& A. H. T. Chisholm, Esq.,
Kuwait Oil Company,
LONDON. 24th September, 1934.
Dear Sirs,

The following documents are enclosed herewith for use in your forthcoming negotiations in Kuwait:

(a) One copy of Draft No. 5 of the Concession.
(b) Draft letter to the Shaikh regarding payments in lieu of petrol.
(c) A further letter regarding the Shaikh's petrol supply which is to be used only in the event of your becoming unable to persuade the Shaikh to accept the draft under (b) above.
(d) A draft of Article 3 of the Concession incorporating provisions for royalty on a 'won and saved' basis.
(e) Arguments to be used with the Shaikh in connection with London Representative and the 'exports and sales' basis.

With regard to the draft Concession, you will have the amendments from Draft 4 translated and an entire Draft 5 printed in Cairo. Kindly send us twelve copies of Draft 5 as soon as you have had it printed.

We are anxious that you should reach a settlement with the Shaikh on the petrol supply question by giving him a letter in terms of the draft under (b) above. If you are unable to secure his acceptance of this draft, you should communicate with us before offering the draft provided in (c) or before incorporating any provisions in the Concession.

You should have these two drafts (b) and (c), translated and typewritten in Cairo.

The draft of Article 3 referred to in (d) above is to be translated and printed, but we impress upon you that in no circumstances is it to be adopted or the Shaikh informed of its contents or existence unless you have communicated with us and we have given our approval. Our anxiety is that you should do everything possible to secure royalty on an 'exports and sales' basis as provided in Draft Concession No. 5.

The arguments regarding the Shaikh's representative in London and the 'exports and sales' basis for royalty are to be used as you think fit; but in order that they may be available for use they should be translated and typewritten as memoranda, in Cairo.

The financial terms which you are authorised to offer to the Shaikh are as discussed in London. They are the same as previously practically agreed upon by the Shaikh and the Company with the exception of the combined royalty and tax payment which should be as follows:

Royalty . . . Rs. 2:13) per ton of petroleum
Taxation exemption . . . 3 annas) exported and sold,
or any other division which may be more acceptable to the Shaikh provided the two items do not exceed Rupees 3 in the aggregate.

In addition to the foregoing, we are still prepared to hand to the Shaikh, after a satisfactory concession is signed by him, the sum of Rs. 25,000 in lieu of inviting him to visit London.

No departure whatever is to be made from these terms without prior consultation with and specific authority from us. We enclose 25 copies of a map of Kuwait for use in connection with the Concession. During the forthcoming negotiations we wish to use a secret code for all confidential cables. We enclose herewith a set of slips for use with the Lombard code, together with a sheet of instructions for making up secret cypher signs, the use of which has been explained to Mr. Chisholm before departure.

Yours faithfully,
FOR KUWAIT OIL COMPANY LIMITED
(Sgd.) H. T. Kemp
Secretary

Enclosures:
P.S. We enclose extract from the Chilean Petroleum Law, 1934, which may be of use to you.

KUWAIT CONCESSION—DRAFT NO. 5

(*Incorporating Amendments of 24.9.34*)

This draft differs from Draft 4 (2nd Version) in the following respects:—

Article 1: The word 'crude' has been inserted between 'ozokerite' and 'petroleum'.
Article 3: Alternative wordings are provided to cover 'exports and sales' and 'won and saved' as the basis for royalty.
Article 3(e): of Draft 4 (free petrol supply to the Shaikh) has been omitted.
Article 4: The word 'London' has been omitted before 'Representative'.
Article 6: New.
Article 7(A): From 'Save as in Article 3' to the end is new.
Article 9: The words 'with the cognisance of the Shaikh' inserted in 1st para.
Article 20: New.

IN THE NAME OF GOD THE MERCIFUL

This is an AGREEMENT made at Kuwait on the in the year 193 corresponding to day of 135 between His Excellency Shaikh Sir Ahmed al-Jabir as-Subah, Knight Commander of the Most Eminent Order of the Indian Empire and Companion of the Most exalted Order of the Star of India, the SHAIKH OF KUWAIT, in the exercise of his powers as Ruler of Kuwait on his own behalf and in the name of and on behalf of his heirs and successors in whom is or shall be vested for the time being the responsibility for the control and government of the State of Kuwait (hereinafter called 'the Shaikh') and the KUWAIT OIL COMPANY LIMITED a Company registered in Great Britain under the Companies Act, 1929, its successors and assigns (hereinafter called 'The Company').

Article 1.

The Shaikh hereby grants to the Company the exclusive right to explore search drill for produce and win natural gas asphalt ozokerite crude petroleum and their products and cognate substances (hereinafter referred to as 'petroleum') within the State of Kuwait including all islands and territorial waters appertaining to Kuwait as shown generally on the map annexed hereto, the exclusive ownership of all petroleum produced and won by the Company within the State of Kuwait the right to refine transport sell for use within the State of Kuwait or for export and export or otherwise deal with or dispose of any and all such petroleum and the right to do all things necessary for the purposes of those operations. The Company undertakes however that it will not carry on any of its operations within areas occupied by or devoted to the purposes of mosques sacred buildings or graveyards or carry on any of its operations except the sale of petroleum housing of staff and employees and administrative work within the present town wall of Kuwait.

The period of this Agreement shall be 75 years from the date of signature.

Article 2.

(A) Within nine months from the date of signature of this Agreement the Company shall commence geological exploration.

(B) The Company shall drill for petroleum to the following total aggregate depths and within the following periods of time at such and so many places as the Company may decide:—

(i) 4,000 feet prior to the 4th anniversary of the date of signature of this Agreement.

(ii) 12,000 feet prior to the 10th anniversary of the date of signature of this Agreement.

(iii) 30,000 feet prior to the 20th anniversary of the date of signature of this Agreement.

(C) The Company shall conduct its operations in a workmanlike manner and by appropriate scientific methods and shall take all reasonable measures to prevent the ingress of water to any petroleum-bearing strata and shall duly close any unproductive holes drilled by it and subsequently abandoned. The Company shall keep the Shaikh and his London Representative informed generally as to the progress and result of its drilling operations but such information shall be treated as confidential.

Article 3.

In consideration of the rights granted by the Shaikh to the Company by this Agreement and of the assistance and protection which the Shaikh hereby undertakes to afford by all means in his power to the Company and its operations employees and property the Company shall pay to the Shaikh the following sums:—

(a) Within thirty (30) days after signature of this Agreement Rupees.

(b) On each anniversary of the date of signature until the Company declares that petroleum has been found in commercial quantities:— Either royalty of Rupees Annas for every English ton of 2,240 lbs. of Kuwait petroleum exported from or sold for consumption in Kuwait during the year ended 3 months prior to the anniversary of the date of signature, Or Rupees , whichever shall be the greater sum.

(c) On each anniversary of the date of signature after the Company has declared that petroleum has been found in commercial quantities:— Either Royalty as defined above, Or Rupees , whichever shall be the greater sum.

Article 4.

On each anniversary of the date of signature of this Agreement the Company shall deliver to the Shaikh a return of petroleum if any on which royalty is payable for the year ended three (3) months prior to such anniversary and a statement of the amount of royalty if any due to the Shaikh for such year, and a report of its operations under this agreement during such year. The Shaikh or his Representative shall have the right to check such returns and statements which, as well as any reports, shall be treated as confidential by the Shaikh with the exception of such figures therein as he may be required by law to publish.

Article 5.

(A) For the purposes of its operations hereunder the Company shall have the right without hindrance to construct and to operate power stations, refineries, pipelines and storage tanks, facilities for water supply including boring for water, telegraph, telephone and wireless installations, roads, railways, tramways, buildings, ports, harbours, harbour works, wharves and jetties, oil and coaling stations, with such lighting as may be requisite and any other facilities or works which the Company may consider necessary and for such purposes to use free of all payments to the Shaikh any stone, sand, gravel, gypsum, clay or water which may be available and may be required for its operations hereunder, provided always that the inhabitants of the State of Kuwait are not prevented from taking their usual requirements of these materials and that the water supply of the local inhabitants and nomad population who may be dependent on the same is not endangered. The Company at its discretion but in consultation with the Shaikh may select the position of any such works. The Company may likewise utilise without hindrance all such means of transportation by land, air and water communication or operation as may be necessary for the effective conduct of its operations hereunder.

But nothing in this Article (5A) shall confer on the Company the right to dispose of stone, sand, gravel, gypsum, clay or water by sale, export or otherwise to any other company or person within or without the State of Kuwait.

(B) The Company shall under normal conditions accept and transmit free of charge on its wireless and telegraph installations such of the Shaikh's messages as will not interfere with the Company's business, and in times of national emergency the Shaikh shall have the full use free of charge of the Company's wireless and telegraph installations and railways for governmental purposes.

(C) The Shaikh's ships shall have the right to use harbours utilised or constructed by the Company, provided that such use in no way hampers the Company or interferes in any way with the safety of its operations of which the Company shall be the sole judge. Any wharves or appurtenances constructed by the Company shall be for its exclusive use.

The Company may use for the purposes of its operations the harbours along the coast of Kuwait but the Company shall not impede or interfere with the subjects of the Shaikh or their right to continue the use of existing harbours, anchorages, wharves and docks along the coast of Kuwait at present utilised by them for their sailing craft and fishing boats.

Article 6.

(A) The Company shall from time to time designate its General Manager or one of its other principal employees in Kuwait as its Chief Local Representative, to represent it in matters relating to this Agreement with the Shaikh.

(B) The Shaikh shall have the right to appoint an Arab conversant with the English language to act as his official Representative and who will represent him in Kuwait in matters relating to this Agreement with the Company and particularly whenever unskilled labour is recruited from among the subjects of the Shaikh the advice and the assistance of this representative shall be available to the Company. The Company shall pay to him monthly the sum of Rupees Eight Hundred (Rs. 800); and such payment shall be a complete discharge of all obligations on the part of the Company in respect to such official Representative for salary or expenses or otherwise.

(C) The Shaikh shall have the right to appoint a person, who may be either a Kuwait, British or American subject, as his Official Representative, to represent him in London in matters relating to this Agreement with the Company. Such representative shall be entitled to receive from the Company current information concerning matters which affect the Shaikh's interests under this Agreement, and to receive notice of and be present at all meetings of the directors of the Company which may be convened to consider any question which has arisen between the Shaikh and the Company under this Agreement. The Company shall pay to such London Representative monthly the sum of Rupees Two Thousand (Rs. 2,000) and such payment shall be a complete discharge of all obligations on the part of the Company in respect to such London Representative for salary or expenses or otherwise.

Article 7.

(A) The Company shall have the right to import water, petroleum, fuel, machinery, motor-cars, and lorries, equipment, plant, timber, utensils, iron work, building materials, food, supplies, medicines, medical supplies, office equipment and household furniture, and all other materials, equipment and goods of whatsoever nature required by the Company and its employees for the purposes of its operations hereunder but not for resale to others, and to export its petroleum and articles previously imported by the Company free of customs or import or export duty and taxes or other charges, but it shall pay on all personal goods, clothing and general merchandise imported by the Company for sale to its employees for their personal use, the ordinary duty in force for the time being in the State of Kuwait. In addition to the payments provided for in Article 3 and in any other articles of this Agreement the Company shall pay to the Shaikh on each anniversary of the date of signature of this Agreement, in lieu of all present or future harbour duties taxes imposts and charges of any kind whether state or local upon or in respect of its operations income profits and property including petroleum, and in lieu of tolls and land surface rent of whatever nature,   annas per ton (2,240 lbs.) of petroleum on which royalty is payable.

(B) The importation by the Company of firearms and other weapons is prohibited except with the written permission of the Shaikh.

(C) If the Company should sell in Kuwait any material or goods previously imported into Kuwait for the purposes of its operations hereunder and no longer required by the Company, the Company shall pay to the Shaikh in respect of such material or goods sold the equivalent of import duty thereon at the rate in force at the time of sale. The duty shall be computed on the price received on sale.

(D) Necessary customs officials at harbours constructed by the Company or additional customs officials required at any other ports utilised by the Company shall be appointed by the Shaikh in consultation with the Company and their salaries which shall not exceed the usual salaries of such officials shall be paid by the Company which shall also provide at its own expense suitable buildings for the accommodation of customs officials at harbours which it has constructed.

Article 8.

(A) The Company shall have the right to purchase at current market rates fuel, water, food, building and constructional materials and other supplies of every kind in connection with its operations hereunder.

(B) The Company shall employ subjects of the Shaikh as far as possible for all work for which they are suited under the supervision of the Company's skilled employees, but if the local supply of labour should in the judgment of the Company be inadequate or unsuitable the Company shall have the right with the approval of the Shaikh which shall not be unreasonably withheld to import labour preference being given to labourers from the neighbouring Arab countries who will obey the local laws. The Company shall also have the right to import skilled and technical employees. Any employee imported by the Company who shall by misconduct cause a breach of peace or public disturbance shall at the request of the Shaikh be dismissed and shall if it is within the power of the Company to do so be sent out of Kuwait. The Company shall pay to the workmen it employs a fair wage, such wage to be decided and stated by the Company at the time the workmen are engaged.

(C) The Company shall provide free of charge medical service for its employees, and the Shaikh and his family shall have the right to such medical service and necessary medical supplies free of charge.

Article 9.

The Shaikh grants to the Company free of cost the unrestricted use and occupation of and surface rights over all uncultivated land belonging to the Shaikh which the Company may need for the purposes of its operations and in particular the Company shall have the right to select in consultation with the Shaikh an area or areas of land chosen by the Company outside the present town wall of Kuwait with exclusive surface rights upon which to erect oil refineries, storage, terminal and shipping facilities and any other works required for the Company's operations; and the Company may with the cognisance of the Shaikh buy or lease for such purposes any lands, houses or buildings with the consent of and on conditions to be arranged with the proprietors thereof but the terms of such purchase or lease shall not be in excess of those ordinarily current in their respective localities.

The Company shall acquire only such land, houses and buildings as are necessary for its operations under this Agreement. The Company shall inform the Shaikh from time to time of the land, houses and buildings which it requires to occupy for its operations; and land houses and buildings previously acquired by the Company from the Shaikh but found no longer necessary for its operations shall be returned by the Company to the Shaikh free of charge.

Article 10.

The Shaikh shall give to the Company and its employees and property all the protection in his power from theft, highway robbery, assault, wilful damage and destruction, and the Company may appoint in consultation with the Shaikh and itself pay trustworthy guards who shall at all times be Kuwait subjects unless the Shaikh permits otherwise to assist in protecting the property of the Company and its employees. The Company shall erect at its own expense suitable buildings for the accommodation of such guards at such places as the Company shall decide.

Article 11.

(A) Before the expiration of the period specified in Article 1 hereof this Agreement shall come to an end either by surrender as provided in paragraph (B) of this Article or in Article 12 or in one of the three following cases:—

(a) If the Company shall fail to fulfil its obligations under Article 2 hereof in respect of geological exploration or drilling.

(b) If the Company shall fail within six (6) months after any anniversary of the date of signature of this Agreement to make to the Shaikh any payments agreed to be due under Article 3.

(c) If the Company shall be in default under the arbitration provisions of Article 18.

In any of the above mentioned cases the Shaikh shall be entitled to terminate this Agreement and all the property of the Company within the State of Kuwait shall become the property of the Shaikh.

(B) In the event of the Company failing to make the declaration provided in Article 3 within 12 years of the date of signature of this Agreement the Company shall at its option either pay to the Shaikh the minimum annual payment provided in Article 3(c) or surrender all rights under this Agreement.

Article 12.

(A) The Company shall have the right at any time after it has drilled the 4,000 feet provided in Article 2(B)(i) or after the expiry of 2 years from the date of signature of this Agreement—whichever shall be the later date—to give the Shaikh one year's notice in advance to terminate this Agreement and the Company shall on expiry of such notice have no further liabilities except to make payment of all monies which may be due to the Shaikh up to the date of termination.

(B) Should this Agreement be terminated by the Company under this Article 12, then:—

(a) If such termination occurs within 35 years from the date of signature of this Agreement all lands granted by the Shaikh and any lands or buildings which the Company may have bought and any houses or buildings constructed by and other immovable property of the Company within the State of Kuwait shall be handed over to the Shaikh free of cost. Producing wells or borings at the time of such termination shall be handed over in reasonably good order and repair.

But

(b) If such termination occurs after 35 years from the date of signature of this Agreement all the movable and immovable property of the Company in the State of Kuwait shall be handed over to the Shaikh free of cost. Producing wells or borings at the time of such termination shall be handed over in reasonably good order and repair.

Article 13.

On the Expiry of this Agreement at the end of the period of 75 years provided in Article 1 or of any extension or renewal of that period all the movable and immovable property of the Company in the State of Kuwait shall be handed over to the Shaikh free of cost. Producing wells or borings at the time of expiry shall be handed over in reasonably good order and repair

Article 14.

The Shaikh hereby agrees that the Company may transfer the obligations and benefits of this Agreement to any Company registered within the British Empire.

Article 15.

(A) Nothing in this Agreement shall be read as restricting in any way the right of the Shaikh to grant to other parties concessions or permits for substances other than petroleum provided that the operations and rights of the Company hereunder are not thereby injuriously affected.

If the Shaikh should at any date subsequent to the date of signature of this Agreement grant to any other parties concessions or permits for substances other than petroleum, the Shaikh undertakes that such concessions shall contain provisions requiring the holders thereof to abstain from damaging impeding or interfering with the property operations and interests of the Company.

Deposits of mineral substances other than petroleum such as gold, silver, copper, lead, potash, sulphur and salt or the like which may be discovered by the Company shall be reported to the Shaikh and shall not be worked by the Company except under a special concession or permit from the Shaikh.

(B) The Company shall use the Shaikh's flag within the State of Kuwait.

Article 16.

Failure on the part of the Company to fulfil any of the conditions of this Agreement shall not give the Shaikh any claim against the Company or be deemed a breach of this Agreement in so far as such failure arises from *force majeure*, and if through *force majeure* the fulfilment by the Company of any of the conditions of this Agreement be delayed the period of such delay shall be added to the periods fixed by this Agreement.

Force majeure as used in this Agreement includes the act of God, war, insurrection, riot, civil commotion, tide, storm, tidal wave, flood, lightning, explosion, fire, earthquake, and any other happening which the Company could not reasonably prevent or control.

Article 17.

The Shaikh shall not by general or special legislation or by administrative measures or by any other act whatever annul this Agreement except as provided in Article 11. No alteration shall be made in the terms of this Agreement by either the Shaikh or the Company except in the event of the Shaikh and the Company jointly agreeing that it is desirable in the interest of both parties to make certain alterations, deletions or additions to this Agreement.

Article 18.

(A) If at any time during the currency of this Agreement any difference or dispute shall arise between the parties hereto concerning the interpretation or execution hereof, or anything herein contained or in connection herewith, or the rights or liabilities of either party hereunder, the same shall, failing any agreement to settle it in any other way, or after consultation with the British Political Agent in Kuwait or the British Political Resident in the Persian Gulf, be referred to two arbitrators, one of whom shall be chosen by each party, and a referee, who shall be chosen by the arbitrators before proceeding to arbitration.

(B) Each party shall nominate its own arbitrator within 60 days after the delivery of a request so to do by the other party failing which its arbitrator may at the request of the other party be designated by the British Political Resident in the Persian Gulf. In the event of the arbitrators failing to agree upon the referee within 60 days after being chosen or designated, the British Political Resident in the Persian Gulf may appoint a referee at the request of the arbitrators or either of them.

(C) The decision of the arbitrators, or in case of a difference of opinion between them the decision of the referee, shall be final and binding upon both parties.

(D) In giving a decision the arbitrators or the referee shall specify an adequate period of delay during which the party against whom the decision is given shall conform to the decision and that party shall be in default only if that party has failed to conform to the decision prior to the expiry of that period and not otherwise.

(E) The places of arbitration shall be such as may be agreed by the parties and in default of agreement shall be London.

Article 19.

The Company shall make all payments that become due to the Shaikh under this Agreement into the Shaikh's account at the   Bank in   and the Bank's receipt shall be a full discharge for the Company in respect to the payment of the sum stated in the Bank's receipt. The Shaikh may from time to time designate in writing another Bank or Banks for the purpose of this Article.

Article 20.

For the purpose of royalty payments the Company shall measure by a method customarily used in good technical practice all petroleum on which royalty is payable and the Shaikh by his representative duly authorised by him shall have the right to observe such measuring and to examine and test whatever appliances may be used for such measuring. Such representative shall comply with all necessary and usual safeguards for the prevention of fire or other accident; and shall make all examinations and tests at such times and in such manner as will cause the minimum of interference with the Company's operations. If upon such examination or testing any such appliance shall be found to be out of order the Company will cause the same to be put in order at its own expense within a reasonable time, and if upon any such examination as aforesaid any error shall be discovered in any such appliance, such error shall if the Shaikh so decide after hearing the Company's explanation be considered to have existed for three (3) calendar months previous to the discovery thereof or from the last occasion of examining the same in case such occasion shall be within such period of three (3) calendar months and the royalty shall be adjusted accordingly. If the Company should find it necessary to alter repair or replace any measuring appliance it shall give reasonable notice to the Shaikh or his representative to enable a representative of the Shaikh to be present during such alteration, repair, or replacement.

The Company shall keep full and correct records of all measurements as aforesaid and the said representative of the Shaikh shall have access at all reasonable times to such records, and shall be at liberty to make extracts from them. Such records shall be treated as confidential by the

Shaikh and his representatives with the exception of such figures therein as the Shaikh may be required by law to publish.

Article 21.

This Agreement is written in English and translated into Arabic. If there should at any time be disagreement as to the meaning or interpretation of any clause in this Agreement the English text shall prevail.

In witness whereof the parties to this Agreement have set their hands the day and year first above written.

ON BEHALF OF THE KUWAIT OIL COMPANY LIMITED.	SHAIKH OF KUWAIT
IN THE PRESENCE OF	IN THE PRESENCE OF

Kuwait Oil Company
Letter to the Shaikh Regarding Free Petrol

Your Excellency,

With reference to the Agreement signed today between Your Excellency and Kuwait Oil Company Limited and to Your Excellency's request for supplies of petrol for your requirements, we have the honour to inform you that, during the continuance of that Agreement, in lieu of supplying petrol or other petroleum for those requirements the Company will pay to Your Excellency within thirty (30) days of the date of signature of the Agreement and on each anniversary of the date of signature until the Company declares that petroleum has been found in commercial quantities the sum of Rupees Five Thousand (Rs. 5,000); but on each anniversary of the date of signature after declaration by the Company that petroleum has been found in commercial quantities or after the expiration of the 12 years mentioned in Article 11(B) of the Agreement, whichever shall be the earlier date, such payment shall be increased to Rupees Ten Thousand (Rs. 10,000).

We should be glad if Your Excellency would acknowledge your acceptance of the terms of this letter.

22nd September, 1934.

Kuwait Oil Company
Alternative Letter to the Shaikh Regarding Free Petrol
(See (c) in letter of 24th September 1934 to
Messrs. Holmes and Chisholm).

Your Excellency,

With reference to the Agreement signed to-day between Your Excellency and Kuwait Oil Company Limited and to Your Excellency's request for supplies of petrol for your requirements, we have the honour to inform you that, during the continuance of that Agreement, within thirty (30) days after signature of the Agreement and on each anniversary of the date of signature until the Company declares that petroleum has been found in commercial quantities, the Company will pay to Your Excellency the sum of Rupees Five Thousand (Rs. 5,000). Between each two consecutive anniversaries of the date of signature after the declaration by the Company that petroleum has been found in commercial quantities, or after the expiration of the 12 years mentioned in Article 11(B) of the Agreement whichever shall be the earlier date, the Company will deliver to Your Excellency free of cost at a suitable storage place within the town wall of Kuwait such quantity of petrol as may be required for the personal use of Your Excellency and your family, not however exceeding in the aggregate in any yearly period Ten Thousand (10,000) English gallons of which not more than Five Thousand (5,000) gallons may be aviation spirit. Any petrol or aviation spirit imported by the Company for free supply to Your Excellency under the terms of this letter shall be free of import or other duties or taxes.

We should be glad if Your Excellency would acknowledge your acceptance of the terms of this letter.

24/9/34.

Article 3.

In consideration of the rights granted by the Shaikh to the Company by this Agreement and of the assistance and protection which the Shaikh hereby undertakes to afford by all means in his power to the Company and its operations employees and property the Company shall pay to the Shaikh the following sums:—

(a) Within thirty (30) days after signature of this Agreement Rupees .

(b) On each anniversary of the date of signature until the Company declares that petroleum has been found in commercial quantities:—
Either Royalty of Rupees for every English ton (2,240 lbs.) of nett crude petroleum won and saved by the Company in Kuwait during the year ending 3 months prior to the anniversary of the date of signature, Or Rupees , whichever shall be the greater sum.

(c) On each anniversary of the date of signature after the Company has declared that petroleum has been found in commercial quantities:—

Either Royalty as defined above, Or Rupees   , whichever shall be the greater sum.

(d) For the purpose of this Agreement and to define the exact quantity to which the Royalty stated above refers, it is agreed that the Royalty is payable on each English ton of 2,240 lbs. of nett crude petroleum won and saved by the Company from within the State of Kuwait, after deducting:
   (i) water, sand or other foreign substances contained in or mixed with the petroleum;
   (ii) petroleum required for use by the Company in its installations and operations in the State of Kuwait;
   (iii) petroleum lost in any process of treatment refining or handling of petroleum that may be carried on within the State of Kuwait;
   (iv) petroleum which may be returned to the underground natural reservoirs through the Company's borings.

Argument to be Presented to Shaikh of Kuwait
(Article 6(C))

While we were in London we discussed many times and very fully with our Directors Your Excellency's wish to have a representative in London during the early years of the concession, and later to have the right to appoint one member to the Board of Directors of the Company, having equal status and privileges with the other members of the Board.

We found that our Directors considered it quite natural, under the circumstances, that Your Excellency should desire to have a representative in London and would be very glad to welcome such a representative and pay to him such a sum monthly as they think would be a liberal allowance to cover his travelling and other expenses and salary.

Our Directors would be quite willing also to supply such London representative currently with information concerning matters which might affect your interests under the concession and to invite him to attend all meetings of the Directors of the Company which may be convened to consider any question which has arisen between Your Excellency and the Company under the Concession.

Our Directors cannot accept, however, that Your Excellency's representative should have full access to the records of the Company, including the Agenda of the Board Meetings and for this they have stated to us several reasons. One is that the records of the Company will naturally and necessarily contain much information regarding materials, construction and other matters which would be of no interest to Your Excellency and information concerning personnel, sales of oil, and other things which in their nature are private and confidential. Another is that if Your Excellency's representative should have the right to full access to the records of the Company and should exercise such right to a great extent, the records of the Company might become confused or lost and might not be readily available when wanted by the Company's officials and employees. In this manner normal conduct of the Company's business at its office in London would be disarranged.

Neither are our Directors able to accept Your Excellency's proposals that you should have the right to appoint a member to the Board of Directors of the Company.

As Your Excellency knows, the Company is a combination of two interests, each having equal rights in all respects in the affairs of the Company. The Board of Directors consists of six members, three of which are appointed by each of the two interests; so that all business transacted by the Board of Directors must receive the approval of directors representing each of the two interests. Your Excellency will fully understand from this explanation that it would be impossible for our Directors to accept the appointment by Your Excellency of a member who would have a deciding vote in the event that one of the interests of which the Company is composed would be in favour of a certain resolution and the Directors representing the other interest in the Company being opposed. In such a situation Your Excellency would, in effect, through a Director appointed by you, be determining the policies and managing the affairs of the Company.

Much as our Directors regret their inability to meet Your Excellency's wishes on this point, they see no possible way, considering the manner in which the Company is constituted, in which it can be done.

Our Directors have, therefore, set forth in Article 6(C) of the new draft concession which we have recently presented to you a provision for a London representative to whom the Company will supply information which we believe will be found entirely adequate for Your Excellency's purposes.

We have discussed this matter at great length with our Directors and the conclusion which they have reached has been arrived at after most careful consideration and study by them; and as the draft provision represents the final views of our Directors we hope it will be acceptable to Your Excellency.

September 20, 1934.

---

118. *The texts of Holmes's cable of 16th and letter of 15th–16th October 1934 from Kuwait to Gulf in London were as follows:*

NOTE 118

From Kuwait 16 October to London
Absolutely strictly confidential THIS TELEGRAM IS FOR STEVENS FULLSTOP Consider that position here very critical in respect of the Kuwait oil concession FULLSTOP Influence from a quarter you would not dream of has been active during our (my) absence therefore do not take any action whatever in connection with Kuwait business until you have received my confidential letter addressed to Stevens through Mrs. Frank Holmes Chelmsford, Essex by airmail due in London about 23rd October—HOLMES

STRICTLY CONFIDENTIAL

KUWAIT
(*Persian Gulf*)
October 15th, 1934.

Guy Stevens, Esq.,
Savoy Hotel,
Strand,
LONDON, W.

Dear Stevens,

I have seen the Shaikh of Kuwait once since our arrival in Kuwait two days ago. The visit took the form of a social rather than a business visit. I stayed two days in Basrah as we arrived there before the 12th October that being the day that the Board wished us to arrive in Kuwait.

Several of the important men of Kuwait have called upon me and several others have sent me messages. The information that the former told me direct and what the latter conveyed to me through a third person seem to be based on very decided conclusions both have arrived at during our absence from Kuwait.

Before proceeding to give you further views on what I gleaned from the above mentioned information of the present aspects of affairs in Kuwait I will tell you of the things I heard in Basrah. The acting British Consul, an old friend of mine, in Basrah (who in real life is the manager of the Basrah branch of the Eastern Bank) invited me to lunch at the Consulate; besides the Consul the local English judge of the Basrah High Court was present making a party with myself of three. I have known the judge for some time but was not over friendly with him as I was not sure of his attitude to me in the role of an opponent of the APOC.

The luncheon was rather prolonged and naturally the question of the Kuwait Oil Concession came up. The judge talked rather freely and in a friendly manner and asked, 'Holmes, do you think for one moment that you and Chisholm will be permitted to secure the Kuwait Oil Concession for a joint Anglo-American group?' I replied 'why not'. He then made the startling statement that a new Company had practically agreed a Concession Agreement with the Shaikh of Kuwait much on the same lines as the Anglo-American group, but with higher initial payments and higher royalty.

When I reached Kuwait my friends told me the same story as the judge with the additional information that the large Oil Company behind the new Company was the APOC and their friends. Further that *Gabriel* the Iraqi lawyer had made several visits to the Shaikh of Kuwait during our absence in England. He also had long and private interviews with the Shaikh of Kuwait and that letters had passed between them at frequent intervals.

Further, Gabriel was acting on behalf of and under instructions of *Khan Bahadur Mirza Mohammed*—a Persian subject turned Iraqi subject—who has been for many years the principal legal adviser (local) to the APOC. He is rated as a very brainy and clever lawyer. He is still the legal adviser to the APOC.

Further, Khan Bahadur Mirza Mohammed recently spoke to my friend Mr. Mohammed Yateem and tried to bribe him to join a Company who could assure his future and offered him alluring terms to quit me and join up with him. Mr. Mohammed Yateem refused, but heard enough from the Khan to realize that the APOC were playing a game that boded ill for those connected with American side of the Kuwait Oil Concession if the scheme proved successful.

Further, I sent for the Shaikh's Private Secretary the most powerful man in the State, but ever short of cash. He is a loyal friend of ours (you will remember the letter I received while in England asking for a loan of Rs. 10,000).

(a) I asked him to tell me exactly what he had heard of Gabriel's Company.
(b) And his opinion as to who was behind it.
(c) And the Shaikh's attitude to the Gabriel's Company.

I will now give his replies, but, he requested me to impress upon my American friends that what he is doing in giving this information to me (Frank Holmes) is very dangerous, and if it were to come out, he would be discharged at once. I promised him that the Americans could be fully relied upon to respect his wishes and that I would guarantee that no harm reached the *Mullah* from that direction.

(a) He told me 'The Trade Arts Company Limited' was the name of the Company that Gabriel was representing.

(b) That Gabriel assured the Shaikh that his Company had most powerful support both official and financial. The Shaikh's Secretary informed me that part of such support—if not the principal part—was that given by the APOC people or their friends.

(c) The Shaikh, the Secretary said, was listening to all Gabriel had to say and the Shaikh had written as recently as yesterday to Gabriel, telling him that as the negotiators of the Kuwait Oil Company Limited had now arrived that he would be hearing from them whether the Kuwait Oil Company Limited accepted in full the terms demanded by the Shaikh in June; if the Company accepted it *in toto* then he could have nothing to say to Gabriel's Company, but if the Kuwait Oil Company did not accept in full and desired further discussion, he would first take the opportunity of hearing the final terms of Gabriel's Company and if satisfied would negotiate a Concession with him. The Shaikh also informed Gabriel that he would require to know who were backing Gabriel's Company.

The Shaikh's Secretary is not certain how much the Shaikh knows of the people behind Gabriel, but the Shaikh does know that Khan is directing Gabriel. The Secretary knows because of Gabriel's efforts to enlist the good-will and active support of (Mulla Saleh) the Secretary on behalf of the new Company.

In the earlier part of this year when Chisholm and I were negotiating I cabled the Company that 'IT WOULD BE UNWISE TO ASSUME AT THIS STAGE THAT THE SHAIKH OF KUWAIT HAS GIVEN UP THE IDEAL OF DEALING WITH OTHER OIL CONCERNS' (Ref. tel No. 13 of 26–3–34). This advice was poohpoohed by both you and naturally by the APOC.

I also cabled later to the Company advising that the terms as per Draft No. 4 be accepted. I gave this advice because of an intangible atmosphere of a baneful influence hovering around our negotiations, whenever our negotiations reached such a stage that early agreement was indicated, Gabriel appeared on the scene and the smooth course of our negotiations was interrupted. I am still much of the same opinion, but I felt in London that you (Americans) were accepting the good intentions and fidelity of the APOC to the extent of 100%. I did not have enough to go on while in England to make it worth while for me to try and make you a little less trustful as to the unspotted integrity and altruistic intentions of your 'friends'.

I was asked by my Arab friends even at that time (June) whether I was satisfied that the APOC were dealing in a straight forward manner with the American group? I could not lay my finger on any one definite action but could not but feel that there was not perfect harmony. Now I am certain where the trouble lies, and to use Mr. Wallace's saying 'I KNOW THE NATURE OF THE ETHIOPIAN IN THE WOODPILE' and I hope you will after you have read this letter.

October 16th, 1934.

Since writing the above the Shaikh of Kuwait made a return call for our call of yesterday. He came at 9 a.m. and stayed till 10:30 a.m. I asked him several questions about local affairs and finally worked up to the point of asking the Shaikh of how affairs stood as between him and Gabriel. He replied that he had been subject to much pressure by the Political during our absence in England, much of which he could not talk to me about. The Shaikh said that he has avoided trying to find out who were behind Gabriel's Company as he wished to retain an open mind until the Kuwait Oil Company Limited had accepted or rejected the terms he agreed to accept just prior to our departure.

He also said that further discussions with the Kuwait Oil Company Limited was useless as the Kuwait Oil Company Limited were looking for more favourable terms than adjoining countries had arranged. He gave us as an example the Standard Oil Company of California's terms to Ibn Saud in the Hasa Agreement. That Agreement has certainly required a deal of explaining away.

If we are to prevent being double crossed my opinion is that the terms as per Draft No. 4 should be accepted by the American group. The Shaikh is not inclined to discuss the Concession further because he wants to secure the best terms and of course the largest initial payment possible and who can blame the Shaikh. He is an Arab Ruler with a clamoring horde of relations ever demanding money from the Shaikh.

The following sentence is difficult to write: I do not believe that any person or persons whether political or otherwise desire the Kuwait Oil Company Limited to secure an Oil Concession from the Shaikh of Kuwait, except those persons long connected with the American group The APOC certainly are praying for a break down in our negotiations and living in hopes of Gabriel Co's success. Your associates are certainly beyond the pale.

There is an old saying 'that those that are not for us are against us'. The English when engaged in a more or less questionable bit of work can always manage to salve their consciences by deluding themselves that what they are doing is for the benefit of their country and therefore a virtue. I think that is what is happening out here. Disraeli once declared 'My Country right or wrong I hope she may always be right but either way still my Country.'

You once spoke much to me about stampeding. I suppose that you referred to my cables to London urging to accept the Shaikh's terms. I was not stampeding then, nor am I now.

The terms are still the same as I recommended acceptance of and judging by the attitude of the Shaikh today it will be difficult to move. I again say that you should now accept the No. 4 Draft terms, after all the terms bear favourable comparison with those areas around the Shaikh's territory. All my Arab friends and advisers consider the situation critical and are convinced that our colleagues are hopeful for a break down in our negotiations and a direct refusal by the Shaikh of our No. 5 Draft terms. So that Gabriel's Company could step in and obtain the Concession or be the cause of endless delay.

Further you should in my opinion force the members of Kuwait Board of Directors to accept the terms as laid down in Draft No. 4. If you do not accept my advice you will play into the hands of those who are not our friends, no matter how long they profess to the contrary.

This letter must be treated as 100% confidential—if the British Government became aware that I had given the game away I would be in a very awkward position. However I am certain of your loyalty and discretion.

October 16th 1934 6 p.m.

Since writing the above I have been in touch with another member of the Council of State, the second in importance to the Secretary but a relation of the Shaikh, he has given us much information and as he was being consulted by the Shaikh, as to how to reply to the enclosed letter from Gabriel. The reply was drafted with our help and has gone off.

Our arrangements made with our two Arab friends is that we are informed and see the contents of letters coming from Gabriel and replies going to Gabriel. By this means we are able to keep abreast of Mr. Gabriel & Co., and check the movements of the Shaikh. The following are the gists of the two letters:—

(A) being from Gabriel to the Shaikh of Kuwait and

(B) being the Shaikh's reply to Gabriel.

A. 'It will be a fault on Your Excellency's part if you will now let the Kuwait Oil Co's representatives resume their negotiations on the basis of their old terms, because now that Company has no right to do so having contradicted their promise to return in September and settle the matter with Your Excellency. And it will be a right action on Your Excellency's part to give a chance to my Company by giving a consideration to their Concession terms, which are agreed by H.M. Government and sanctioned to negotiate with Your Excellency.'
Received by the Shaikh 16/10/34

B. 'Replied by the Shaikh on 17–10–34. We are sorry to inform you that H.M. Government did not communicate with us about your Company's desire neither prior nor after the arrival of the Kuwait Oil Co's negotiators, now they are on the spot and ready for negotiations. We cannot give consideration to your application until their negotiations with ourselves have come to an end with unsatisfactory result.'

One thing I would like to mention is that no word should reach the Shaikh direct from an American Director of the Kuwait Oil Company Limited through the Political Agent of Kuwait. This has happened (no doubt with quite harmless intent) as Major Davis saw Col. Dickson at Fenchurch Street when latter was seeing his wife off to Kuwait and sent the following message through Dickson to the Shaikh—'Tell the Shaikh of Kuwait that though I have not had the pleasure of meeting His Excellency that he (Major Davis) would see and put right any point the Shaikh of Kuwait objected to in the Draft Concession about to be presented to the Shaikh.' The Shaikh of Kuwait was surprised that 'one of Major Holmes own Directors should use the Political Agent' as a medium of sending messages direct to him (the Shaikh).

This all means little but one never knows how the Shaikh will view these outside messages. All messages should come through the correct channel if at all. The position here is such that I am confronted with a most difficult task, more difficult than at any time since we have been working in Kuwait. I have added to my trouble perfidy and deceit and defection from such a quarter.

I have not given the slightest hint to Chisholm that I know of the perfidy of his crowd, nor do I until I hear from you.

Do not fail to advise me of the arrival of this letter into your hands and please treat this as confidential to all apart from your own people. My wife will ring up and tell you that the letter for you has arrived at 'Millhill' so please notify me of its safe arrival.

With best wishes,

Yours sincerely,
(Sgd.) Frank Holmes

---

119. *Mirza Mohammed (1885-1974) died recently in Basrah, Iraq; since becoming an Iraqi subject he has changed his name to Muhammad Ahmad. In 1934, when he still retained his original Persian nationality, his full name was Khan Bahadur Mirza Muhammed, C.I.E., LL.B.*

Because of the references to him in Holmes's letter of 15th–16th October 1934, Mr Chisholm wrote to Muhammad Ahmad from Kuwait on 19th June 1971 asking if, among his recollections of the Kuwait concession negotiations in and before 1932–34 (he was legal adviser in Basrah to APOC from 1921 to 1945) he recalled a company named Traders Ltd. as having been connected with the negotiations at any time. In his reply dated 26th June 1971, Muhammad Ahmad said he could remember no such company but detailed various recollections of the negotiations from 1922 onwards (he had been with Sir Arnold Wilson at his interview with Shaikh Ahmad on 2nd June 1923; see page 6 and Note 18).

On 26th July 1971 Mr Chisholm again wrote to Muhammed Ahmad, asking him for his comments on the statements in Holmes's letter of 15th–16th October 1934 that at that time Gabriel was acting for Traders Ltd. 'on behalf of and under instructions of' himself, and also that he had spoken to Mohammed Yatim 'and tried to bribe him' to quit Holmes and join up with him. In Muhammad Ahmad's reply, dated 30th August 1971, he wrote 'I was astonished to read about the false reports purporting my employment of Yatim for the interests of Traders Ltd. and I hasten to repudiate them most strongly', and 'as regards Traders Ltd. I have heard this name only this year' (i.e. from Mr Chisholm's letter to him of 19th June 1971). This letter from Muhammad Ahmad also contained various personal recollections concerning the 1932–34 negotiations and mentioned that his colleague Gabriel who 'was the attorney for' Shaikh Ahmad, had died several years ago.

There can be no doubt that the statements in Holmes's letter of 15th–16th October 1934 connecting Muhammad Ahmad (then Mirza Muhammed) with Traders Ltd. were untrue. As detailed later in this Record, all that letter's statements connecting APOC with Traders Ltd. were also untrue.

---

120. *See Note 96.*

---

121. *The negotiators sent the following report to KOC on 17th October 1934:*

KUWAIT
17th October 1934

Messrs. Kuwait Oil Company Limited
LONDON
Dear Sirs,

Further to our letter dated Cairo 6th October, we beg to submit the following report.

We arrived at Gaza on 8th October where we remained until a.m. on the 10th. Through the good offices of Mr. Cross (Manager Imperial Airways, Cairo) we were then able to get passages to Basrah for ourselves and a certain amount of luggage. The balance of luggage, including the radio set for presentation for the Shaikh, had to be left in Gaza for forwarding by a later plane.

We arrived in Basrah p.m. on 10th October, and confirm despatch of our telegram of 11th and receipt of your No. 26 in reply. In accordance with the latter we proceeded to Kuwait on 13th October, as advised in our telegram dated 14th.

We called on the Shaikh on 15th October, and were cordially received. He returned our calls on the 16th, and we dined with him the same evening.

As regards the question of the Company's Chief Local Representative, we understand from the Political Agent that the Shaikh has informed him that the solution put forward by the India Office will be approved by him, if and when he agrees to grant the concession to our Company. This advice corresponds with the information contained in your telegram No. 27, which we have just received.

We had our first business interview with the Shaikh this morning, when we handed him new draft No. 5. He stated that he would discuss it with us after he had studied it, but this is unlikely to be before next week, as the Political Resident and family are due to arrive here today, and will occupy the Shaikhs attention until Sunday.

A. H. T. Chisholm
Frank Holmes

---

122. *The negotiators' report, 24th October:*

KUWAIT
24th, October 1934.

Messrs. Kuwait Oil Company Limited.
Britannic House,
Finsbury Circus,
London, E.C.2.
Dear Sirs,

Further to our letter dated 17th October, we have to confirm receipt of your telegram No. 28 and despatch of our No. 34 in reply. The general situation on our return showed no appreciable change since our departure. The Shaikh remains most cordial to ourselves and repeats that he would prefer our Company to have his Concession rather than any other. He still states however that other companies are anxious to deal with him if we do not comply with his demands.

There is still no development to report, neither has the Shaikh yet indicated what his reaction is to be to our draft No. 5 as presented to him on 17th October. Yesterday he wrote asking us for a copy of draft No. 4 so that he could compare it with draft No. 5. Mr. Chisholm wished to discuss with Major Holmes the possibility of avoiding this request, but Major Holmes sent the draft to the Shaikh, being of the opinion that the least delay in so doing might be extremely prejudicial.

Our luggage, including the Shaikh's radio set, has arrived in Kuwait, and the latter will shortly be installed by an electrical engineer who is kindly being provided by APOC Abadan.

Col. Dickson has resumed his duties as Political Agent as from 18th October.

Yours faithfully.
A. Chisholm   Frank Holmes

---

123. *The negotiators' report, 31st October:*

KUWAIT,
PERSIAN GULF.
31st, October 1934.

Messrs. Kuwait Oil Company Ltd;
Britannic House,
Finsbury Circus,
London, E.C.2.
(England.)
Dear Sirs,

Further to our letter of 24th October, we regret that we have as yet no development in our negotiations to report.

The Shaikh indicated that he would see us today to discuss our latest draft, but he has now put us off until Saturday 3rd November. We hope to telegraph you some news of the situation after seeing him then. So far we have been able to gather no indication whatsoever of his intentions.

Yours faithfully.
A. Chisholm Frank Holmes

---

124. *Details of Holmes's private interview with the Shaikh were cabled to Stevens in London by Holmes on 25th October:*

CONFIDENTIAL This is for Stevens FULLSTOP I have had at my house two hours interview with Shaikh of Kuwait this morning STOP Shaikh wrote to Chisholm and Holmes requesting copy of the Kuwait Oil Concession draft including terms as Shaikh presented to us in June which later formed our #4 Draft STOP Chisholm objected to give Shaikh a copy but we (I) sent Shaikh a copy as our important Arab friends insisted that an impasse would be result if Shaikh's request be refused STOP Before coming to early lunch with me he had compared essential articles in #4 with #5 Draft STOP We then discussed for long the two vital clauses firstly one dealing with Director second dealing with won and saved clause STOP I explained in detail to Shaikh the advantage of sold or exported but Shaikh said he had promised Ibn Saud that he would follow the same line as latter has in the Hassa Oil Concession STOP With regard to Director Shaikh agrees to accept London Representative for the whole period of Concession in lieu of Director so that point is overcome STOP You will understand that this has been agreed privately and that he will make some show of resistance when we approach Shaikh jointly officially STOP Question of Gabriel's company was broached Shaikh of Kuwait told me that he had no intention to permit the English to stab Americans in the back and that I could assure Stevens that he could rest assured that Shaikh would not separate himself from his American supporters STOP Shaikh told me to tell you that as he had to yield Director question and to refuse Gabriel's application Stevens should help him by yielding won and saved clause STOP Shaikh was very obdurate during the whole interview but at the same time evincing desire to settle the outstanding questions with me at this private meeting before having official discussion with Chisholm present STOP Seeing Mohamed Yateem and we have at last obtained promise from Shaikh that he will not deal with Gabriel's company is it worth while to raise any question with Kuwait Oil Company about Gabriel's company it may be better not to raise it but as you decide STOP I would like your reaction to the won and saved question bearing in mind that the Shaikh considers he has to yield much in the Director and Gabriel's company STOP You will appreciate that I had to tackle Shaikh privately on the Gabriel question in order to check our friends I am now satisfied that I have done so STOP All that passed between Shaikh and myself at above mentioned meeting is private and confidential STOP Only Shaikh Holmes Yateem were present at lunch and later discussion—HOLMES

Holmes' information of 25th October was cabled on to Gulf in New York by Stevens in London on 2nd November as follows:

Firstly He had private interview with Shaikh morning Oct. 25th
Secondly Shaikh agreed to accept London representative for the whole period concession in lieu of director
Thirdly Shaikh said with reference to Gabriel's company he had no intention permit the English to stab Americans in the back and Holmes could assure Stevens that Shaikh would not separate himself from his American supporters but Shaikh asked Holmes to tell Stevens that as Shaikh had yielded on director question Stevens should help Shaikh by yielding won and saved clause
Fourthly Shaikh said he promised Ibn Saud that he could follow won and saved provision as in Hassa concession
Fifthly Holmes suggested that under the circumstances it might be better not to raise any question with Kuwait Oil Co. Ltd about Gabriel's company FULLSTOP
Reply to Holmes dated Oct. 26th expressed agreement with fifthly above FULLSTOP

---

125. *Gulf (New York) expressed incredulity at Holmes's theory to its Pittsburgh head office:*

New York, N.Y.,
November 1, 1934

PRIVATE
Mr. F. A. Leovy,
Pittsburgh, Pa.
Dear Sir,

Pursuant to Mr. Hamilton's telephone conversation with Mr. Bothwell of this morning, you will find herewith copies of the correspondence referred to in the cable received from Major Davis dated October 29th and our cable to London which went out yesterday—copies of which cables were forwarded to you in last night's mail.

After talking to Mr. Bothwell we cabled Mr. Stevens in London that for our advance information he cable us briefly the substance of Holmes' message of October 25th and the replies made thereto of October 26th and 29th. After this message was despatched there was received here by steamer mail Major Davis' private letter to Mr. Wallace of October 24th, a copy of which is enclosed.

Naturally we are very much 'up in the air' here as to what this all means. Even though Stevens cables the substance of the cable communications with Holmes in all probability we will not get his full reaction to this situation until the 'Aquitania' mail arrives next week.

We cannot believe that the APOC would take the steps suggested by the information which has come to Major Holmes. Apparently subsequent communications from Kuwait have thrown some doubt on the rumour or else Major Holmes now has information that the APOC is in no wise implicated in the efforts being made by third parties to obtain an oil concession from the Sheikh. However, pending further information from London, any opinion we have here is largely conjecture.

Yours very truly,
Wm. T. Wallace

---

126. *Chisholm and Holmes reported their meeting with the Shaikh on 3rd November to London by cable on 4th November:*

To London 4th November 1934:
At SOK's request visited him yesterday for first time since 17th October fullstop Stated he had studied draft 5 and was disappointed to note its failure to meet his demands as expressed in draft 4 fullstop He was preparing written reply showing his demands and alternatively certain new demands if he accepts draft 5 in full fullstop We found his main objection to draft 5 is exports-and-sales basis royalty so have discussed that point with him today fullstop His view generally is that as won-and-saved basis was offered by both APOC & EGS it is unreasonable for KOC to renounce it now as unsuitable for best exploitation Kuwait oil fullstop Gather his letter will include demand for won-and-saved basis as before with allowance for return to reservoir of certain percentage of production fullstop Amount royalty not yet discussed as overshadowed by question of basis fullstop Are now awaiting his letter.

---

127. *On 28th December 1934 the Political Agent in Kuwait wrote as follows to the Political Resident in Bushire:*

SECRET — *POLITICAL AGENCY,*
D.O. No. C–377 — *KUWAIT*
The 28th December 1934.

The Hon'ble Lt.-Colonel T. C. Fowle, C.B.E.,
The Political Resident in the Persian Gulf,
BUSHIRE.
Mr dear Fowle,

I am glad to say the Oil negotiations are through at last, for they have caused me no little worry and anxiety, especially the last stages.

2. As you will have seen from my telegram and other correspondence on the subject, the Shaikh signed the Concession in the Drawing Room of the Agency at 11 A.M. on Sunday the 23rd December and Holmes and Chisholm followed suit. I acted the part of official witness etc.

3. There is no doubt in my mind that the Kuwait Oil Company had a very narrow escape, and that the American side of the Company (Gulf) were very nearly the victims of which I believe to have been a barefaced attempt to 'double cross' them, on the part of their allies and friends, the British side (APOC Ltd.)

4. If you will refer to my Secret D.O. C–297, dated 17th November [see Note 136] and my telegram No. 338 of 13th December, you will see that in November I was well on the track of the hostile Oil concern, which was operating from Basra (and which later turned out to be Traders Ltd.) and

every fresh piece of evidence that I have since been able to get hold of has tended to confirm my suspicion that the APOC were the hidden hand behind 'Traders'.

5. Of course the APOC have covered their tracks well, and one will never, I suppose, be able to prove anything against them, which in any case would be undesirable. That very strong suspicion attaches to them, however, will be seen from the enclosed 'Note', which I send you, the contents of which I have gone to great pains to get hold of both from Basra, Baghdad and London.

6. The Shaikh of course has been to blame from our point of view for his stupid excess of secrecy, but from his own point of view (and don't forget he is a Nejdi) he has played a supremely clever game, and has forced the Kuwait Oil Company in the end to grant him all he wanted.

7. I take a little credit to myself for having 'saved the Shaikhs bacon' to some extent with His Majesty's Government, by getting through to you on 17th November my Secret D.O. No. C–297, which I trust not only put His Majesty's Government on their guard against the APOC, but to a certain extent showed that the Shaikh did make some sort of belated effort to keep me informed of what was going on, though not until I had got most of the facts myself.

8. After the Concession was signed, I tackled Holmes judiciously about Traders Ltd., and found out enough to satisfy me that he had been suspicious as far back as May 1934, that something seriously wrong with the internal affairs of the Kuwait Oil Company, especially when he received a wire from London telling him and Chisholm to get hold of K. B. Mirza Mohamed of Basra, and entrust him with all their secret Arabic translation work.

Holmes who says he refused to act as ordered, did not apparently begin to find out about the full nature of the 'plot' until he and Chisholm returned from London early in October last. From that date, however, although he did not say this to me, I feel sure he got well on to the track of Traders and kept his American principals in London fully informed of every move.

9. Chisholm, I am confident now, knew all about Traders Ltd. as far back as May 1934, also, and though I say this with the utmost reserve, I feel sure he was made to act the part of liaison officer between some one high up in the APOC in London, and Frank Strick & Co., of Basra, including Mirza Mohamed and Gabriel. In part proof of this, I may mention that since his return from London in October he has made some 6 trips to Basra and on each occasion visited Frank Strick's offices, and those of Mirza Mohamed and Gabriel.

10. I will go further and say that on the night of June 17th 1934, and actually when my wife and I broke our journey in Baghdad whilst *en route* to the United Kingdom, Chisholm secretly it seems met the following persons (Maude Hotel) also in Baghdad:—

(a) K. B. Mirza Mohamed,
(b) Mr. Gabriel,
(c) Hajji Abdulla Williamson,
(d) The local APOC Agent in Kuwait (an Arab),
(e) Shaik Salim el Hamud Al Subah, cousin of the Shaikh of Kuwait and a very pro-APOC personage.

At this meeting Chisholm, I have reason to believe, propounded the forthcoming campaign of Traders Ltd., and discussed plans and methods of procedure.

11. It is a curious fact that ever since my arrival back from England (16th October) Chisholm from being the very open and delightful friend of the family that he used to be, all of a sudden adopted a close and decidedly aloof attitude. I now put this down to the fact that he had received 'his orders' in London, and not only 'disliked' playing a dual role, but definitely felt he was playing a 'dirty' game, and was ashamed of himself. He certainly was continually at loggerheads during this period with Holmes, and whatever progress Holmes seemed to make in the negotiations he, Chisholm, appeared to want to spoil.

12. Actually when the Oil Concession was finally signed (on 23rd December) Chisholm showed a curious lack of enthusiasm, in my house, just as if he were a 'beaten man'. Holmes on the contrary acted the part of 'conquering hero'. According to Holmes—though I say this with reserve—Chisholm volunteered the information to him (after they had gone home from signing) that he had all along been aware of Traders activities in Basra but had not liked to say so to Holmes, lest he get 'worried' about it. He also passed the cryptic remark that 'some people in this world had to do lots of things by order which greatly went against the grain of a decent minded person'. From this you will gather that right up to 23rd December neither mentioned a word to the other about the existence of Traders, presumably because for his part Holmes believed Chisholm to be acting the traitor and Chisholm because he felt guilty.

13. I give the above facts at length in order to show you what has been going on out here, and the unpleasant atmosphere that has surrounded and hampered the last stages of the Oil negotiations.

14. Can it be wondered, I ask, that the Shaikh who (as he told me a few days before he signed the Concession) believed that he saw in Traders Ltd., the hand of the APOC, decided to go all out and play with them to suit his own hand. You remember that I reported this to you when you were last here (22nd December).

15. The last stages of the Oil negotiations were I now know marked by an intensive propaganda from Traders Ltd. among the leading people of Kuwait, and it is fully admitted now that Gabriel distributed large sums of money in the Town, on behalf of Traders Ltd., so that pressure should be brought on the Shaikh, to decide the right way. Gabriel employed sub-Agents to pay out the money, one of whom was the APOC local Agent himself, I understand. It is even said—though I can hardly believe this—that in two cases certain individuals received Rs. 5000/- apiece, and two others Rs. 10,000. Certain it is now that Gabriel and his backers were confident of success right

up to the end, especially after Traders in London had made contact with His Majesty's Government.

16. In conclusion, I enclose, with my attached note, a cutting which appeared in the Iraq TIMES of December 18th 1934 and which curiously supports what I have said about the APOC. The relative portion reads as follows:—

> Lord Greenway, *President of the APOC Ltd.*, died at Ramsey, Hampshire, yesterday at the age of 77. Lord Greenway was one of the founders of the Company, and was also *senior partner in Shaw, Wallace, & Co.* who have large business connections in Iraq, India and Ceylon.

17. I trust that what I have written will not be taken in any way as suggesting that I want to 'down' anyone, or get anyone into trouble. My sole desire is to show that the suspicions, which I propounded in my D.O. C.–297 of 17th November were not empty ideas based on mere hearsay, but had more behind them than met the eye.

Yours sincerely,
Sd. H. R. P. Dickson.

NOTE ON ***'TRADERS** (OIL) CO Y LTD'*.
Known in Basra under the name of 'FINE
ARTS TRADERS (OIL) LTD)'.

A. *The persons acting in the interest of, as well as the various ramifications of the above Company.*

(i) *Lord Greenway* (*died 17th December*).
- (a) President of the APOC Ltd.
- (b) Senior partner of Shaw, Wallace & Co. Ltd. of London and India.
- (c) Partner in R. G. Shaw & Co. of London.

Note. A lawyer named Mr. Everet is legal adviser in London for Lord Greenway, and also for (b) and (c) Companies above and, I now understand, also for '*Traders Ltd.*'.

(ii) *R. G. Holmes* (*of London*).
- (a) A director of Messrs R. G. Shaw & Co. (Stockbrokers, 5 Drapers Gardens, London EC.2), which Company is intimately connected with Shaw Wallace and Co., Ltd., who in turn are intimately connected with Frank Strick & Co., Ltd of Basra and London.
- (b) The APOC Ltd., and their daughter Coy. the Raffidain Oil Co of Iraq are distributing Company to Frank Strick & Co., of London and Basra, and have closest business connections with them and with Shaw Wallace.

(iii) *K. B. Mirza Mohamed, C.I.E.*
- (a) Is a lawyer practising in Basra and is legal partner to Mr. Gabriel (mentioned below).
- (b) He is the legal attorney of Frank Strick and Co. of Basra.
- (c) He is legal Adviser in Basra to the APOC Ltd.
- (d) There is every reason to believe he personally drew up the 'Traders Ltd.' draft Oil Concession, sent to Shaikh of Kuwait by hands of Mr. Gabriel.

(iv) *Mr. Gabriel*
- (a) An Armenian lawyer of Basra and legal partner of K. B. Mirza Mohamed (see above).
- (b) He (Mr. Gabriel) is known to have been directed by Mirza Mohamed in everything pertaining to the Oil Concession which he tried to get out of the Shaikh.
- (c) He is legal adviser to the Shaikh of Kuwait and looks after his gardens and other interests in Iraq.

'*B*' *Chairmen & Directors of* '*Traders Ltd.*'

Lord Glenconner. Nephew of Lady Oxford and Asquith. Lord Glenconner's sister is believed to be married to present Minister of Agriculture.

Sir Richard Redmayne, K.C.B. Uncle to Lady Oxford and Asquith and so related to Lord Glenconner. Has much influence.

H. H. Holmes. Director of R. G. Shaw & Co of London, and brother of R. G. Holmes, who is a director of
- (a) Shaw, Wallace & Co. Ltd.
- (b) Frank Strick & Co. (Basra) Ltd.
- (c) R. G. Shaw & Co.

Mr. Chas: Tennant
Col. Edward Tennant, DSO, MC. } (a) Both related to Lord Glenconner.
(b) One is a member of the firm of R. G. Shaw & Co. stockbrokers.

NOTE. Mr. W. T. Shaw, the senior member of R. G. Shaw & Co., is the son of the 'Shaw' of Shaw, Wallace & Co., Ltd. of London and India.

Dickson's letter of 28th December 1934 was forwarded from Bushire to the India Office in London, where it was received on 28th January 1935 by Mr (later Sir J. G.) Laithwaite, the official (see page 71) who had received Mr Hunting and Lt-Col Bovill on 28th November 1934 when they first disclosed to H.M.G. the existence, ownership and Kuwait activities of Traders Ltd. The Political Agent's suggestion of an APOC connection with Traders Ltd. was so clearly absurd, in view of all the facts known to the India Office about the latter company (which were only to reach Bushire and Kuwait a month later) that his letter was not acknowledged, marked for non-circulation and filed.

It was only in 1970, thirty-six years later, during researches for this Record in the London archives of the British Foreign and Commonwealth Office, that this letter first came to the notice of The British Petroleum Co. (formerly APOC) and Mr Chisholm. To prevent its misleading future readers of the archives, a note has now been attached to the letter stating that its suspicions of a connection between Traders Ltd. and APOC were unfounded, and referring to the file in the same archives describing Traders Ltd.'s origins, ownership and dealings with the India Office from 28th October 1934 onwards. The details given in the letter as supporting the Political Agent's suspicions of pro-Traders Ltd. activities between June and December 1934, of Mr Chisholm and pro-APOC Arab personages are all without any foundation in fact, as Mr Chisholm and the others concerned, Mr Yusuf Alghanim of Kuwait (25) and Mr Muhammed Ahmad of Basrah (177), have recently attested.

The Political Agent's letter, like its predecessor of 17th November 1934 (see page 71 and Note 136), evidently reflected Holmes's 'APOC plot' theory of 16th October 1934 (see page 66) regarding Traders Ltd. For so experienced an H.M.G. official to have become convinced that Traders Ltd. could possibly be a creation of the then H.M.G.-controlled APOC designed to double-cross the H.M.G.-favoured KOC is an outstanding example of Holmes's persuasive skill.

See also Addendum page 250

---

128. *See Note 124.*

---

129. *The following was the text of the Shaikh's letter to the negotiators:*

Kuwait.
28th Rajab 1353
6th November 1934

No. R/4/1801

Dear Friends Major Holmes and Mr. Chisholm.
After compliments;

With reference to the two drafts of the Kuwait Oil concession agreement which you have recently sent us, and which have been noted, we have also noted your statements regarding the proofs and reasons of your directors for having altered several articles that were agreed upon at the interview which you had with us in our palace at Bayan prior to your journey to England in the month of June last.

On having thoroughly investigated the alterations that appear in the last draft of the concession agreement we have come to the conclusions that further discussions of any sort will not result to the interest of either of us. Therefore as we desire to bring our discussions to an end, and to end matters definitely, we have decided to put up the following two suggestions either of which is acceptable to us.

We draw your attention particularly to have your directors warned that our demands include all the terms you find shown in our suggestions (A) and (B). We have written down each one separately. There are a few more points of little importance, shown in the two drafts of the agreement, but these are mostly agreeable to us.

Yours sincerely
Sd/ Ahmad Al Jaber As Subah

The terms desired to be inserted in the agreement.

(A) The royalty to be paid on crude oil 'won and saved' in Kuwait, after deducting waters, and other foreign substances, likewise the company's normal requirements of oil for its operations and works within the territory of Kuwait.

Payments as per our suggestion (A) should be as follows:—

(1) Payment on signature Rs. 475,000
(2) Yearly rent ,, 95,000
(3) Royalty on each English ton Rs. 3/- on crude oil production
(4) Minimum royalty per year Rs. 250,000
(5) For taxation exemption As4 per each ton of crude oil on which royalty is payable.

As regards the question of the director whom we wish to appoint we shall discuss that if and when your directors have agreed to our suggestions.

(B) Royalty to be paid on the basis of export and sales; and payments as per this our suggestion (B) should be as follows:—

(1) Initial payment Rs. 600,000
(2) Yearly rent ,, 100,000
(3) Royalty on each English ton Rs. 3/3 on exports & sales
(4) Minimum royalty per year Rs. 300,000
(5) For taxation exemption As.3 per each English ton.

Besides the above the company shall have the right to return a quantity of the crude oil (partly treated) to its original source to the maximum of $7\frac{1}{2}\%$ and this maximum quantity is to be estimated on the monthly production of crude oil after extracting from it the water etc.
The questions of the director and the Chief local representative shall be dealt with, as shown in our suggestions (A) above, if and when your directors have agreed to our suggestion (B).

---

130. *See Note 139.*

---

131. *See Note 117.*

---

132. *See Note 136.*

---

133. *See Note 117.*

---

134. *Mr (later Sir George) Rendel recorded his conversation of 19th November 1934 with Lord Lloyd as follows:*

KOWEIT OIL

CONFIDENTIAL

Lord Lloyd rang me up this morning to say that he was becoming increasingly disturbed about the oil position in the Gulf, and that he feared that he would have to raise the whole matter publicly in the near future. As I knew, he felt extremely strongly that oil exploitation in the Gulf should, as far as possible, be kept under purely British control, and he regarded the intervention of American interests with the utmost misgiving. It will be remembered that at his speech at the Central Asian Club Dinner on October 18th Lord Lloyd delivered a strong attack on the Anglo-Persian Oil Company for having missed the opportunity of making sure that all the oil in the Persian Gulf was under British control, and was very critical of H.M. Government for not having influenced the Anglo-Persian Oil Company more strongly in this direction, and for having acquiesced in American interests getting a footing in the Gulf.
2. Lord Lloyd then went on to say that before taking the whole matter up publicly he wished to inform me most confidentially of the position as regards the Sheikhdom of Koweit (as distinct from the Neutral Zone). He said that I was no doubt aware that an entirely British group was now competing against the Anglo-Persian–Gulf Oil of California Combine. I told him that I had not myself heard of this, but that the Foreign Office were only somewhat indirectly involved, and that it was possible that the information was still on its way. Lord Lloyd seemed somewhat surprised, and in these circumstances he said he would be glad if I would keep his name out of the affair as far as possible. The information had been given to him in confidence, and he did not wish to be quoted, but at the same time he felt bound to pass it on to me confidentially, so that there might be no misunderstanding about his own position if he eventually raised the matter publicly, and he would indeed be glad if I would make a note of what he said so that it might be on record. He then went on to tell me that the new interest, which he was informed was wholly British, was represented by Mr. P. L. Hunting and was known as the Hunting Group, or 'Traders Limited'. Mr. Hunting was a director of the B.O.D. (Mosul Oilfields Ltd.), but the B.O.D. were not concerned in this matter, and Mr. Hunting was acting independently. The following were, however, associated with him in 'Traders Limited': Lord Glenconner, Charles Tennant, Son and Company, Limited, Sir Richard Redmayne, Mr. Edward Tennant of Hohler and Company, Mr. H. H. Holmes, and several others.
3. It appears that Mr. Hunting has now informed Lord Lloyd that his group have concluded an agreement with the Sheikh of Koweit (the signed and sealed text of which, in English and Arabic, has apparently been shown to Lord Lloyd by Mr. Hunting) by which the Sheikh grants an oil concession over the whole of his territory to the Hunting Group 'subject to the approval of H.M. Government'. Lord Lloyd was unable to tell me anything about the terms of concession except that the document is extremely long and complicated. But he clearly anticipated that if the Sheikh had a free hand the concession would go to the Hunting Group and not to the Anglo-Persian Gulf Oil Combine. He was inclined to take the line that it was now up to H.M. Government to ensure that the concession went to a purely English concern instead of to one which was at least half American. Moreover, Lord Lloyd, who gave me the impression of being now definitely hostile to the APOC, explained that he thought that for political reasons it was most undesirable

that all the oil in the Gulf, or rather all the British control over Gulf oil, should be concentrated in the hands of a single group, more especially as it was unlikely that that group would hurry to develop it.

G. W. Rendel

19th November, 1934.

---

135. *Mr Rendel recorded the conversation which he and Mr Laithwaite of the India Office had with Lord Lloyd on 27th November 1934 as follows:*

KOWEIT OIL

Conversations with Lord Lloyd

Lord Lloyd rang me (Mr. G. W. Rendel) up this afternoon (27/11/34) with reference to his previous (19/4/34) conversation with me and asked whether I had any further information as to the oil situation in Koweit. I said that, as a result of his previous communication, we had made discreet enquiries, but had been unable to obtain any information as to any concession having been given to any British group such as he described. Lord Lloyd said this was very odd as the concession had actually been signed and he had been given a copy. Lord Lloyd admitted, however, after some further discussion, that the copy of the agreement that he had before him was undated, and that the Hunting group, though they had clearly given him to understand that it had been signed and concluded, had not categorically said so in writing. Lord Lloyd eventually said that he would obtain further particulars as to the concession from Mr. Hunting and ring me again.

2. Lord Lloyd was anxious to know whether the negotiations between the Anglo-Persian Oil Company–Gulf Oil group and the Sheikh of Koweit had made any progress. I said that I believed that they were well advanced.

3. Lord Lloyd also enquired whether any progress had been made as regards the negotiations which he understood were going on between British and American oil interests in connexion with the Bahrain oil concession. He had understood that there had been a proposal on foot which might have resulted in transfer of the Bahrain concession to British interests in return for the grant to American interests of increased marketing facilities, especially in the Far East. The American Bahrain Oil Company had recently tried to market their oil in Japan, but had experienced difficulties. Lord Lloyd was under the impression that the result of the negotiations might be much affected by what happened as regards the solution of these difficulties. I told Lord Lloyd in reply that this question was one which was being dealt with entirely between the various oil interests concerned by direct negotiations carried on in their own way, and that I was therefore very much in the dark as to the situation, about which I feared I could give him no information.

4. Lord Lloyd had already explained that he would almost certainly feel it his duty to raise the whole question of our oil policy in the Persian Gulf in the House of Lords at an early date, as he felt most strongly that the virtual exclusion of purely British interests from this area was in the nature of a national disaster.

5. Immediately after this conversation I asked Mr. Laithwaite to come over, and having explained the position to him I rang up Lord Lloyd again and asked him to discuss the matter over the telephone direct with Mr. Laithwaite, since he was much better informed than I as to the situation in Koweit. The greater part of the remainder of this record, which embodies what passed in Mr. Laithwaite's telephone conversation with Lord Lloyd (to which I listened in on my second telephone receiver) has been dictated by Mr. Laithwaite himself.

6. Lord Lloyd said that in the first place he did not wish to ask for any confidential information, but he felt that in the interest of British interests in the Gulf, which he considered of great importance, it would be his duty, unless some steps could be taken to break up the American block in the Gulf, to raise the matter in the House of Lords. He had felt it his duty to give Mr. Rendel the information recorded in E6972 and now to raise the matter again with a view to avoiding any suggestion by the Government, if he had to attack them in public, that he had not laid all his cards on the table and given them the fullest possible information. In his view the Government ought to have taken steps, before allowing matters to go as far as they had in Koweit, to make quite sure that there was no British company which could be induced to interest themselves in that area.

7. Mr. Laithwaite said in reply that the information which Lord Lloyd had given me had come as a complete surprise. We had no information that any negotiations had taken place between the Sheikh and the group mentioned by him. The group in question did in fact appear to contain individuals of some substance and closely associated with the British Oil Development Company, but if they were in fact competing as a British group, they must have known that the Sheikh was bound by special arrangements to His Majesty's Government as regards oil, and that their proper course was to approach the Sheikh only through His Majesty's Government or the proper political authorities.

8. As regards a further point which Lord Lloyd had raised, he was probably well aware that the Koweit Oil Company, which represented the fusion of Anglo-Persian Oil Company and American interests, had come into being only after lengthy endeavours by the Anglo-Persian Oil Company to capture this concession for themselves. The terms which they were in a position to offer as independent competitors appeared to have been capped by American interests. Mr. Laithwaite thought

it was fair to say that there had really been no alternative British group to the Anglo-Persian Oil Company at any earlier stage of the negotiations. Traders Limited were now apparently coming forward as competitors on a British basis, but Lord Lloyd would appreciate in the first place that it would be necessary to conduct close investigations into the financial standing and bona fides of any group which came forward. Secondly, very substantial financial offers were understood to have been made to the Sheikh of Koweit by the Koweit Oil Company and it would be difficult to persuade the Sheikh to accept any less favourable offer from another company, even if its composition was British. Thirdly, negotiations between the Sheikh and the Koweit Oil Company were, so far as we were aware, now at a very advanced stage indeed.

9. Lord Lloyd said that he had no personal interest in Traders Limited. He was solely concerned with British prestige and British interests in the Gulf. He was supporting Traders Limited because, so far as he knew, they were British and they had asked for his help on that ground. If for any reason they had made an error in approaching the Sheikh direct he much regretted it and could only think that it was due to uncertainty as to the attitude of the Government Departments, who were regarded as being active supporters of the Anglo-Persian Oil Company. But he trusted that any such error would not prejudice them if they were now to make an official approach to His Majesty's Government. As regards the Sheikh and his interests, these could not, he thought be paramount. We had done much for Koweit in the past, and the deciding factor in a question such as the present could not be whether the Sheikh would benefit to the extent of a few thousand pounds more or less: it must be British interests. He knew that we could if we wished bring pressure to bear on these small Persian Gulf Sheikhs and make them do what we wanted. As regards the state of the negotiations between the Sheikh and the Koweit Oil Company, he begged us, even if it were now the eleventh hour, to lend a sympathetic ear to Traders Limited if they approached us. Would we be prepared to discuss matters with them, and if so ought they to approach the Foreign Office or the India Office?

10. Mr. Laithwaite said that he did not think we could entirely accept his views as to the financial position of the Sheikh. It would be most difficult for us to say to the Sheikh, 'You must break off 'negotiations with the Company with which, with the approval of His Majesty's Government 'you have now been negotiating for a considerable time, and which has offered substantial financial 'inducements to you, and you must commence negotiations with an alternative group which has 'indeed the great advantage, from the point of view of the Government, of being British, but 'which may not be able to give you the same price for your concession'. We should find it very difficult to convince the Sheikh of the validity of such a proposition. As for Traders Limited, we should be prepared to discuss matters with them if they had any positive proposition to make. But they must realise that matters were now at a most advanced stage in the discussion at Koweit.

11. If they approached the Government they should approach the India Office, which, while it concerted its action in oil questions with the Foreign Office, was now the department primarily responsible for oil questions in this area, in consultation with the other interested departments.

12. Lord Lloyd said that he could not say whether Traders Limited would be prepared in fact to communicate with a Government department. But he would himself at once approach them about it, and only hoped that if, as a result, they should approach the India Office, we should deal with them sympathetically. Mr. Laithwaite said that we should certainly give them a sympathetic hearing within the limits which he had indicated above.

13. Lord Lloyd again deplored the way in which His Majesty's Government had, according to him, capitulated to America throughout this affair. He said that one of the most deplorable and scandalous pages in the history of the business and one which would produce a very bad effect if he had to refer to it in public, was that the peak of the negotiations took place during Mr. Mellon's tenure of the United States Embassy here. He then indulged in a mild diatribe against our general subservience to the United States, and explained that his essential object was to redress the balance in the Persian Gulf in favour of the true interests of this country.

14. Lord Lloyd subsequently again rang up and spoke to Mr. Laithwaite. He said that in the light of the conversation he had got in touch with Traders Limited and strongly advised them to put themselves right with the Government and get into touch with the India Office.

15. He made the further point that he had for some time been restrained by the Foreign Office only with difficulty from ventilating the whole of this question, of which he took so serious a view. Was it entirely fair that when he had held his peace because he was advised that it was in the best interests of the country, Government departments should be pressing on in the Gulf with steps designed to hand the concession over to a syndicate which, even if 50% British, was likely in practice to be under a decisive American influence. Was it too late even at this stage, when he had shown that there was a purely British group in the field, to close the negotiations down? Mr. Laithwaite said that he doubted whether there was any very material difference, except as regards small points, between the position of the Koweit concession when Lord Lloyd had first spoken to the Foreign Office last May, and the position today. The essential features of the concession had been decided by last spring. Negotiations for a concession had now almost reached finality and could not be abruptly broken off. It was not only as though it were a question of a new concession for which a British company were competing. Mr. Laithwaite would, however, put his point on record.

16. Lord Lloyd more than once hinted in the course of the conversation that he would probably feel compelled to raise the matter in the House of Lords in the near future. If so he thought that our weakness towards the Americans would create a deplorable impression. He might even feel

obliged to say that we had begged him to hold his peace and taken advantage of his silence to push forward our particular policy.

---

136. *The texts of the Political Resident's telegram of 22nd November and the Political Agent's letter to him which the telegram reports, are as follows:*

TELEGRAM 5871

From Political Resident in the Persian Gulf to
Secretary of State for India.

Dated Bushire, 22nd November, 1934.

IMPORTANT. Addressed to Secretary of State for India and repeated to the Government of India, copy by mail Koweit. Paragraph No. 2 of your telegram of the 19th instant, 2743. In a letter yesterday, Political Agent, Koweit, reports that Sheikh had of his own accord informed Political Agent that in August an Iraqi approached him with draft oil concession offering terms financially and otherwise more favourable than those offered by Koweit Oil Company. Replied that company must first approach His Majesty's Government and secure their approval.

2. Sheikh did not divulge name of the company or of agent but Political Agent suspects that company may be Frank Strick and Company of London and Basrah and has reason to believe that agent was lawyer, Gabriel.

3. If Frank Strick are concerned there is presumably another group behind or associating with them.

4. Political Agent's letter follows by post.

POLITICAL AGENCY,
KUWAIT.
The 17th November, 1934.

The Hon'ble Lt.-Colonel T. C. Fowle, C.B.E.,
Political Resident in the Persian Gulf,
BUSHIRE.

Subject:— 'OIL'

My dear Fowle,

Since my return from leave last month (on 18th Oct) I have been quietly trying to find out something about the intrigues of a strange Oil concern or Company, inimical to the Kuwait Oil Company, which (and I say this guardedly at present) as far as I can make out has been operating from BASRA during the past Summer, and continues to be active.

2. If you recollect in my Confidential Express Letter No. C–146, dated the 24th April 1934 and paragraph 6(C). I mentioned the fact that the Shaikh had recived a telegram from a rival Oil concern which at the time was thought to have come from the B.O.D. Company by Holmes and Chisholm. The purport of that telegram was that the strange Company offered to the Shaikh to go one better in every respect than the Kuwait Oil Company, and was obviously sent with a view to prevent the Shaikh coming to a favourable decision with the Kuwait Oil Company (as it was thought he would).

3. The Shaikh, as you know, admitted at the time to me that he had received the telegram in question, and said that it purported to come from an all British concern, but he would not tell me the name.

4. I believe later enquiries from London satisfied Holmes and Chisholm that the B.O.D. had nothing to do with the telegram, nor were the I.P.C. suspected. I myself came to the conclusion that it had probably been sent at the instigation of the Shaikh himself in a rather childish attempt to try and force the Kuwait Oil Company to bid higher than they were doing.

5. I believe I have now got on the track of the senders of the telegram, but of course I say this with the greatest reserve, seeing that my evidence is only of a circumstantial nature at present.

It was Messrs: Frank Strick & Co., of Basra (and London), working through their Lawyer in Basra, K. B. Mirza Mohamed, C.I.E.

6. I have also practically proved that friend Mirza Mohamed is in secret partnership with Mr. Gabriel the Shaikh's lawyer (thereby also hangs a tale where the Shaikh's gardens are concerned), and that before I returned from England the former sent Mr. Gabriel on 20th August last to Kuwait, to see the Shaikh. Mr. Gabriel was armed with a completely drawn up and printed Oil Agreement in which he offered the Shaikh every single thing he wanted including money, royalty, director, Chief Local Representative, etc., etc. The figures of course being better in every respect than those offered by the Kuwait Oil Company.

7. The Shaikh was, I understand, very gratified that another Company had come forward to bid against the Kuwait Oil Company, but when he was asked to sign, his native caution came to his aid and he said No, and told Gabriel that the Company he represented must first go to His Majesty's Government in London, be 'vetted' and then come back with His Majesty's Government's blessing.

8. The Shaikh himself and 'quite off his own bat' after pledging me to secrecy, told me the story and said that the completely drawn up concession had been brought him from Iraq and by

an Iraqi, that he had been pressed hard to sign. He declined to say however who the person was who brought the papers to him, nor would he divulge the name of the Company.

9. I have not of course discussed a word of the above with either Major Holmes or Mr. Chisholm of the Kuwait Oil Company, but from one or two remarks which the former has passed I should not be surprised if he was not himself well on the track of the enemy.

10. One thing I have gathered, and that is that Holmes and Chisholm are not working as smoothly as they should be these days. The reason may be (and I only suggest this, mind you) that Holmes has found out about the rival concern, and puts down the whole business to a plot on the part of the APOC to 'double cross' the Kuwait Oil Company, and may also feel that Chisholm is not as ignorant of the business as he pretends to be.

11. The following facts are significant, I think:

(a) F. Strick & Co., are connected with the APOC and receive an annual sum of money from the latter for 'good will' etc.

(b) Mirza Mohamed is legal adviser to the APOC Ltd. and is also lawyer to F. Strick & Co.

(c) Mirza Mohamed and Mr. Gabriel are partners in the same firm, though the world believes the latter to be working with a Turk named Arteen Effendi.

(d) Mirza Mohamed recently became an Iraq subject.

12. Is it beyond the bounds of possibility that the APOC smarting under the fact that they were unable themselves to get an Oil Concession out of the Shaikh because of the Americans, for appearances sake leagued themselves with these selfsame Americans, and while outwardly agreeing to work with them under the name Kuwait Oil Company secretly set themselves the task of hindering progress from the start with the deliberate intention of preventing the Shaikh from giving the Oil Concession to the Kuwait Oil Company, and at the psychological moment and when negotiations with the Kuwait Oil Company had reached deadlock they had ready a new and secret subsidiary company to throw into the breach, and so walk off with the prize. This Company, of course, would in no way be connected with the APOC, as far as the world was concerned, it would be all British, and be registered in London and Iraq.

13. I gather also that Mirza Mohamed and Mr. Gabriel are already directors in the new concern.

14. I send you the above for what it is worth and I hope you will not think my suggestions about the APOC being behind Strick too fantastic or unfair.

Yours sincerely,
Sd. H. R. P. Dickson

---

137. *See Note 118.*

---

138. *See Note 127.*

---

139. *Mr Hunting's letter of 28th November 1934 on behalf of Traders Ltd., and its Appendices A, B and C, are given here in full, preceded by Mr Laithwaite's record of his meeting with Mr Hunting (later Sir Percy, 1885–1974) and Lt-Col Bovill on 28th November 1934:*

Koweit Oil
Note of Interview at India Office on 28th November 1934 with
Mr. P. L. Hunting and Lieutenant-Colonel Bovill of Traders Ltd.

After discussion with Mr. Wakely, Mr. Walton and I saw Mr. Hunting and Lieutenant-Colonel Bovill, representing Traders Limited, on 28th November.

2. Mr. Hunting at the opening of the meeting produced an official letter of 28th November from Traders Limited to the Secretary of State for India intimating that the Sheikh of Koweit had declared his readiness to grant an oil concession to Traders Ltd., covering the whole of Koweit, subject to the approval of His Majesty's Government. The Company's official letter enclosed a draft of the concession in the terms stated to have been accepted by the Sheikh, together with a translation by Mr. Gabriel, the Sheikh's lawyer, of an Arabic letter, stated to be the Sheikh's autograph, from the Sheikh intimating his willingness to grant a concession to the Company, subject to the approval of His Majesty's Government. A list of the persons and companies associated with the application now put forward by Traders Limited was also enclosed.

3. The Company's representatives were informed that the application they had put forward would clearly require consideration and that we would submit the matter. In reply to a question as to the period over which they had been negotiating with the Sheikh they stated that negotiations had been proceeding since last April. When reminded of the Sheikh's treaty obligations in regard to the grant of concessions and of the fact that their approach should in the first place have been made to His Majesty's Government or the political authorities concerned, they answered that had they adopted that course they would not to-day be so far advanced as they were. Mr. Hunting

offered no explanation of this remark. (Presumably he may have meant that—as Lord Lloyd alleged—His Majesty's Government supposed attitude towards the Anglo-Persian Oil Company made his company think that they had better try to confront us with a *fait accompli* so far as possible) They were, Mr. Hunting said, well aware that the terms embodied in their concession were more favourable than those which had so far been offered by the Kuwait Oil Company and they doubted if the Kuwait Oil Company were so far advanced as they might imagine.

J. G. Laithwaite
29th November 1934

Cunard House,
Leadenhall Street,
London, E.C.3.
28th November, 1934.

The Secretary-of-State for India,
India Office,
Whitehall, S.W.1.
Sir,

I have the honour to inform you that, subject to the approval of His Britannic Majesty's Government, His Excellency Sir Ahmad al Jabir as Subah, K.C.I.E., C.S.I., Shaikh of Kuwait, has granted a Concession to Messrs. Traders Limited of Newcastle-upon-Tyne.

The concession covers the whole of Kuwait territory, and is to explore, prospect, search and drill for petroleum and its products, natural gas, etc.

Messrs. Traders Limited of Milburn House in the City and County of Newcastle-upon-Tyne, England, is a Company of British invested capital registered in Great Britain under the Companies Act 1929, and is entirely owned and controlled by Messrs. Hunting & Son, Ltd. and their associated companies: ship owners, oil merchants and brokers of Newcastle and London.

On behalf of Messrs. Traders Limited, I have the honour to request that formal and official approval of His Britannic Majesty's Government be given to the grant of this concession and that His Excellency the Shaikh be informed accordingly.

For the information of His Majesty's Government I have the honour to attach the following appendices:—

Appendix 'A' List of Companies and gentlemen associated with this application, giving certain relevant details.

Appendix 'B' Copy of terms agreed between His Excellency Sir Ahmad al Jabir as Subah, Shaikh of Kuwait, and Messrs. Traders Limited.

Appendix 'C' English translation of autograph letter granting the concession, subject to the approval of His Britannic Majesty's Government, from the Shaikh of Kuwait to Mr. J. Gabriel, Advocate of Basrah, local representative of Messrs. Traders Limited, who arranged details on the spot. The whole letter is in the Shaikh's own handwriting and it is signed by him.

Any correspondence in connection with this matter should please be addressed to:—

P. Ll. Hunting, Esq.,
c/o Messrs. E. A. Gibson & Co. Ltd.,
88, Leadenhall Street,
London. E.C.3.

I have the honour to be,
Sir,
Your obedient Servant,
P. Ll. Hunting
DIRECTOR OF TRADERS LIMITED

The Secretary-of-State for India.

APPENDIX A

List of Companies and Gentlemen associated with this Application

LORD GLENCONNER—Head of Messrs. Charles Tennant, Sons & Co. Ltd., Merchants of Mincing Lane, E.C.3.

MESSRS. CHAS TENNANT, SONS & CO. LTD.—Merchants.

SIR RICHARD REDMAYNE, K.C.B.—Formerly Chief Inspector of Mines.

COLONEL EDWARD TENNANT, D.S.O., M.C.—Senior member of Messrs. Hohler & Co. Ltd., Bill brokers of Cornhill, London.

H. H. HOLMES, ESQ.—Managing Director of Messrs. Berry Wiggins & Co. Ltd.

MESSRS. BERRY WIGGINS & CO. LTD.—Oil refiners of Kingsnorth-on-the-Medway and Manchester.

HUNTING & SON LTD. AND THEIR ASSOCIATED COMPANIES.—Tank Steamship Owners, Oil Merchants and Brokers.

APPENDIX B

In the Name of God the Merciful and Compassionate

AGREEMENT made on this day of one thousand nine hundred and thirty four, corresponding to one thousand three hundred and fifty three Hajri, *BETWEEN HIS EXCELLENCY SHAIKH SIR AHMAD AL-JABIR AS-SUBAH*, Knight Commander of the most Eminent Order of the Indian Empire and Companion of the Most Exalted Order of the Star of India, the Prince and independent ruler of the State of KUWAIT (hereinafter called the Prince) and *TRADERS LIMITED* of Milburn House in the City and County of Newcastle upon Tyne England, a Company of British invested capital registered in Great Britain under the Companies Act 1929, its successors and assigns (hereinafter called the Company) *WHEREBY* the Prince grants and the Company accepts the Kuwait Oil Concession as hereinafter described under the following terms and conditions:

Article 1

SUBJECT to the sanction of His Britannic Majesty's Government the Prince hereby grants to the Company *THE* exclusive rights to explore, prospect, search and drill for, produce, extract and render suitable for trade, Asphalt, Ozokerite, Petroleum and their products, derivatives and cognate substances (hereinafter termed 'oil') and also natural gas and its products and derivatives and the right to carry away and sell the same within the State of Kuwait including all islands and territorial waters appertaining to Kuwait as shown generally on the map annexed hereto, together with the exclusive ownership of all oil and natural gas produced and won by the Company within the State of Kuwait for a period of *SEVENTYFIVE YEARS* beginning from the date of finally signing this agreement. The term 'finally signed or final signature' shall for the purpose of this agreement mean signed by both parties and sanctioned by His Britannic Majesty's Government and for the purpose of this agreement the boundaries of the State of Kuwait shall be those recognised and laid down by His Britannic Majesty's Government and the Prince at the date hereof .

Article 2

THE Company undertakes to pay in advance regularly and continually to the Prince, his successors, assigns or legal representatives a yearly deadrent of the sum of *Rs. 500,000* (*FIVE HUNDRED THOUSAND RUPEES*) commencing (subject to Article 3(A) hereof) on and from the expiration of a period of one year as from the date of finally signing this agreement. The payment of the deadrent is to be continued until such time (hereinafter referred to as 'the commencement of regular exports') as the oil actually produced and exported and/or reserved for exportation amounts to a total of 166,666 tons (One Hundred and Sixtysix Thousand Six Hundred and Sixtysix Tons) in any one year of this concession.

Article 3

(A) *THE Company* undertake to pay the Prince forthwith on the date of final signature of this agreement a sum of *Rs. 500,000* (*FIVE HUNDRED THOUSAND RUPEES*) being the deadrent due for the first year following the first immediate anniversary of the date of final signature of this agreement. No further payment shall become due and payable in respect of deadrent until the second anniversary of the date of final signature of this agreement.

(B) *THE Company* may make all payments that become due to the Prince under this agreement into the Prince's account at any bank that may be nominated from time to time by him (the Prince) and the Bank's receipt shall be a full discharge for the Company in respect of such payments.

Article 4

FROM and after the commencement of regular exports or reserves for exportation the Company shall pay to the Prince a royalty of *Rs. 3* (*THREE RUPEES ONLY*) for every English ton of Lbs. 2240 of oil exported or reserved for exportation from Kuwait territory provided that the aggregate yearly sum of such Royalty shall be not less than the sum fixed for the yearly deadrent under Article 2 hereinabove. Payments due under this Article shall be accounted for and become payable by the Company as follows:— *Rs. 500,000* (*FIVE HUNDRED THOUSAND RUPEES*) on each anniversary of the date of final signature of this agreement and the balance of any within three calendar months thereafter.

(*NOTE:* Article 4 originally read as follows:—

FROM and after the commencement of regular exports the Company shall pay to the Prince a Royalty of *Rs. 3* (*THREE RUPEES ONLY*) for every English ton of Lbs. 2240 of oil exported from Kuwait territory, etc.)

Article 5

(A) *THE Company* shall pay to the Prince *Annas 2* (*TWO ANNAS*) for every English ton of oil exported or reserved for exportation from the territory of Kuwait in lieu of Customs, Excise, Port and Harbour and all other duties on the Company's products, machineries, implements and all such goods, dry or liquid materials including (*inter alia*) medical and sanitary stores, medicines, drugs, appliances and instruments for the use and benefit of the employees of the Company or otherwise required in connection with the business of the Company, which may be exported from or imported into the territory of Kuwait for the purpose of prospecting, searching for, winning, exploiting, refining and exporting oil or natural gas; provided always:—

(i) *THAT* the Company when wishing to dispose of by sale within the State of Kuwait any goods which have been imported free from Customs Import Duty under this Article shall pay the Customs Duty therefor in accordance with the Customs tariff prevailing at the time of and when selling such goods as aforesaid within the territory of Kuwait, provided however that in the case of sale of used or damaged or surplus goods, the payment to the Prince of a sum equal to 4% (Four Percent) of the gross price received on sale of such goods shall be accepted by the Prince in lieu of Customs duty otherwise payable thereon.

(ii) *THAT* all other goods and provisions shall be subject to the usual Customs duty but no discrimination shall be exercised in relation to any goods or provisions imported by the Company.

(B) *THE Company* is precluded from importing into Kuwait territory any kind of firearms and ammunitions without the previous written sanction of the Prince.

(C) *THE Company* shall be at liberty to re-export free of duty any goods or materials imported free of duty under the provisions of this agreement.

(D) *THE Company* shall build at their own expense habitable residences and offices for the Customs, Port and Harbour officers, guards and watchmen whom the Prince may appoint for functioning and discharging their duties at the Company's out-station or on the Company's premises. The said residences and offices shall be similar in type to those provided for the local employees of the Company of similar grade and status and shall be passed as adequate by the resident medical officer of the Company, and the Company shall pay to the Prince the wages for such additional officers, guards and watchmen and their travelling expenses to and from their stations.

(*NOTE:* Reference ARTICLE 5. Para. (A.): The words 'or reserved for exportation' have been inserted in the final draft agreement. *ARTICLE 5. Para.* (*D.*): The words 'and the Company shall pay to the Prince the wages for such additional officers, guards and watchmen and their travelling expenses to and from their stations' have been inserted in the final draft agreement.)

Article 6

IN the event of the Company neglecting to commence the geological survey for the exploration of oil within one year from the date of finally signing this agreement and/or in the event of the Company's deliberate failure to commence exploration and exploitation within five successive years from the date of finally signing this agreement, the Prince shall have the option of either cancelling this agreement or continuing to demand and receive from the Company the yearly deadrent due under Article 2 hereinabove or the equivalent thereof as provided by Article 4 hereof.

Article 7

IN the event of this agreement becoming void by optional cancellation under Article 6 herebefore or by deliberate failure under Article 3 herebelow, all lands, buildings and all fixtures thereof and all ports, docks, harbours and landing stages and all factories, refineries, fixed machineries, railways, railroads and all the oil wells and borings, pumps and pipes and all other immovables and fixtures shall automatically and unconditionally vest in the Prince free of all costs and charges, whereby the Prince shall become the rightful owner thereof, but the Company shall have the right to remove all other property of the Company in Kuwait and to export the same free of export duty, but only upon the same terms and conditions as are laid down in Article 25 hereof.

Article 8

IF the Company deliberately fails to pay on the due date the deadrent under Article 2, or the Royalty under Article 4, or the duty under Article 5 hereinbefore, the Company shall be liable to pay to the Prince the amounts so due within thirty days, with interest thereon at London bank rate from the date of maturity, and in the event of continuous default of any due payments in full for a period of six successive months from the date of maturity, the Prince shall have the option either to prevent the Company from exploiting and exporting oil until full payment of the amounts due are made or cancel this agreement forthwith.

Article 9

THE Company shall have full right to enter upon, occupy and use free of charge and cost such State or Government lands as it may require them for the purposes of this agreement and in particular for the purpose of erecting, building, constructing thereon all kinds of buildings, abodes, residences, refineries, harbours and suchlike buildings, railways, tramways, roads and ports, docks and landing stages and laying pipelines, telegraph, telephone and wireless communications and other plant or installations as may be necessary or advantageous provided always:—

(a) *THAT* such occupation or use of land does not affect, except as hereinafter mentioned, the interest of the population of the Kuwait territory nor does it deprive them from their acquired rights by traditional use and custom.

(b) *THAT* the Prince shall have the right to a reasonable extent to make use of the harbours, ports, docks and landing stages constructed by the Company for his own ships and boats provided the Company's operations are not hindered by such use.

(c) *THAT* the Company shall not without the consent of the Prince construct any railway to a gauge exceeding two feet.

(d) *PRIVATELY* owned land and legal rights in land shall be acquired by agreement between the Company and person concerned, or failing agreement the Prince will regard such lands or rights as being required for a work of public utility, and will acquire them accordingly to the law for time being in force and at the expense in all things of the Company. Provided that in fixing the value of such lands, no regard shall be had to the purpose for which they may be used by the Company and provided also that the lands so acquired by the Prince be registered in the name of the Prince but placed free of charge at the disposal of the Company during the period of this agreement. The Prince further agrees and undertakes that in any such expropriation proceedings the Prince will so far the law permits instruct his Government to act in accordance with the requirements of the Company as if they were the agents of the Company, and generally will in accordance with the provisions of Article 10 hereof, use his powers as Prince to prevent any unreasonable obstruction on the part of other persons.

Article 10

THE Prince shall give all geological and other information in his power and shall grant all possible facilities and assistance to the Company for the proper and good working of the project and in particular will use his powers as Prince to prevent any unreasonable obstruction on the part of other persons, and shall as reasonable and practicable protect the Company's rights, properties, employees and interests against theft, highway robbery, assault, wilful damage and destruction provided always that:—

(a) *The Prince* for the purpose of this Article shall have the right to nominate and appoint chief guard and guards or watchmen at the Company's expense, but so that the wages or salaries paid to such guards and watchmen shall not exceed the wages or salaries paid by the Company to persons of similar status and grades.

(b) *That* the Company shall erect at its own expense buildings in accordance with Article 5(d) hereof for such guards and watchmen and at such place as the Company decide.

(c) *That* the Company shall use always within the State of Kuwait the Prince's flag.

(d) *That* the Company shall duly close any unproductive holes drilled by it and subsequently abandoned.

(e) *The Prince* shall prohibit the erection of buildings, tents or any other erections for human occupation in areas which the Company may declare to be dangerous by reason of its operations under this agreement and shall cause prompt measures to be taken to remove any such buildings, tents or other erections and any persons disobeying the said prohibition.

Article 11

THE Prince shall have the full right from time to time to nominate and appoint and at his pleasure to dismiss:—

(1) *A person* to represent him at the administrative board of the Company in London. The said person shall be a director of the Company and shall enjoy the same rights and privileges and shall receive the same emoluments from the Company as the other Directors.

(2) *A person* to represent him at the head office at Kuwait for a salary not exceeding Rs. 1000 (One Thousand rupees) per month payable by the Company for examination inspection and checking of the Company's accounts and generally keeping in touch with the works of the Company and its schemes and projects which it intends to put into practical effect, such as building railways, railroads, erecting wireless, telegraph and telephone stations, pipelines and telegraph and telephone lines etc, within the territory of Kuwait.

(3) *A person* to be responsible for the supply and distribution of good and competent labourers to the Company and for organisation and fixing their wages, always provided that the rate of wages fixed shall not exceed the ordinary rates payable in Kuwait for labour of similar grade or class. The salary of such a person shall not exceed Rs. 800 (Eight hundred rupees) per menuem payable by the Company.

(4) *If* requested by the Company to do so for a reasonable cause, the Prince will remove any person so appointed by him and will appoint some other person in place of the person so removed. (NOTE: *Clause 4* originally reads as follows:— *If* requested in writing by the Company to do so, the Prince will remove any person so appointed by him and will appoint some other person in place of the person so removed.)

Article 12

THE Company shall not employ labourers other than Kuwaity subjects except that:—

(i) The Company may employ experts and craftsmen if persons are not available among Kuwaity subjects whose attainments are in accordance with the standard required by the Company.

(ii) *The Company* may also employ others when no sufficient numbers of Kuwaity labourers are available in Kuwait territory provided that:—

(a) *the* Prince's written sanction is obtained for employment of foreign labourers.

(b) *that* such labourers shall be subject to the laws and orders of the State.

Provided always that any of the Company's British officers and employees shall be subject to deportation after due trial of his case by His Britannic Majesty's representative in Kuwait for the charge of misconduct or any other act harmful to the life, property, interests and/or reputation of Kuwait subjects, and/or harmful to the public peace and order.

Article 13

(1) *THE Company* shall have full right to acquire by purchase in the markets of Kuwait territory foodstuffs, water and fire fuel for its employees and building and constructing materials and other supplies of all kinds necessary for its operations provided it adheres to the standard prevailing prices at the time.

(2) *THE Prince* shall as much as circumstances may permit render every facility to the Company in the exercise of its rights under this Article.

Article 14

(A) *THE Company* shall for the purpose of its operations in general construct its tanks, refineries etc. at places outside and away from the town walls of Kuwait by a distance of not less than one mile. The Company shall have the option to erect dwelling houses for its employees within the walls of and in the town of Kuwait.

(B) *THE Company* shall accept and transmit free of charge on its wireless and telegraph or telephone installations the Prince's messages provided that the operations of the Company are not unduly hampered thereby, and in time of national emergency, of which the representative of His Britannic Majesty's Government in Kuwait shall be the sole judge, the Prince shall have free of charge the full use of the Company's wireless, telegraph and telephone installations, and railways, tramways, water pipelines, tanks etc. for Governmental purposes, paying however to the Company reasonable compensation for all damage done thereby to the property of the Company.

(*NOTE:* In Article 14 (B), the words 'free of charge' have been inserted after 'the Prince shall have' and the last part of the paragraph originally read as follows:— 'paying, however, to the Company fair and reasonable compensation for the said usage and for all damage done thereby to the property of the Company'.)

Article 15

FROM and after the commencement of regular exports the Company shall set aside for sale to the public of the State of Kuwait such quantities of petroleum as are reasonable sufficient to meet their needs. Petroleum consumed under this Article shall be exempted from all royalties and duties provided for by Article 2, 3 and 4 hereinbefore, provided that the sale price shall not be higher than the prices prevailing at the time in the markets of oil producing neighbouring states. No petroleum supplied under this Article may be exported except with the consent both of the Prince and of the Company.

Article 16

THE Prince grants the Company the right of occupy and/or to use the royal buildings and houses free of charge, but in other respects on terms to be agreed upon hereafter, provided that:—

(a) *Such* buildings and/or houses are not under the occupation of the State Government.

(b) *Or* under the occupation of any of the Royal family members.

Article 17

THE Company shall employ and keep permanently a competent European doctor and pharmacist for the control and inspection of the general health in the Company's camps and stations. The Company shall provide the Prince and his family with the medical services and attendances of such a European doctor and pharmacist and all reasonable medicines and drugs prescribed by him free of charge.

Article 18

THE Company shall have the right to acquire, take and use free of charge any quantity of stone, sand, gravel, gypsum, clay or other materials or water which may be available on the Government and State lands and waters for the Company's operations, provided that the taking of such materials

(a) *Does* not expose the land to floods or inundations or give it a trench like aspect so as to attract robbers to hide therein.

(b) *Does* not destroy or deface the public roads or passages and/or lands adjacent to such roads and passages.

(c) *The Company* shall not prevent the Kuwait public from acquiring taking and/or using such materials, except so far as is necessary to protect the Company's property from damage or to prevent interference with their rights.

Article 19

(A) *THE Prince* shall be at liberty to grant to other parties concessions of mines of such substances other than mines of oil and cognate substances, provided that

(i) *The Company* shall have the right of first refusal, but this preference does not in any way mean that the Company's terms and conditions shall be less than those offered by other parties.

(ii) *Due* to the Company's refusal the grant of the concession shall not affect injuriously the operations and rights of the Company.

(B) *THE Company* shall inform the Prince of any mine or water wells other than oil and deliver to him any and every historical or ornamental antiquities which may be traced accidentally and/or during the drilling for oil or wells. The Prince in particular and his subjects in general shall have the full benefit of and the right to use all water wells, water fountains found or dug by the Company but not so as to hamper or interfere with the operations and rights of the Company.

Article 20

THE Company shall from and after the commencement of regular exports deliver to the Prince for his use continually during each one year 10,000 gallons (Ten thousand gallons) of Petroleum free of charge at any of its petroleum depots, but so that not more than 10,000 gallons (Ten thousand gallons) in all may be called for by the Prince under this Article.

Article 21

WITHIN three calendar months after each anniversary of the date of final signature of this agreement, the Company shall deliver to the Prince a *return* of Petroleum, if any, on which Royalty is payable for the year ended prior to such anniversary, and a statement of the amount of Royalty due to the Prince for such year, and a full report of its operations under this agreement and the results thereof. The Prince shall have the right always and at all times to check such returns and statements which as well as any report shall be treated as confidential by the Prince with the exception of such figures therein as he may be required by law or traditional customs to publish.

Article 22

THE Company undertakes not to carry out any of its operations or build, erect or construct on and/or within the areas occupied by or devoted to the purpose of Mosques, sacred buildings, graveyards, religious and/or historical shrines.

Article 23

THE Company shall be exempted from all kinds of government, Municipal and port taxes impositions duties fees charges etc. which may be payable on its returns, profits, buildings and properties.

Article 24

NO failure or omission on the part of the Company to carry out or perform any of the stipulations, covenants or conditions of this agreement shall give the Prince any claim against the Company or be deemed a breach of this agreement insofar as the same arises from 'force majeure', and if through 'force majeure' the fulfilment by the Company of any of the conditions of this agreement be delayed, the period of such delay together with such period as may be necessary for the restoration of any damage done during such delay shall be added to the periods fixed by this agreement.

Article 25

THE Company shall have the right at any time to abandon the undertaking upon giving three calendar months previous notice in writing to the Prince of its intention to do so. No such notice shall be given, however, until at least three years have expired from the final signature of this agreement. Upon the expiration of such notice, this agreement shall absolutely determine. Upon such determination, all buildings and structures of a permanent nature shall become the property of the Prince without payment, but the Company shall have the right to remove all other property of the Company in Kuwait and to export the same free of export duty provided that the Prince, upon so notifying the Company at any time during the currency of the said notice, may purchase the said property or any part thereof at a price equal to the replacement value thereof at that date less depreciation, which price shall be fixed by agreement or by arbitration under Article 26 hereof, and provided that the Company shall have no right to remove any of the said property or to receive any sum by way of price thereof until all amounts due to the Prince up to the date of such determination have been paid or fully set off by the said price.

Article 26

IN the event of difference or dispute arising out of this agreement between the contracting parties, the Prince and the Company undertake to refer the point in difference or in dispute to Arbitration in Kuwait. Each party to nominate and appoint his own Arbitrator within sixty days from the date of receipt by him of the intimation given by the other party calling for nomination of the Arbitrator. Should there arise a difference of opinion between the two Arbitrators, an Umpire shall be nominated by the Arbitrators within sixty days from the date of their appointment. In the event of failure to nominate and appoint the Arbitrator or the Umpire within the stipulated period, the British Political Resident in the Persian Gulf may on the request of either part, nominate and appoint an Arbitrator or Arbitrators and/or Umpire as the case may be. The award given by the Arbitrators unanimously or by the Umpire shall be final and binding on both parties.

Article 27

THE Company shall have full liberty and right to assign, transfer, sell and let the Concession granted to it by this Agreement to any firm or firms, Company or Companies of British invested capital registered in England, such assignees or transferees shall have the same obligations and enjoy the same privileges provided for by this Agreement. The term 'British invested capital' in this agreement shall mean having the whole or at least fifty percent of its invested capital British and registered in England. The Company shall also have the right to form any subsidiary Company or Companies of British invested capital for the purpose of taking over and working the said concession or any portion thereof or all or any of the rights hereby granted to the Company, or for the purpose of carrying on any part of the business of the Company as a separate concern.

Article 28

IN the event of difference in the interpretation of any of the text of this agreement, the English version shall prevail.

Article 29

WHEREVER mentioned in this agreement the Title 'The Prince' shall mean and include The Prince, His Government, Successors, Assigns and/or his Legal Representatives and the expression 'the Company' shall mean and include Traders Limited, its successors, and/or assigns as the context may require or admit.

Article 30

EXCEPT as otherwise herein provided the penalty for any breach of this agreement shall be damages which shall be fixed by agreement or under Article 26 hereof.

Article 31

ALL references to time herein contained shall be construed as referring to the calendar and system of measuring time used in Great Britain.

Article 32

WHEREVER in this agreement it is mentioned that any matter is subject to the consent approval or sanction of either party such approval shall not be unreasonably withheld and the decision shall not in the absence of any provision in this convention to the contrary be delayed for more than thirty days.

Article 33

IN this agreement if not inconsistent with the subject or context the following definition shall apply:—

Writing shall include printing and lithography and any other mode or modes of representing or reproducing words in a visible form.

Words importing the singular number only shall include the plural number and vice versa.

Words importing the masculine gender only shall include the feminine gender.

Words importing persons shall include corporations.

Article 34

(A) *ON* signature by both the contracting parties and on obtaining the sanction of His Britannic Majesty's Government this agreement shall become final and binding on the Prince, his Government, Successors, Assigns, Heirs and Legal Representatives.

(B) *ON* the termination of the concession period of Seventy-five years this agreement shall cease to operate and the Company undertakes to quit and invest in favour of the Ruler and/or the Government of Kuwait free of charge all its operations, buildings, harbours, refineries, tanks, railways, tramways, telephones and telegraph and pipelines and all its immovable properties, rights and interests in general in reasonably good and working condition.

IN WITNESS whereof the Prince has hereunto set his hand and affixed his Royal Seal and Traders Limited have caused their common seal to be affixed hereto this day of One thousand nine hundred and thirty four.

(*NOTE:* The last sentence 'reasonably good and working condition' has been inserted in this final agreement.)

APPENDIX C

2nd September, 1934.

My dear Agob,

After compliments and enquiry about your health, I received two copies of the Concession, retained one and returned you the other that you may despatch same to the Company and inform them to communicate with His Britannic Majesty King George's Government in London and produce to them the copy of the Concession; that if His Majesties Government assent to it they may please telegraph their assent to the Political agent in Kuwait.

We have accepted the conditions of the Concession and I will affix my signature thereunto on hearing the result from the Political agent in Kuwait to the effect that His Majesties Government have sanctioned the agreement of the Concession and assented thereto, provided that you undertake for us that the Company agrees to the points undermarked in red pencil in the arabic version of the exact text of the Concession.

Wishing your existence,

Yours sincerely,
AHMED ALSABAH
True translation:
(Sgd.) J. GABRIEL

---

140. *The Political Agent's three letters to Shaikh Ahmad, all sent on 14th December 1934, were as follows:*

No. C–341

After Compliments,

I have the honour to address Your Excellency on the following important matter:—

His Majesty's Secretary of State for India, London, has directed me by telegram through the Hon'ble the Political Resident in the Persian Gulf, to inform Your Excellency that he has received a letter dated the 28th November 1934 from a certain firm calling themselves Messrs: Traders Ltd. of Newcastle-upon-Tyne, England, in which the latter state that your Excellency *has granted* an Oil concession to them covering the whole of Kuwait territory, subject to the approval of His Majesty's Government, and further request that the formal and official approval of His Majesty's Government be given *to the grant* of such concession, and that Your Excellency be informed accordingly.

2. As enclosures to their letter Messrs: Traders Ltd. also sent the following three appendices:

(a) A list of the Companies and Gentlemen associated with their application.
(b) Copy of the 'Terms' agreed upon between Your Excellency and the aforesaid firm.
(c) English translation of an autograph letter *granting the oil concession* to Messrs: Traders Ltd., said to have been sent by Your Excellency to Mr. Gabriel, Advocate of Basra.

3. His Majesty's Secretary of State for India directs me to ascertain from Your Excellency without delay whether you confirm or otherwise the information given by the firm in question, namely Messrs: Traders Ltd.

4. I would request therefore that Your Excellency be so good as to give me a very early reply to His Majesty's Secretary of State's question, as I have been asked to send a telegraphic answer to London and to Bushire.

Usual Ending.

No. C–342

After Compliments,

Reference my confidential letter No. C–341 of today's date regarding the statement of Messrs: Traders Ltd., that Your Excellency had granted them an Oil concession and that you had sent a letter to this effect to Mr. Gabriel, a copy of which was sent to His Majesty's Secretary of State for India by the said Company.

2. I am directed by His Majesty's Government to remind Your Excellency that on 23rd April 1934 I visited Your Excellency at 'Bayan' and read out to Your Excellency (word for word) telegram No. 963 dated 17th April from His Majesty's Secretary of State for India, in which amongst other things I was instructed '*to remind Your Excellency that you were not at liberty to grant any Oil whatsoever to any one without the prior consent of His Majesty's Government*', and further '*that His Majesty's Government would expect Your Excellency to consult them before you even opened any negotiations with any Company*'.

3. I am further directed by His Majesty's Government to draw your attention to the fact that on the 14th June 1934 Your Excellency gave '*your promise*' to Messrs: Holmes and Chisholm, negotiators of the Kuwait Oil Company '*that for three months as from that date, you would not receive or consider offers for an Oil concession from any party other than themselves*,' which promise was officially communicated by the Representatives of the Kuwait Oil Company to this Agency in their letter dated 16th June 1934, and by the Political Agent to the Hon'ble the Political Resident on the 18th June (telegram No. 215), and by the Hon'ble the Political Resident to His Majesty's Government in London on 20th June (telegram No. 568).

4. I am therefore instructed by His Majesty's Government to ask Your Excellency to explain the circumstance under which you secretly carried on negotiations with Messrs: Traders Ltd., and also addressed your letter dated 2nd September 1934 to Agob Effendi (Mr. Gabriel) Advocate of Basra, in face of His Majesty's Government's warning as stated in Paragraph (2) above, and in face of Your Excellency's 'guarantee' to the Kuwait Oil Company negotiators as stated in paragraph (3) above.

5. Lastly I would request that Your Excellency give me as early a reply as possible as His Majesty's Government desire a telegraphic reply.

Usual Ending.

No. C–343

After Compliments,

Reference my letters No. C–341 and No. C–342 of today's date, as well as my verbal conversation with Your Excellency regarding the communication of Messrs: Traders Coy. Ltd., to His Majesty's Secretary of State, and also regarding the letter of 2nd September which Your Excellency wrote to Agoob Effendi (Mr. Gabriel) at Basra.

I am directed by the Hon'ble the Political Resident (under orders of His Majesty's Government) to request you to ask Agoob Effendi (Mr. Gabriel) by telegraph to come to Kuwait at once and bring with him the original of the autograph letter dated 2nd September 1934, which Your Excellency sent him, in order that I may compare same with the translation which was made by Mr. Gabriel for Messrs: Traders Ltd., and which was forwarded to His Majesty's Government by the latter Company.

Usual Ending.

---

141. *The Political Agent's telegram of 16th December 1934 to London and Bushire is given below, followed by the text of the Shaikh's three letters to the Political Agent and their enclosures:*

TELEGRAM

From Political Agent, Kuwait to Political Resident in the Persian Gulf and repeated to Secretary of State for India.

Dated Kuwait, 16th December, 1934.

347. Addressed to Political Resident in the Persian Gulf, repeated to Secretary of State for India.

Secretary of State for India telegram 2954, 12th December. Saw Sheikh yesterday. He admitted writing letter 2nd September to Gabriel, copy of which he gave me (it tallies with Gabriel's translation), but denied that this is any way meant actual 'grant of' oil concession as stated by Traders, Ltd. He denied that he had acted improperly by His Majesty's Government or broke (? promise to) Koweit Oil Company. Following were facts according to the Sheikh

(a) In April, Traders telegraphed to him offer (? for, omitted) oil concession. He informed Koweit Oil Company and the Political Agent.

(b) In spite of several determined advances by Traders, Limited, he refused all discussion till expiry time limit given to the Koweit Oil Company negotiators which was August 30th, and not September 14th as stated by Chisholm and wired you by Watts.

(c) On September 1st, and as Koweit Oil Company had done absolutely nothing to meet his demands, he allowed Gabriel show him for the first time draft concession offered by Traders, Limited. He did not tell Watts as he feared to talk in presence of interpreter.

(d) On September 2nd he gave Gabriel letter in which expressed approval and readiness to sign under certain conditions.

(e) As Traders Limited made no further move he informed Gabriel on December 12th, that he was about to come to terms with Koweit Oil Company who now offered to agree to all his final demands.

(f) On December 15th, he telegraphed to Gabriel that he had arrived at agreement with the Koweit Oil Company and further conversations must cease.

2. Sheikh promised to give an explanation and information in writing called for in (a) (b) your telegram 1205 to Secretary of State for India.

Translation of a confidential letter No. R–4/1862, dated the 8th Ramadhan 1353, corresponding to the 15th December 1934, from His Excellency the Ruler of Kuwait to the Political Agent, Kuwait:

After Compliments,

I have, with pleasure, received your letter No. C–343, dated the 7th inst (14.12.34) and understood the order issued by the Hon'ble the Political Resident in the Persian Gulf, under instructions of His Majesty's Government, that I should ask Agoob Effendi to come to Kuwait and bring with him the original of the letter dated 2nd September 1934, which I sent to him so that you may compare same with the translation he made for Messrs: Traders Co.

I beg, therefore, to enclose copies of two letters, which I sent to the said Agoob, one dated the 2nd September 1934, and the other 5th Ramadhan 1353 (12.12.34) from which you will be able to know the manner in which negotiations were carried on, in respect of the above mentioned company.

I may further mention that today, I telegraphed to the said Agoob, regarding my coming to terms with the Kuwait Oil Company Ltd.

Usual Ending.

Translation of a letter dated 2nd September 1934 (22nd Jamad Auwal 1353) from His Excellency the Ruler of Kuwait to Mr. Gabriel, Advocate, Basra:

After Compliments, Etc.

My dear Agoob Effendi,

I have received two copies of the concession. One I have retained, and the other I have sent to you, so that you may dispatch it to the Company and advise them to communicate with His Majesty King George's Government in London, and submit the said copy to them. Should His Majesty's Government agree to the same, they may telegraph to the Political Agent, Kuwait, to this effect. I myself accept the terms of the concession and shall affix my signature to them when I hear the result from the Political Agent, Kuwait, namely that His Majesty's Government have approved the concession agreement and had agreed to it, provided always that you give me your word that the Company shall agree to the points marked with red pencil on the Arabic text of the Concession.

Translation of a letter dated 5th Ramadhan 1353 (12.12.34) from His Excellency the Ruler of Kuwait, to Mr. Gabriel, Advocate, Basra:

After Compliments,

I have received your letter No. 343 dated the 28th Sha'ban (6.12.34) and have understood what you stated regarding the contents of the letter received from your company.

I should therefore inform you that Messrs: Holmes and Chisholm's Company have agreed to all my demands. I have been so long alleging pretexts to them for my delaying of their question, that I have become verily ashamed of the abundance of my promises. All this time I have been waiting for your Company to make a move, but up to date, unfortunately, I have not received from them any thing of a satisfactory nature that I can depend on. It is not I, therefore, who have failed your Company by delay. Actually the matter has not as yet been quite settled with Messrs: Holmes and Chisholm's Company, but it is on the point of being so finally settled. I cannot now refuse them at this stage. You will have to excuse me, after all.

You state you have got some references to make, which you prefer not to expose by writing, but will personally disclose them to me: and that on account of press of business you could not proceed (come to Kuwait). Permit me to say that your business (i.e. looking after my date gardens) concerns you more than anything else. You will Insha Allah do your best in the administration of same. There is no longer need for you to come now.

Usual Ending.

Translation of a letter No. R–4/1863, dated the 9th Ramadhan 1353, corresponding to the 16th December 1934, from His Excellency the Ruler of Kuwait, to the Political Agent, Kuwait:

After Compliments,

I have, with pleasure, received your letter No. C–341, dated 7th inst (14.12.34), and have understood what you have stated with regard to the contents of the telegram emanating from His Majesty's Secretary of State for India in London, and dealing with the letter dated 28th November, which he received from a certain firm calling themselves Messrs. Traders Co. Ltd. and with what the said company had stated, namely that I had granted to them an oil concession covering the whole of the Kuwait territory, subject to the formal and official approval of His Majesty's Government to the grant of such a concession. That also the said company had enclosed in their letter three appendices, namely:

(a) A list of the companies and gentlemen associated with their application.
(b) A copy of the 'terms' agreed upon between me and the said company.
(c) An English translation of a letter emanating from me, granting an oil concession to the said Company, which was forwarded through Mr. Gabriel, Advocate of Basra.

And that finally His Majesty's Secretary of State for India had directed you to ascertain from me whether I confirmed or otherwise, the aforesaid statement.

I accordingly beg to lay before you the real facts: During negotiations with Messrs: Holmes and Chisholm and at one of our first meetings, Mr. Gabriel, the Lawyer, was in Kuwait. He informed me then that there was a British firm, which desired to take the Kuwait Oil concession and on terms which would satisfy me. I told him (Mr. Gabriel) then that negotiations were in progress with Messrs: Holmes and Chisholm, and that they had asked my leave to depart in order to consulting with their company on certain points, (in London) and that a decision had been arrived at between them and myself (to postpone negotiations) until the end of August 1934, during which period I was not at liberty to open negotiations with anybody else. That when that period expired, if the Company he (Mr. Gabriel) represented was a purely British concern, he could communicate with them, but on the one condition that they should apply, through their local representative (in England) to His Majesty's Government for the issue of an order to me, as without such an order I could not see my way to grant any concession.

After Messrs: Holmes and Chisholm had left (for England) and after the time limit which I promised had expired, namely on 30th August 1934, the said Mr. Gabriel brought me two copies of the terms of the concession (of Traders Ltd.) I examined these terms and found them suitable. I next wrote on the 2nd September to Mr. Gabriel a copy of which, giving exact wording I forwarded to you under my letter No. R.4/1862, dated 8th inst (15.12.34). Was there, I ask, anything in the wording of that letter from me to justify the Company in assuming (as they have done) that I had actually granted them the concession?

Usual Ending.

Translation of a letter No. R–4/1864, dated the 9th Ramadhan 1353, corresponding to the 16th December 1934, from His Excellency the Ruler of Kuwait, to the Political Agent, Kuwait.

After Compliments,

I have, with pleasure, received your letter No. C–342, dated the 7th inst. (14.12.34), and have understood what you stated therein, namely that you had been directed by His Majesty's Government to ask me to explain the circumstances under which I secretly carried on negotiations, which were contrary to the warning you conveyed to me at 'Bayan', under instructions of His Majesty's Government, and to the promise I gave to Messrs Holmes and Chisholm which the latter communicated to you in their letter dated the 16th June 1934.

I beg to say that God willing (Insha'Allah) I will never go against an order of His Majesty's Government and that I shall always obey the behests of His Majesty's Government, in every respect. The nature of the negotiations that had taken place, were as outlined in my letter to you, No. R–4/1863, dated the 8th inst (15.12.34). Indeed at the time I wished to inform Major Watts of these negotiations, but they were of a very secret nature and as he did not know Arabic, and discussions with him had to be carried on by means of an interpreter, I kept the knowledge of them to myself, for fear of being betrayed. On your return (from England) we both were kept very busy for sometime, with date-garden questions, but soon afterwards I informed you verbally and very confidentially of everything. I trust you informed His Majesty's Government at the time of what I told you.

As regards my promise to Messrs: Holmes and Chisholm that it was for a period of three months with effect from the 14th June 1934, as they presumed to be the case in their letter (which you showed me). I admit that I gave a promise to the said gentlemen, but I carefully told them that the period would expire at the end of August 1934. It was I who disclosed to them that I had received a telegram from a British firm in London, which desired to negotiate for a concession. This also you heard from me. Further, I warned them that should they fail to settle the question with me within this period, (i.e. by 30th August 1934) I should be compelled to open negotiations with the above mentioned new Company. These are the real facts.

I have never thought, nor would I think, of going against His Majesty's Government's orders, because I am quite sure that His Majesty's Government have got no other intention or wish than the peace and welfare of ourselves and our country. We are always most grateful to His Majesty's Government.

Usual Ending.

142. *Shaikh Ahmad's letter of 9th December to the negotiators was as follows:*

No. R/4/1853 — 2nd Ramadhan 1353
9th December 1934

Dear Friends Major Holmes and Mr. Chisholm,

After compliments,

We draw your attention to our letter No. R/4/1801 of 28th Rajab corresponding with 6th November 1934 with regard to the observations and the few alterations that we mentioned to you as desirable to be inserted in the draft agreement you presented to us after your return from London. We have accordingly completed (in the attached draft of the concession agreement) all the observations, alterations and cancellations required for insertion and which you will find to be in accordance with what was agreed upon with you at 'Bayan' in last June.

And we hereby inform you that we desire to add a sub-clause to article 9(B) which you will find attached in its respective place in the same article.

And you have informed us at our last interview with you that your Company has decided to accept financial terms as shown in our suggestion (A) and therefore the Company has decided to work the oil basis on the foundation of royalty of won-and-saved.

Therefore we cannot allow the insertion of sub-clauses '3' and '4' in article 3(d) as they are part of our suggestion 'B' which you have rejected, and chosen for your work the basis of the won-and-saved, suggestion (A).

REGARDING OUR RIGHT TO APPOINT A DIRECTOR TO THE COMPANY'S BOARD IN LONDON

In accordance with discussions with you we forego our right as shown above but on condition that we appoint a representative in London in lieu of our foregoing this. In addition the salary of this representative should be increased from Rs. 2000/- to Rs. 2250/- because he shall in future bear the expenses during his movements under our directions from London to Kuwait for instructions or to discuss our relations with the Company.

And we have also decided to insert a conditional clause to article 6(d) as shown below. The insertion of this clause we find is absolutely essential for the protection of our own interests and that of our state and to prevent any disputes that may arise between us and the Company.

> 'If any dispute arises as to the accuracy of the Company's accounts in connection with the amount of royalty and/or other payments due for payment to the Shaikh under this agreement the Shaikh shall have the right to appoint (after consulting H.M.G.)) a recognised firm of accountants to check the Company's registers on his behalf either in Kuwait or in London as is found necessary and he shall pay to this firm of accountants all the expenses they happen to incur for this business and the Company shall have to allow the recognised firm of accountants appointed by the Shaikh all that is necessary to facilitate the work of checking the Company's books and registers and to afford them all facilities to obtain a final check of what is due to the Shaikh. And the Shaikh shall keep secret all the information he obtains apart from the matter under dispute for which the check has been made.'

So after you present to us a copy printed and translated & after we find it corresponding with the copy mentioned and which is in our possession then you may inform your Company that we shall sign it and grant the concession in accordance with these terms.

---

143. *Stevens of Gulf in London relayed to New York Holmes's cable to him of 12th December as follows:*

/2 Dec. 12th Your/1 from Stevens Private cable received from Holmes today though somewhat mutilated is substantially as follows:
Quote Strictly confidential for Stevens I had personal interview with British Political Agent Kuwait he had received from His Majesty's Government additional information by telegram as follows:
Quote Following particulars Traders Limited sponsors all are so powerful that HMG cannot decline to accept their application because Traders offer much more above Kuwait Oil Co. Stop HMG has been told that question is to be asked in Parliament as to the reason why American capital is being encouraged in preference to British capital in Kuwait stop HMG cannot do otherwise than agree to consider Traders application in view of Shaikh of Kuwait's letter to Traders and because influential Stop B.P.A.K. cabled HMG time has arrived when Kuwait Oil Co. must be informed officially of Traders application activity Unquote

---

144. *The Political Resident's telegram of 14th December to the India Office was as follows:*

From Political Resident in the Persian Gulf to
Secretary of State for India.
Dated Bushire Sub, 14th December, 1934.
1228. Addressed to Secretary of State for India, copy sent by sea mail to Government of India,

and repeated to Koweit, for information. Political Agent, Koweit telegraphs that he is suspicious that Traders, Limited, represents efforts by Anglo-Persian Oil Company to block success of Koweit Oil Company.

---

145. *KOC's telegram of 13th December 1934 to the negotiators authorising them to accept the Shaikh's proposals and sign the agreement:*

From London
Date 13th December 1934
No. 39

We accept proposal(s) of the SOK and authorise you to sign concession accordingly namely draft No. 5 with financial terms as specified reference our telegram No. 32 and alterations as specified reference your telegram No. 41 reference your telegram No. 45 fullstop

You should communicate our acceptance to the SOK forthwith by letter fullstop

No attempt should be made to discuss insertion sub-clause D.3 and 4 of alternative article 3 since our purpose is to accept forthwith the terms upon which SOK stated his letter he would sign and grant concession fullstop

How do you intend to prepare copies of concession for signature and when do you expect concession will be signed fullstop

Express directors' compliments to the SOK in suitable terms and their satisfaction with conclusion agreement.

---

146. *The negotiators wrote to Shaikh Ahmad on 14th December as follows:*

14 December 1934

Your Excellency,

After compliments.

We have the honour to refer to Your Excellency's letter No. R/4/1853 of 9th December.

We beg to inform you, on behalf of the Kuwait Oil Company, that we have accepted all the amendments proposed by Your Excellency in that letter. In accordance with Your Excellency's request we are having the necessary copies of the concession documents printed and translated, showing all the amendments required by Your Excellency and also the financial terms A as shown in Your Excellency's letter No. R/4/1801 of 6th November 1934. On completion of these documents we shall be happy to present them for Your Excellency's signature.

The Directors of the Kuwait Oil Company in London, in authorising us to inform Your Excellency as above, desire to present their respectful compliments to Your Excellency, and to express their great satisfaction at the Agreement thus come to between Your Excellency and themselves.

We ourselves also wish to express to Your Excellency our happiness at this favourable outcome of these negotiations, during which we have enjoyed so much kindness and consideration from Your Excellency.

Your Excellency's obedient servants & friends
for Kuwait Oil Co. Ltd.

---

147. *The negotiators confirmed the agreement arrived at concerning payments to the Shaikh for travel expenses and petrol supplies as follows:*

23rd December, 1934.
16th Ramadhan, 1353.
CH/1

H.E. Shaikh Sir Ahmad al Jabar As Subah,
K.C.I.E., C.S.I.,
KUWAIT.

Your Excellency,

After compliments.

The Directors of the Kuwait Oil Company have heard with great pleasure of the possibility of Your Excellency shortly making a visit to London. The Directors much look forward to such an opportunity of meeting Your Excellency in person.

On the occasion of the signing of the Agreement so happily concluded today between Your Excellency and the Kuwait Oil Company Ltd., and as a mark of the personal respect and esteem in which they hold Your Excellency, the Directors propose to pay into Your Excellency's Bank the sum of Rupees twentyfive thousand (Rs. 25,000) which they hope may contribute towards the expense of a visit to London by Your Excellency.

We have the honour to be,
Your Excellency's obedient servants
and friends,
(sgd.) F. Holmes A. Chisholm
for KUWAIT OIL COMPANY LIMITED

NO.CH/
23rd December, 1934.
16th Ramadhan, 1353.

H.E. Shaikh Sir Ahmad al Jaber as Subah,
K.C.I.E., C.S.I.,
KUWAIT.

Your Excellency,

After compliments.

With reference to the Agreement signed today between Your Excellency, and the Kuwait Oil Company Limited and to Your Excellency's request for supplies of petrol for your requirements, we have the honour to inform you that, during the continuance of that Agreement, in lieu of supplying petrol or other petroleum for those requirements the Company will pay to Your Excellency within thirty (30) days of the date of signature of the Agreement and on each anniversary of the date of signature until the Company declares that petroleum has been found in commercial quantities the sum of Rupees Five Thousand (Rs. 5000) but on each anniversary of the date of signature after declaration by the Company that petroleum has been found in commercial quantities or after the expiration of the 12 years mentioned in article 11(B) of the Agreement, whichever shall be the earlier date, such payment shall be increased to Rupees Ten Thousand (Rs. 10,000).

We should be glad if Your Excellency would acknowledge your acceptance of the terms of this letter.

We have the honour to be,
Your Excellency's obedient servants and friends.
(Sgd.) F. H. A. Chisholm
for KUWAIT OIL COMPANY LIMITED

---

148. *Stevens relayed to Gulf in New York Holmes's telegram to him of 16th December as follows:*

/6 Dec. 17th from Stevens: Private telegram received from Holmes today
Quote for Stevens: Have your telegram Dec. 15th Stop Received Telegram from Kuwait Oil Co. accepting Shaikh's draft on Friday stop Immediately advised Shaikh by official letter Friday evening of Kuwait Oil Co.'s acceptance also advised British Political Agent Kuwait Dec. 15th stop Have telegraphed today Kuwait Oil Co. strongly advocating that His Majesty's Government be asked to inform British Political Resident Persian Gulf that Kuwait Oil Co. has accepted Shaikh's demands in full stop Holmes interviewed Shaikh and Shaikh agreed to telegraph Gabriel that he had granted Kuwait concession to Kuwait Oil Co. and he was not prepared to discuss with Gabriel further stop Both urgent telegram and confirming letter went to Gabriel Dec. 15th stop T. C. W. Fowle is still inclined argue about 6 if he is able hence desire for His Majesty's Government to telegraph British Political Resident Persian Gulf that demands accepted by Kuwait Oil Co. stop Since writing above British Political Agent Kuwait called and showed to me telegram that His Majesty's Government had sent BPRPG suggesting that Kuwait Oil Co. and Shaikh could not assume their combined agreement *ipso facto* carries compliance with His Majesty's Government stop Shaikh called my house and stated Shaikh is ready to sign on receipt of copy printed concession stop Fowle is expected to arrive here Dec. 17th *Unquote*.

---

149. *The negotiators had cabled KOC in London on 3rd June 1934 from Kuwait as follows:*

No. 30 As requested by Shaikh we interviewed him yesterday. Shaikh informed us that if your reply to his demands does not constitute full acceptance as in draft number 4 and financial terms he does not propose to discuss concession further with us until September. He definitely instructed us in that case to return to London forthwith and stated that in our absence he will undertake not to entertain or discuss any offers from other parties. He says there is no need to return before September or later if circumstances justify. We asked his reason for this move and he said it was:

1. to avoid prolongation of negotiations into hot weather which uncomfortable for him and for us, further he has other pressing business attend to which our negotiations are holding up; and
2. because he is convinced points now in question can only be settled by us personally in London, system exchanging letters and telegrams having proved unsatisfactory to both parties.

In our opinion Shaikh's attitude confirms what we have telegraphed reference our telegram No. 27. Shaikh is most anxious to come to an agreement with Kuwait Oil Company but feel you do not appreciate his position which can only be rectified by verbal discussion with us. Owing to our long standing relations with Shaikh personally he believes we will advocate his reasonable demands. He has therefore broken off negotiations with us and refuses to further talk on details concession until we have represented his views and his reasons to you and brought back your final reply with full powers to conclude. If Shaikh's instructions as above are not complied with it will render our position impossible therefore it is our opinion that failing full acceptance draft number 4 you do not present any new suggestions until you have discussed position and Shaikh's ideas with us in London.

Early instructions requested.

---

150. *The meeting of British Government departments in London on 19th December 1934 to discuss the granting of the concession to KOC was minuted as follows:*

KOWEIT OIL

A meeting was held at the India Office on 19th December to discuss the applications of the Kuwait Oil Company and of Traders Ltd. for the Koweit oil concession.

Present:	
Mr. Walton Mr. Laithwaite Mr. Crombie	India Office
Mr. Starling	Petroleum Department
Mr. Rendel Mr. Johnstone Mr. Malcolm	Foreign Office
Mr. Seal Commander Dickson	Admiralty

Mr. Walton circulated copies of a confidential report which had been obtained from Sir Henry Strakosch on the subject of Traders Ltd., briefly recapitulated the history of the case as disclosed in the correspondence which had been circulated to the interested Departments, and explained the reasons for which circulation had not been made until 17th December.

Mr. Starling pointed out that in view of the issues raised in this matter which it was considered involved more than a local question, and might have already have been raised in Parliament, he was instructed to say that his higher authorities were surprised that the Petroleum Department had not been informed of the intervention of Lord Lloyd at an earlier date. This course had been suggested by Mr. Rendel in his note of the 29th November. He reminded the meeting that the Petroleum Department had not agreed with some of the other Departments at an earlier stage in the Koweit Oil negotiations, before the APOC had become associated with American interests, a step which he felt sure the Company would contend had been forced upon them, as a result of the policy of H.M.G. as the only means of securing for British interests a share in any Koweit oil concession which might be granted. In February and March, 1933, the Secretary for Mines had written personally to the Secretaries of State for the Colonies and India urging that it was a mistake to treat the Americans so favourably, and pressing the claims of the APOC. The Petroleum Department's line all along had been to support the APOC. They appreciated that British oil interests must in certain directions work in with other big oil interests, but they regarded the APOC as entitled to special consideration in the Gulf. The instructions which he had now received from his Minister were to press at this meeting that in view of what had happened in the earlier negotiations, it could not now be reasonably contended that some other Company, even a hundred per cent. British concern, had a prior right to the APOC. The non-official approach which Traders Ltd. had made to the Shaikh was very similar to what had happened in the case of the Eastern and General Syndicate at Bahrein. In that case an apparently British concern obtained the concession only for HMG to discover shortly afterwards that it had assigned its rights to a purely United States Company. He wished to stress the fact that as long ago as the 5th March last the Kuwait Oil Company had signed an Agreement with H.M.G. covering the safeguards which were considered necessary to protect British interests and had been given to understand that if they came to terms with the Sheikh on a basis which H.M.G. could approve, as being in the Sheikh's interests, their approval would be given to the grant of the concession.

Mr. Walton said that the whole position had been explained in general terms a day or two before to the Kuwait Oil Company who had been unable to understand why, after their long negotiations and now that they had reached an agreement with the Sheikh, H.M.G. could not approve it at once. The question was now for decision by higher authority and the provisional view of the India Office was that H.M.G. should approve the Sheikh's grant of a concession to the Kuwait Oil Company. Their reasons were:—

(1) The Sheikh's wishes: He had apparently changed his mind since September and now proposed to accept the Kuwait Oil Company's concession. This was of much importance.

(2) The manner in which Traders Ltd. had handled the question: They had acted wrongly in approaching the Sheikh direct and especially in not informing H.M.G. They had let almost three months elapse before approaching the India Office after receiving the Sheikh's letter of 2nd September.

(3) A comparison of the merits of the two applicants: The information furnished by Sir Henry Strakosch, which the Petroleum Department might perhaps be able to supplement, seemed to show that although Traders Ltd. were financially solid they did not apparently have that intimate knowledge and experience of exploiting oil concessions which the APOC and the Gulf Oil Company were known to have. It did not seem so probable that Traders Ltd. would work the concession themselves as that the Kuwait Oil Company would do so. It seemed more probable that they would transfer it to another Company or throw it back on the Sheikh after a short

attempt to work it. It seemed better to go to a Company which was willing and able to work the concession if there was any oil there.

(4) The Commitments of H.M.G. to the Kuwait Oil Company: H.M.G. were already very closely committed to the Kuwait Oil Company. H.M.G. might have to face criticism whatever action they took; but if they turned down the Kuwait Oil Company the criticism from the British and American interests behind it would be, to say the least of it, pungent.

Mr. Rendel suggested that it was very important to be as far as possible consistent in these intricate negotiations. We had committed ourselves by our past policy to acquiescing in allowing the Sheikh himself to grant a concession subject to certain essential conditions being fulfilled. H.M.G. should therefore make it quite clear on this occasion that these conditions having been fulfilled their action was limited to approving the Sheikh's grant of a concession. It seemed important so far as possible to place the responsibility for the final decision on His Excellency.

Mr. Walton pointed out that H.M.G. had never committed themselves to letting the Kuwait Oil Company have the concession if a competitor appeared in the field. Sir Findlater Stewart had drawn attention to the fact that Traders Ltd. had in their possession a letter from the Sheikh which on the face of it was a definite grant of a concession and on which, if the Sheikh were a private individual, there might conceivably even be a question of an action in the Courts, except for the fact that the grant was *subject to the approval of H.M.G.* He suggested that on that letter they would be in a strong position to criticise H.M.G. if the latter now failed to take the merits of the competitors into consideration.

Mr. Rendel suggested that the Sheikh's subsequent letter of 12th December was relevant in this respect, and must be regarded as definitely superseding the letter of September 2nd.

Mr. Walton agreed but said that the main point in the letter of 2nd September was that the Sheikh put the onus of approval on H.M.G. Sir Findlater Stewart would like to be satisfied that on merits the Kuwait Oil Company was the better potential concessionaire, but he felt that if the concession went to the latter and H.M.G. were criticised they would not be able to depend solely on the Sheikh's having chosen, but would have to take their stand on the merits of the case as well.

Mr. Rendel suggested that the line to be taken by H.M.G. in answering criticism might be as follows:— This matter of the Koweit Oil concession had a very long history. At first there had been two companies competing for a concession. H.M.G. had then laid down, and committed themselves to, a certain policy namely that, subject to certain essential safeguards for British interests and for the Sheikh himself, there should be a fair field and no favour and that the Sheikh himself should decide. He reminded the meeting that correspondence had passed on this subject (e.g. with the U.S. Government) which was accessible to persons outside H.M. Service. H.M.G. could base their defence on the fact that they had consistently followed the policy which they had then publicly laid down. It was true that the Sheikh's letter of 2nd September purported to give a concession to the Company subject to certain conditions, but it was obvious from that letter that the Company should have got into immediate touch with H.M.G. Instead of this the Company's first approach had been more than two months later, when they had allowed Lord Lloyd to telephone through a circuitous channel and only privately and confidentially an obscure and inaccurate hint. Even before Lord Lloyd had begun his private and confidential messages the Sheikh himself in October has resumed his negotiations with the Kuwait Oil Company. H.M.G. had been kept entirely in the dark by Traders Ltd., and in these circumstances it had been impossible for them, in view of their past policy, not to commit themselves to the Kuwait Oil Company, who had accepted their conditions and were on the point of reaching an agreement with the Sheikh.

Mr. Seal suggested that it was precisely this policy to which H.M.G. had committed themselves in 1932 which Lord Lloyd would criticise.

Mr. Rendel suggested that, in order to attack the policy of H.M.G. Lord Lloyd would have to take the line that H.M.G. had refused a fair opportunity to compete to Traders Ltd. an ostensibly British company. H.M.G.'s policy, however, had been laid down long before Traders Ltd. entered the field and there had been no information of the existence of this Company until H.M.G. were completely committed to the Kuwait Oil Company owing to the latter's acceptance of H.M.G.'s conditions. It was true that in 1932 there was a purely British Company in the field, namely the APOC; but he did not think that Lord Lloyd would take the line that H.M.G. ought to have supported the APOC at that time, since his attitude to the APOC was now definitely hostile. What Lord Lloyd would say was that H.M.G. should have given the concession to Traders Ltd., his own protégés. Against this attack we had an extremely good defence, namely, that Lord Lloyd had been much too late in bringing them forward.

Mr. Walton said that the fact that the Sheikh had now chosen the Kuwait Oil Company, coupled with the policy which H.M.G. had laid down in 1932, was certainly a powerful argument against Traders Ltd.

Mr. Rendel agreed. H.M.G. might reasonably say that they had had no opportunity of changing that policy in the absence of another purely British competitor for the concession. Had Lord Lloyd produced his purely British company earlier, H.M.G. might have had an opportunity of reconsidering their policy, though this did not necessarily mean that they would have done so.

Mr. Starling suggested that if the concession were to be given to Traders Ltd. H.M.G. would then have to meet the argument that there had once been a purely British company in the field, namely, the APOC, and that they had failed to support it.

Mr. Rendel suggested that the immediate point at issue was to decide how to meet the criticism which might be expected from Lord Lloyd that H.M.G. had unjustifiably sacrificed an important British interest in allowing the concession to go to the Koweit Oil Company instead of to Traders Ltd. But it would be quite compatible with this for H.M.G. to produce other arguments later in defence of their past policy.

Mr. Laithwaite agreed that it was a good argument that in the absence of any declared competition with the Kuwait Oil Company there had been neither reason nor opportunity for H.M.G. to reconsider the policy laid down in 1932. If Traders Ltd. had approached H.M.G. last April, H.M.G., if satisfied as to their qualifications etc., would have had to allow them to compete for a concession. Traders Ltd. had in fact postponed their application until 28th November, by which date (on 26th November) the Kuwait Oil Company had reached agreement with the Sheikh save on certain unimportant 'final points'.

Mr. Rendel suggested that the possibility ought also to be considered of Lord Lloyd taking the line that in April the Sheikh had informed H.M.G. that there was a purely British group in the field, but that in spite of this H.M.G. had not slowed down the Kuwait Oil Company's negotiations. He might say that when his protégés obtained their concession from the Sheikh on 2nd September, they had not understood that they ought to have approached H.M.G. at once, and he might insinuate that when they did ultimately do so on 28th November, H.M.G. had put up the Sheikh to refusing their offer. It would be necessary to be ready with replies to these arguments.

Mr. Starling said that the answer to this was that in response to a definite enquiry the Sheikh had declined last April to give the name of the British company which had made him an offer. Further the Sheikh had been told at that time that he must consult H.M.G. before entering into negotiations with any group.

Mr. Walton referred back to the point that the Sheikh had put the onus of making a decision upon H.M.G. The position was that the Sheikh had informed first Traders Ltd. and then the Kuwait Oil Company that he would grant them concessions.

Mr. Seal suggested that the offer which the Sheikh had made to Traders Ltd. on 2nd September could not be regarded as open for acceptance indefinitely, and that the Sheikh was perfectly justified in informing the Company on 12th December that the offer was now withdrawn.

Mr. Walton said he agreed with this up to a point, but that Lord Lloyd could make a good deal of capital out of the letter of 2nd September. That was why Sir Findlater Stewart suggested that H.M.G. might have to base their case partly on the qualifications of the two companies. At any rate it would be well to have in reserve an argument on these lines. He asked the Petroleum Department whether H.M.G. could say that prima facie, on a comparison of the two applicants, Traders Ltd. were not so likely as the Kuwait Oil Company to make a good concessionaire?

Mr. Starling said that the Petroleum Department felt clear on this point, but some of the reasons which influenced the Department in reaching its conclusion might be difficult to disclose in a discussion in Parliament. Traders Ltd. might, for example, come forward with still further associated companies. From the company's articles of association there was nothing to show that they were competent to do much except obtain concessions.

Mr. Walton suggested that we could not leave out of account the interests backing the Company.

Mr. Laithwaite said that the difficulty would be to meet the argument in Parliament that without considering in detail the status of Traders Ltd., their ability to work concessions, etc., H.M.G. had approved the grant of the concession to the Kuwait Oil Company. It was important to find as much justification for this course as possible. Apparently the financial position of the Company was sound. But were they equally sound technically? Were we in a position (without a detailed investigation, which would be a lengthy process) to say at once that they did not compare with the Kuwait Oil Company from this standpoint?

Mr. Starling said that looking at the two companies it could be said that prima facie the Kuwait Oil Company was undoubtedly the stronger group of the two. Traders Ltd. were not equipped to search for, exploit or sell oil in large quantities. Huntings, being a tanker shipping company, might sell a little as bunker oil, and Berry Wiggins, a comparatively small concern, might take small quantities for refining. Traders Ltd. had produced no evidence that they had the necessary qualifications for working a concession. With regard to the B.O.D. connection of Traders Ltd., Lord Glenconner, who was one of the Directors of the Company, was associated with the notorious Mr. Rickett in an enterprise known as the Anglo-Iraq Transport Company which had a distinctly B.O.D. flavour. The Foreign Office had supplied a report on this concern in which it was suggested that it was an extremely unsavoury business, of the sort where British people came in, made money, and then got out again as quickly as possible. Mr. Rickett was still a large shareholder in the B.O.D. He also referred to a report of an interview given a few months ago to Colonel Bovill and Mr. Hunting as representing the B.O.D. at the Admiralty (he had Mr. Seal's permission to mention this) when they told a fantastic story about the enormous quantity of oil which had been discovered in the area west of the Tigris.

Mr. Seal intervened to say that they had asked the Admiralty not to pass this information on to other Department, not even the Petroleum Department whom they regarded as wholly identified with the APOC interests; he therefore asked that the matter should be treated as confidential. *Mr. Starling* continuing said it was difficult to know why these gentlemen, who were also acting for Traders Ltd., should attempt to give information of this character to a Government Department especially as although the B.O.D. had had their concession for more than two years as yet they had failed to discover oil in commercial quantities and had had considerable difficulties in finding the money to pay their last dead rent of £125,000 (gold). These facts seemed clearly to justify H.M.G. in regarding the Kuwait Oil Company as a definitely better concern than Traders Ltd. and their associates.

Mr. Rendel suggested that there seemed to be three matters for decision:—

(1) Whether H.M.G. ought now definitely to approve the Sheikh's grant of a concession to the Kuwait Oil Company;
(2) In the event of their doing so how to meet the expected criticism from Lord Lloyd;
(3) The actual machinery of carrying out their decision.

The first two had already been discussed. With regard to (3) he still thought that H.M.G. should put the onus of making a decision on the Sheikh.

Mr. Walton agreed that H.M.G. might do that in the first instance, but said that Sir S. Hoare in considering the matter would attach great importance to the question whether Traders Ltd. were definitely not likely to be as good a concessionnaire as the Kuwait Oil Company. Were we in a position to say this?

Mr. Starling said he thought we were. Besides the points which he had mentioned a most important point was that there was nothing to show that Traders Ltd. would be able to market their oil if they got the concession.

Mr. Rendel pointed out that it was urgent to get a decision on the matter if possible before Christmas. The Sheikh was a slippery individual. He was now ready to clinch with the Kuwait Oil Company, but there was a danger that if any delay took place he might change his mind again. In view of this it was important that H.M.G. should not be responsible for delaying matters. It was impossible for H.M.G. definitely to go against the Kuwait Oil Company, and he agreed that they should decide in their own minds in favour of the latter, but he thought it preferable, and Sir Robert Vansittart, with whom he had briefly discussed the whole question, had agreed, that H.M.G. should not themselves veto the application of Traders Ltd., but should fix the responsibility on the Sheikh, which could easily be done in view of His Excellency's recent communications to that Company.

Mr. Walton said that the India Office would have to write letters both to Traders Ltd. and to the Kuwait Oil Company. These letters would require very careful drafting.

Mr. Rendel suggested that the India Office might remind the Kuwait Oil Company that the understanding reached earlier in the year had been that H.M.G. would raise no objection to the Company obtaining a concession from the Sheikh provided certain conditions were fulfilled. These conditions had now been complied with and the Sheikh was now ready to sign the concession. Therefore in view of their earlier understanding with the Company, H.M.G. had no objection to the Sheikh's granting the concession to the Company. The letter to Traders Ltd. presented more difficulty, and he suggested that the India Office might explain that the policy of H.M.G. had for some years been to allow the Sheikh a free choice in the matter of granting oil concession, subject to certain specific conditions. These conditions were

(1) that any Company should communicate with H.M.G. before entering into negotiations and that the Sheikh should not enter into negotiations with any company without prior consultation with H.M.G.;
(2) that the essential interests of H.M.G. were properly safeguarded;
(3) that the interests of the Sheikh were also properly safeguarded.

Neither the Sheikh nor Traders Ltd. had informed H.M.G. of their negotiations. The Company had failed to communicate to H.M.G. the Sheikh's letter of 2nd September, despite the fact that the first paragraph of that letter, with its request for a telegraphic reply, made it clear that it was intended that the Company should at once get into touch with H.M.G. On the other hand another Company (which appeared to be the only one in the field) had complied with all the conditions mentioned above, and had, in view of this, been given to understand that H.M.G. would raise no objection to their obtaining a concession from the Sheikh if he would grant it. The Sheikh now informed H.M.G. that he was ready to grant a concession to this Company, and in view of their earlier understanding with the Company H.M.G. had given their approval to the Sheikh's decision and were unable at this late stage to reconsider the matter.

Mr. Laithwaite suggested that they might also say that even in so far as the Company had received the letter of 2nd September the Sheikh had denied that this letter was in any way intended to represent the definite grant of a concession. In any event the Sheikh said that he had heard nothing from the Company since 2nd September and there was nothing to show that the Company had in fact accepted the conditions put forward by the Sheikh.

Mr. Seal pointed out that the weakest link in the argument was that the Sheikh's letter of 12th December did not antedate the application received from Traders Ltd., but he agreed that H.M.G. could, if necessary, answer effectively any insinuation that H.M.G. had put the Sheikh up to writing his letter of 12th December to the Company, since it was clear from the correspondence that the Sheikh had not seen the Political Agent or the Political Resident before writing that letter.

After further discussion it was agreed that the India Office should draft letters to Mr. Hunting and to the Kuwait Oil Company. Advance copies of the drafts would be circulated to other departments and the India Office would simultaneously submit them to the Secretary of State for India with the results of the meeting.

Copies of the two letters [*see Note 153*] as eventually concurred in by other Departments concerned and despatched on December 21st to the Kuwait Oil Company and to Traders Ltd. respectively, are now attached (Annex).
INDIA OFFICE
9th January, 1935.

151.

TELEGRAM

To London
Date 20th December 1934
No. 48
Completed documents arrived from Cairo yesterday fullstop ten copies despatched to you from Cairo 17th December fullstop we have checked them and they are ready for signature fullstop we should like to sign Saturday but await H.M.G. approval for which we understand P.R. has also telegraphed fullstop documents will be signed by Shaikh ourselves and P.A.

152.

20th December 1934

H.E. Shaikh Sir Ahmad Al Jaber As Subah,
K.C.I.E., C.S.I.,
Ruler of Kuwait.
Your Excellency,

After compliments.

We have the honour to inform Your Excellency that we received the printed concession documents by aeroplane yesterday, and immediately sent two copies to Your Excellency as you requested.

We have now carefully studied these documents and find that they are exactly similar to that enclosed with Your Excellency's letter to us No. R/4/1853 of the 9th December.

We should be glad to learn from Your Excellency when it is your wish that the formalities of signing should take place.

Your Excellency's obedient
servants and friends
Frank Holmes A. Chisholm
for KUWAIT OIL COMPANY LIMITED

No. R/4/1871 *Kuwait*, 13th Ramadhan 1353.
20th December 1934.

(*TRANSLATION*)

Dear Friends Major Holmes and Mr. Chisholm,

After compliments; I have received your letter of today's date and noted what you conveyed through same, and accordingly I have to acknowledge that yesterday I received the two printed copies of the Kuwait Oil Concession Agreement, and wish to inform you that on completion of my examination of these copies I shall communicate with you about fixing a suitable date for the formalities of signing the document, as soon as possible.

May God protect you.

Yours sincerely.
sd/ Ahmad Al Jabir As-Subah

---

153. *The India Office's letters of 21st December to KOC and Traders Ltd. were as follows:*

India Office to Koweit Oil Company

December 21, 1934.

Gentlemen,

I AM directed by the Secretary of State for India to refer to your interview at the India Office on the 17th December, in which you reported that agreement had now been reached between you and his Excellency the Sheikh of Koweit on the subject of the terms of an oil concession in respect of Koweit, and that his Excellency was prepared to sign the concession in your favour immediately, subject to the approval of His Majesty's Government.

2. It was understood in the earlier discussions between your company and His Majesty's Government that no objection would be raised to the grant of a concession by the Sheikh to your company, subject to certain specified conditions. These conditions have been fulfilled, and having regard to the understanding in question and to the fact that the Sheikh has now decided to grant the

concession to your company, I am directed to inform you that His Majesty's Government have approved the Sheikh's decision and to say that it is assumed that you will now take the necessary steps to complete the signature of your agreement with his Excellency.

I am, &c.
J. C. WALTON

India Office to Mr. P. L. Hunting, Traders Ltd.

December 21, 1934.

Sir,

I AM directed by the Secretary of State for India to state that he has had under consideration your letter of the 28th November on the question of the grant of an oil concession in Koweit to Messrs. Traders Limited.

2. On receipt of your letter of the 28th November, which was the first intimation which he received from your company that they were interested in the matter, the Secretary of State took steps to ascertain from the Sheikh of Koweit the position in regard to his relations with Traders Limited. The Sheikh has now stated, in reply to this enquiry, that the letter of the 2nd September to Mr. Gabriel, of which a copy was enclosed in your letter under reference, was not in any way intended to represent the grant of a concession to your company by him. He added that, as he had heard nothing further from your company subsequently to the date of that letter, he informed Mr. Gabriel on the 12th December that he was about to come to terms with the Koweit Oil Company, and that on the 15th December he telegraphed to Mr. Gabriel that he had done so and that further conversations with your company must cease.

3. I am to explain that His Majesty's Government have for some considerable time been guided in regard to the question of the grant of oil concessions in Koweit by certain definite principles which they felt bound to adopt in view of the special circumstances of the case, and which are well known to the sheikh and to the various applicants who have from time to time approached His Majesty's Government in the matter. The effect of these principles is that the responsibility for deciding as to the grant of an oil concession shall rest with his Excellency, subject to the following conditions:—

(a) That the sheikh (and similarly any applicant for a concession) should consult with His Majesty's Government before entering into negotiations for a concession, and that His Majesty's Government should be kept fully informed of all negotiations in connexion with the grant of any concession.

(b) That arrangements should be made to ensure that the interests of His Majesty's Government should be specifically safeguarded in the event of any concession being granted.

(c) That the interests of the sheikh himself and of his principality should be sufficiently safeguarded in any concession granted.

4. As you are aware, the Koweit Oil Company have for a considerable period been in negotiation with the Sheikh. Before initiating those negotiations, the Koweit Oil Company placed themselves in touch with His Majesty's Government and were informed of the conditions set out above. Having undertaken to satisfy those conditions, they were informed that no objection was seen to their initiating negotiations with his Excellency, and they were also given to understand, long before it was known that your company were in the field, that, in the event of their reaching agreement with the sheikh as to the terms of a concession within the limits indicated above, the approval of His Majesty's Government to the grant of such concession would not be withheld.

5. In the case of the negotiations between Messrs. Traders Limited and the sheikh, no approach was made to His Majesty's Government by your company, and no information that they desired to secure a concession in respect of this area was received by His Majesty's Government until the 28th November, although from the papers submitted with their application on that date it appeared that discussions with the sheikh had reached an advanced state by the 2nd September. Despite the request in the opening paragraph of the sheikh's letter of the 2nd September (which, as explained above, the sheikh states that he did not in any way regard as constituting a grant of a concession to your company) that your company should place themselves in communication with His Majesty's Government and obtain a telegraphic reply, a delay of almost three months took place before any communication was made to His Majesty's Government by your company, by which time the negotiations between the sheikh and the Koweit Oil Company had been completed except for one or two minor points.

6. Having regard to the understanding with the Koweit Oil Company referred to in paragraph 4 above, to the fact that the conditions referred to in paragraph 3 have been satisfied by that company, and that the sheikh, in the exercise of his choice, has elected to grant a concession to that company, and to the communication which his Excellency states he has made to Mr. Gabriel, the Secretary of State now directs me to say that His Majesty's Government are not prepared to interfere with the decision which the sheikh has reached. They have therefore informed his Excellency, in accordance with the understanding previously reached with the Koweit Oil Company, that they concur in his granting the concession to that company.

I am &c.
J. C. WALTON

154. *Exchange of correspondence between Mr Hunting of Traders Ltd., the India Office and Shaikh Ahmad, December 1934–July 1935:*

The Under Secretary of State for India,
Political Department,
India Office,
Whitehall,
London, S.W.1.

31st December, 1934.

Sir,

With reference to my application of 28th November, 1934, regarding the grant of a Concession in Kuwait to Messrs. Traders Limited, I have the honour to acknowledge receipt of your reply thereto dated 21st December, 1934, under No. P.Z. 7815/34.

In view of all the correspondence etc. in my possession, I cannot, on behalf of Messrs Traders Limited, accept the ruling of the Secretary of State in this matter as being final. The whole question is now in the hands of my Legal Advisers and I shall communicate with you again in due course.

I am, Sir,
Your obedient servant,
P. Ll. Hunting
For and on behalf of
TRADERS LIMITED

PLH/WIN
The Under Secretary of State for India,
Political Department,
India Office,
Whitehall,
London, S.W.1.

21st January, 1935.

Sir,

I have the honour to request that you will be kind enough to acknowledge receipt of my letter dated 31st December, 1934.

I have the honour to be, Sir,
Your obedient servant,
For and on behalf of
TRADERS LIMITED.

India Office
Whitehall,
London, S.W.1.
22nd January, 1935.

P. Ll. Hunting, Esq.,
Messrs Traders Limited,
Cunard House,
Leadenhall St.,
E.C.3.

Sir,

I am directed by the Secretary of State for India to acknowledge the receipt of your letter of 21st January, and of your letter of 31st December, reference PLH/WIN. A formal acknowledgement of your letter of 31st December was addressed to you on the 2nd January.

I am, Sir,
Your obedient Servant,
J. C. WALTON

Your Ref: P.Z. 7815/34
Our Ref: PLH/WGL

24th July, 1935.

The Under Secretary of State for India,
India Office,
Whitehall,
LONDON S.W.1.

Dear Sir,

With reference to previous correspondence regarding the grant to Messrs: Traders Limited of a concession to drill for oil in Kuwait, and with particular reference to my letter of 31st December last.

I have the honour to forward for your information copy of a letter sent by Messrs. Traders Limited to the Shaikh of Kuwait.

I have the honour to be,
Sir,
Your obedient Servant,
P. Ll. Hunting
For and on behalf of
TRADERS LIMITED

PLH/WGL 24th July, 1935.

His Excellency Shaikh Sir Ahmad Al-Jabir As-Subah,
K.C.I.E., C.S.I.,
Hans Crescent Hotel,
Hans Crescent,
LONDON, S.W.1.

Your Excellency,

With reference to the concession to drill for oil in the territory of Kuwait granted to Messrs. Traders Limited by your letter of 2nd September, 1934, and since alleged to have been given to the Kuwait Oil Company by Your Excellency; in view of the original correspondence and direct concluded negotiations that took place with your Excellency through our accredited agent, as informed His Britannic Majesty's Government on 31st December, 1934, it is not the intention of Messrs. Traders Limited to accept your Excellency's decision to grant this concession to Messrs. The Kuwait Oil Company.

I have to add that Messrs. Traders Limited have placed the matter in the hands of their legal advisers with a view to proceeding being taken to contest the validity of this concession.

I have the honour to remain,

Yours faithfully,
For and on behalf of
TRADERS LIMITED

Copy for information:—
The Under Secretary for India,
Whitehall,
London, S.W.1.
Reference Traders Limited's letter to the India Office dated 31st December, 1934.

---

155. *Eighteen copies of the Concession Agreement document were signed on the 23rd December 1934, six being leather-bound copies with maps annexed (as mentioned in Article 1 of the Agreement) and twelve paper-bound. Of these Shaikh Ahmad retained one leather-bound copy and three paper-bound, the negotiators four leathcr-bound and seven paper-bound, and the Political Agent one leather-bound (which was registered and recorded in the Agency archives) and two paper-bound. Each leather-bound copy had two maps of Kuwait annexed, one small-scale showing Kuwait's position in the Gulf, the other larger-scale (eight miles to the inch).*

The Concession Agreement document was printed in English and Arabic (as mentioned in Article 21 of the Agreement) in parallel columns. The English text, with the annexed maps, was as follows:

IN THE NAME OF GOD THE MERCIFUL. This is an AGREEMENT made at Kuwait on the 23rd day of December in the year 1934 corresponding to 16th day of Ramadhan 1353 between His Excellency Shaikh Sir Ahmad al-Jabir as-Subah, Knight Commander of the Most Eminent Order of the Indian Empire and Companion of the Most Exalted Order of the Star of India, the SHAIKH OF KUWAIT in the exercise of his powers as Ruler of Kuwait on his own behalf and in the name of and on behalf of his heirs and successors in whom is or shall be vested for the time being the responsibility for the control and government of the State of Kuwait (hereinafter called 'the Shaikh') and the Kuwait OIL COMPANY LIMITED a Company registered in Great Britain under the Companies Act, 1929, its successors and assigns (hereinafter called 'the Company').

Article 1

The Shaikh hereby grants to the Company the exclusive right to explore search drill for produce and win natural gas asphalt ozokerite crude petroleum and their products and cognate substances (hereinafter referred to as 'petroleum') within the State of Kuwait including all islands and territorial waters appertaining to Kuwait as shown generally on the map annexed hereto, the exclusive ownership of all petroleum produced and won by the Company within the State of Kuwait the right to refine transport sell for use within the State of Kuwait or for export and export or other wise deal with or dispose of any and all such petroleum and the right to do all things necessary for the purposes of those operations. The Company undertakes however that it will not carry on any of its operations within areas occupied by or devoted to the purposes of mosques sacred buildings or graveyards of carry on any of its operations except the sale of petroleum housing of staff and employees and administrative work within the present town wall of Kuwait.

The period of this Agreement shall be 75 years from the date of signature.

Article 2

(A) Within nine months from the date of signature of this Agreement the Company shall commence geological exploration.

(B) The Company shall drill for petroleum to the following total aggregate depths and within the following periods of time at such and so many places as the Company may decide:—

(i) 4,000 feet prior to the 4th anniversary of the date of signature of this Agreement.

(ii) 12,000 feet prior to the 10th anniversary of the date of signature of this Agreement.

(iii) 30,000 feet prior to the 20th anniversary of the date of signature of this Agreement.

(C) The Company shall conduct its operations in a workmanlike manner and by appropriate scientific methods and shall take all reasonable measures to prevent the ingress of water to any petroleum-bearing strata and shall duly close any unproductive holes drilled by it and subsequently abandoned. The Company shall keep the Shaikh and his London representative informed generally as to the progress and result of its drilling operations but such information shall be treated as confidential..

Article 3

In consideration of the rights granted by the Shaikh to the Company by this Agreement and of the assistance and protection which the Shaikh hereby undertakes to afford by all means in his power to the Company and its operations employees and property the Company shall pay to the Shaikh the following sums:—

(a) Within thirty (30) days after signature of this Agreement Rupees Four Hundred and seventy five thousand (Rs. 475,000).

(b) On each anniversary of the date of signature until the Company declares that petroleum has been found in commercial quantities:—
EITHER Royalty of Rupees Three (Rs. 3) for every English ton (2,240 lbs.) of Kuwait petroleum won and saved by the Company in Kuwait during the year ending 3 months prior to the anniversary of the date of signature.
OR Rupees Ninety five thousand (Rs. 95,000)
whichever shall be the greater sum.

(c) On each anniversary of the date of signature after the Company has declared that petroleum has been found in commercial quantities:—
EITHER Royalty as defined above, OR Rupees Two Hundred and fifty thousand (Rs. 250,000) whichever shall be the greater sum.

(d) For the purpose of this Agreement and to define the exact product to which the Royalty stated above refers, it is agreed that the Royalty is payable on each English ton of 2,240 lbs. of nett crude petroleum won and saved by the Company from within the State of Kuwait—that is after deducting water sand and other foreign substances and the oil required for the customary operations of the Company's installations in the Shaikh's territories.

Article 4

On each anniversary of the date of signature of this Agreement the Company shall deliver to the Shaikh a return of petroleum if any on which royalty is payable for the year ended three (3) months prior to such anniversary and a statement of the amount of royalty if any due to the Shaikh for such year, and a report of its operations under this agreement during such year. The Shaikh or his Representative shall have the right to check such returns and statements which, as well as any reports shall be treated as confidential by the Shaikh with the exception of such figures therein as he may be required by law to publish.

Article 5

(A) For the purposes of its operations hereunder the Company shall have the right without hindrance to construct and to operate power stations, refineries, pipelines and storage tanks, facilities for water supply including boring for water, telegraph, telephone and wireless installations, roads, railways, tramways, buildings, ports, harbours, harbour works, wharves and jetties, oil and coaling stations, with such lighting as may be requisite and any other facilities or works which the Company may consider necessary and for such purposes to use free of all payments to the Shaikh any stone, sand, gravel, gypsum, clay or water which may be available and may be required for its operations hereunder, provided always that the inhabitants of the State of Kuwait are not prevented from taking their usual requirements of these materials and that the water supply of the local inhabitants and nomad population who may be dependent on the same is not endangered. The Company at its discretion but in consultation with the Shaikh may select the position of any works. The Company may likewise utilise without hindrance all such means of transportation by land, air and water communication or operation as may be necessary for the effective conduct of its operations hereunder.

But nothing in this Article (5A) shall confer on the Company the right to dispose of stone, sand, gravel, gypsum, clay or water by sale, export or otherwise to any other company or person within or without the State of Kuwait.

(B) The Company shall under normal conditions accept and transmit free of charge on its wireless and telegraph installations such of the Shaikh's messages as will not interfere with the Company's business, and in times of national emergency the Shaikh shall have the full use free of charge of the Company's wireless and telegraph installations and railways for governmental purposes.

(C) The Shaikh's ships shall have the right to use harbours utilised or constructed by the Company, provided that such use in no way hampers the Company or interferes in any way with the safety of its operations of which the company shall be the sole judge. Any wharves or appurtenances constructed by the Company shall be for its exclusive use.

The Company may use for the purposes of its operations the harbours along the coast of Kuwait but the Company shall not impede or interfere with the subjects of the Shaikh or their right to continue the use of existing harbours, anchorages, wharves and docks along the coast of Kuwait at present utilised by them for their sailing craft and fishing boats.

Article 6

(A) The Company shall maintain in the region of the Persian Gulf a Chief Local Representative to represent it in matters relating to this Agreement with the Shaikh. The Shaikh has the right to select on the first occasion the chief Local Representative in consultation with His Majesty's Government.

(B) The Shaikh shall have the right to appoint an Arab conversant with the English language to act as his Official Representative and who will represent him in Kuwait in matters relating to this Agreement with the Company and particularly whenever unskilled labour is recruited from among the subjects of the Shaikh this Representative shall be consulted and advise the Company regarding any such recruitment. The salary of the Representative shall not be less than Rupees Eight hundred (Rs. 800/-) per month, and such salary shall be paid by the Company monthly to the Representative as from the date of his appointment by the Shaikh.

(C) The Shaikh shall have the right to appoint—from the effective date of this Agreement—a Representative in London to represent the Shaikh in all matters relating to this Agreement with the Company in its London Office and such Representative shall have full access to the production records of the Company including the agenda of the Board meetings and shall be entitled to attend the Board's meetings at which the Shaikh's interests are discussed. The salary of such Representative shall not be less than Rupees Two thousand two hundred and fifty (Rs. 2250) per month which shall be paid to the Representative by the Company and not by the Shaikh. The salary of such Representative shall be paid either in London or Bombay as requested by him. Travelling and general expenses of the Representative shall be defrayed from the above mentioned sum of Rupees Two thousand two hundred and fifty (Rs. 2250).

(D) If any any time during the currency of this Agreement any dispute shall arise regarding the accuracy of the accounts of the Company in connection with the amount of the Royalty and/or other payments due to the Shaikh under this Agreement, the Shaikh shall have the right to appoint in consultation with His Majesty's Government—a registered firm of Auditors to examine the books of the Company, on behalf of the Shaikh, at Kuwait and/or in London as he may consider necessary. All expenditure incurred in connection with such auditing shall be paid by the Shaikh.

The Company shall provide the registered firm of Auditors appointed by the Shaikh the necessary facilities to enable them to check the books and registers of the Company and to render every assistance to enable the Auditors to thoroughly examine such accounts and in every way to assist them safeguard the interests of the Shaikh.

The Shaikh shall regard as confidential all information supplied in connection with all such auditing with the exception of such items as may have an actual bearing on the dispute or are connected with it.

Article 7

(A) The Company shall have the right to import water, petroleum, fuel, machinery, motorcars and lorries, equipment, plant, timber, utensils, iron work, building materials, food, supplies, medicines, medical supplies, office equipment and household furniture, and all other materials, equipment and goods of whatsoever nature required by the Company and its employees for the purposes of its operations hereunder but not for resale to others, and to export its petroleum and articles previously imported by the Company free of customs or import or export duty and taxes or other charges, but it shall pay on all personal goods, clothing and general merchandise imported by the Company for the personal use of its employees or for resale to them, the ordinary duty in force for the time being in the State of Kuwait. Saving as in Article 3 and in this Article provided, the Company, its operations, income, profits and property including petroleum shall be exempt and free during the period of this Agreement from all present or future harbour duties, import duties, export duties, taxes, imposts and charges of any kind whether state or local, tolls, and land surface rent of whatever nature; and in consideration thereof the Company shall in addition to the payments provided for in Article 3 pay to the Shaikh on each anniversary of the date of signature of this Agreement four annas (annas 4) per ton (2,240 lbs.) of petroleum on which royalty is payable.

(B) The importation by the Company of firearms and other weapons is prohibited except with the written permission of the Shaikh.

(C) If the Company should sell in Kuwait any material or goods previously imported into Kuwait for the purposes of its operations hereunder and no longer required by the Company, the Company shall pay to the Shaikh in respect of such material or goods sold the equivalent of import duty thereon at the rate in force at the time of sale. The duty shall be computed on the price received on sale.

(D) Necessary customs officials at harbours constructed by the Company or additional customs officials required at any other ports utilised by the Company shall be appointed by the Shaikh in consultation with the Company and their salaries which shall not exceed the usual salaries of such officials shall be paid by the Company which shall also provide at its own expense suitable buildings for the accommodation of customs officials at harbours which it has constructed.

Article 8

(A) The Company shall have the right to purchase at current market rates fuel, water, food, building and constructional materials and other supplies of every kind in connection with its operations hereunder.

(B) The Company shall employ subjects of the Shaikh as far as possible for all work for which they are suited under the supervision of the Company's skilled employees, but if the local supply of labour should in the judgment of the Company be inadequate or unsuitable the Company shall have the right with the approval of the Shaikh which shall not be unreasonably withheld to import labour preference being given to labourers from the neighbouring Arab countries who will obey the local laws. The Company shall also have the right to import skilled and technical employees. Any employee imported by the Company who shall by misconduct cause a breach of peace or public disturbance shall at the request of the Shaikh be dismissed and shall if it is within the power of the Company to do so be sent out of Kuwait. The Company shall pay to the workmen it employs a fair wage, such wage to be decided and stated by the Company at the time the workmen are engaged.

(C) The Company shall provide free of charge medical service for its employees, and the Shaikh and his family shall have the right to such medical service and necessary medical supplies free of charge.

Article 9

The Shaikh grants to the Company free of cost the unrestricted use and occupation of and surface rights over all uncultivated land belonging to the Shaikh which the Company may need for the purposes of its operations and in particular the Company shall have the right to select in consultation with the Shaikh an area or areas of land chosen by the Company outside the present town wall of Kuwait with exclusive surface rights upon which to erect oil refineries, storage, terminal and shipping facilities and any other works required for the Company's operations; and the Company may with the cognisance of the Shaikh buy or lease for such purposes any lands, houses or buildings with the consent of and on conditions to be arranged with the proprietors thereof but the terms of such purchase or lease shall not be in excess of those ordinarily current in their respective localities.

(A) The Company shall acquire only such land, houses and buildings as are necessary for its operations under this Agreement. The Company shall inform the Shaikh from time to time of the land, houses and buildings which it requires to occupy for its operations; and land houses and buildings previously acquired by the Company from the Shaikh but found no longer necessary for its operations shall be returned by the Company to the Shaikh free of charge.

(B) The Shaikh shall retain for himself the right to grant—in consultation with the Company—to another Company or Companies operating petroleum areas within territories adjoining the Kuwait borders the right to lay down pipelines and to permit such Company or Companies to construct and erect within the Kuwait territory and across same, the necessary buildings and machinery required for the transport in transit or passage over Kuwait territory of crude oil to a suitable site within the State of Kuwait considered convenient for the loading of the said crude Oil.

Article 10

The Shaikh shall give to the Company and its employees and property all the protection in his power from theft, highway robbery, assault, wilful damage and destruction, and the Company may appoint in consultation with the Shaikh and itself pay trustworthy guards who shall at all times be Kuwait subjects unless the Shaikh permits otherwise to assist in protecting the property of the Company and its employees. The Company shall erect at its own expense suitable buildings for the accommodation of such guards at such places as the Company shall decide.

Article 11

(A) Before the expiration of the period specified in Article 1 hereof this Agreement shall come to an end either by surrender as provided in paragraph (B) of this Article or in Article 12 or in one of the three following cases:

(a) If the Company shall fail to fulfil its obligations under Article 2 hereof in respect of geological exploration or drilling.

(b) If the Company shall fail within six (6) months after any anniversary of the date of signature of this Agreement to make to the Shaikh any payments agreed to be due under Article 3.

(c) If the Company shall be in default under the arbitration provisions of Article 18.

In any of the above mentioned cases the Shaikh shall be entitled to terminate this Agreement and all the property of the Company within the State of Kuwait shall become the property of the Shaikh.

(B) In the event of the Company failing to make the declaration provided in Article 3 within 12 years of the date of signature of this Agreement the Company shall at its option either pay to the Shaikh the minimum annual payment provided in Article 3(c) or surrender all rights under this Agreement.

Article 12

(A) The Company shall have the right at any time after it has drilled the 4,000 feet provided in Article 2(B) (i) or after the expiry of 2 years from the date of signature of this Agreement—whichever shall be the later date—to give the Shaikh one year's notice in advance to terminate this Agreement and the Company shall on expiry of such notice have no further liabilities except to make payment of all monies which may be due to the Shaikh up to the date of termination.

(B) Should this Agreement be terminated by the Company under this Article 12, then:—

(a) If such termination occurs within 35 years from the date of signature of this Agreement all lands granted by the Shaikh and any lands or buildings which the Company may have bought and any houses or buildings constructed by and other immovable property of the Company within the State of Kuwait shall be handed over to the Shaikh free of charge. Producing wells or borings at the time of such termination shall be handed over in reasonably good order and repair, but

(b) if such termination occurs after 35 years from the date of signature of this Agreement all the movable and immovable property of the Company in the State of Kuwait shall be handed over to the Shaikh free of cost. Producing wells or borings at the time of such termination shall be handed over in reasonably good order and repair.

Article 13

On the Expiry of this Agreement at the end of the period of 75 years provided in Article 1 or of any extension or renewal of that period all the movable and immovable property of the Company in the State of Kuwait shall be handed over to the Shaikh free of cost. Producing wells or borings at the time of such expiry shall be handed over in reasonably good order and repair.

Article 14

The Shaikh hereby agrees that the Company may transfer the obligations and benefits of this Agreement to any Company registered within the British Empire.

Article 15

(A) Nothing in this Agreement shall be read as restricting in any way the right of the Shaikh to grant to other parties concessions or permits for substances other than petroleum provided that the operations and rights of the Company hereunder are not thereby injuriously affected.

If the Shaikh should at any date subsequent to the date of signature of this Agreement grant to any other parties concessions or permits for substances other than petroleum the Shaikh undertakes that such concessions shall contain provisions requiring the holders thereof from damaging impeding or interfering with the property operations and interests of the Company.

Deposits of mineral substances other than petroleum such as gold, silver, copper, lead, potash, sulphur and salt or the like which may be discovered by the Company shall be reported to the Shaikh and shall not be worked by the Company except under a special concession or permit from the Shaikh.

(B) The Company shall use the Shaikh's flag within the State of Kuwait.

Article 16

Failure on the part of the Company to fulfil any of the conditions of this Agreement shall not give the Shaikh any claim against the Company or be deemed a breach of this Agreement in so far as such failure arises from *force majeure*, and if through *force majeure* the fulfilment by the Company of any of the conditions of this Agreement be delayed the period of such delay shall be added to the periods fixed by this Agreement.

Force majeure as used in this Agreement includes the act of God, war, insurrection, riot, civil commotion, tide, storm, tidal wave, flood, lightning, explosion, fire, earthquake and any other happening which the Company could not reasonably prevent or control.

Article 17

The Shaikh shall not by general or special legislation or by administrative measures or by any other act whatever annul this Agreement except as provided in Article 11. No alteration shall be made in the terms of this Agreement by either the Shaikh or the Company except in the event of the Shaikh and the Company jointly agreeing that it is desirable in the interest of both parties to make certain alterations, deletions or additions to this Agreement.

Article 18

(A) If at any time during the currency of this Agreement any difference or dispute shall arise between the parties hereto concerning the interpretation or execution hereof, or anything herein contained or in connection herewith, or the rights or liabilities of either party hereunder, the same shall, failing any agreement to settle it in any other way, or after consultation with the British Political Agent in Kuwait or the British Political Resident in the Persian Gulf, be referred to two arbitrators, one of whom shall be chosen by each party, and a referee, who shall be chosen by the arbitrators before proceeding to arbitration.

(B) Each party shall nominate its own arbitrator within 60 days after the delivery of a request so to do by the other party failing which its arbitrator may at the request of the other party be designated by the British Political Resident in the Persian Gulf. In the event of the arbitrators failing to agree upon the referee within 60 days after being chosen or designated, the British Political Resident in the Persian Gulf may appoint a referee at the request of the arbitrators or either of them.

(C) The decision of the arbitrators, or in case of a difference of opinion between them the decision of the referee, shall be final and binding upon both parties.

(D) In giving a decision the arbitrators or the referee shall specify an adequate period of delay during which the party against whom the decision is given shall conform to the decision and

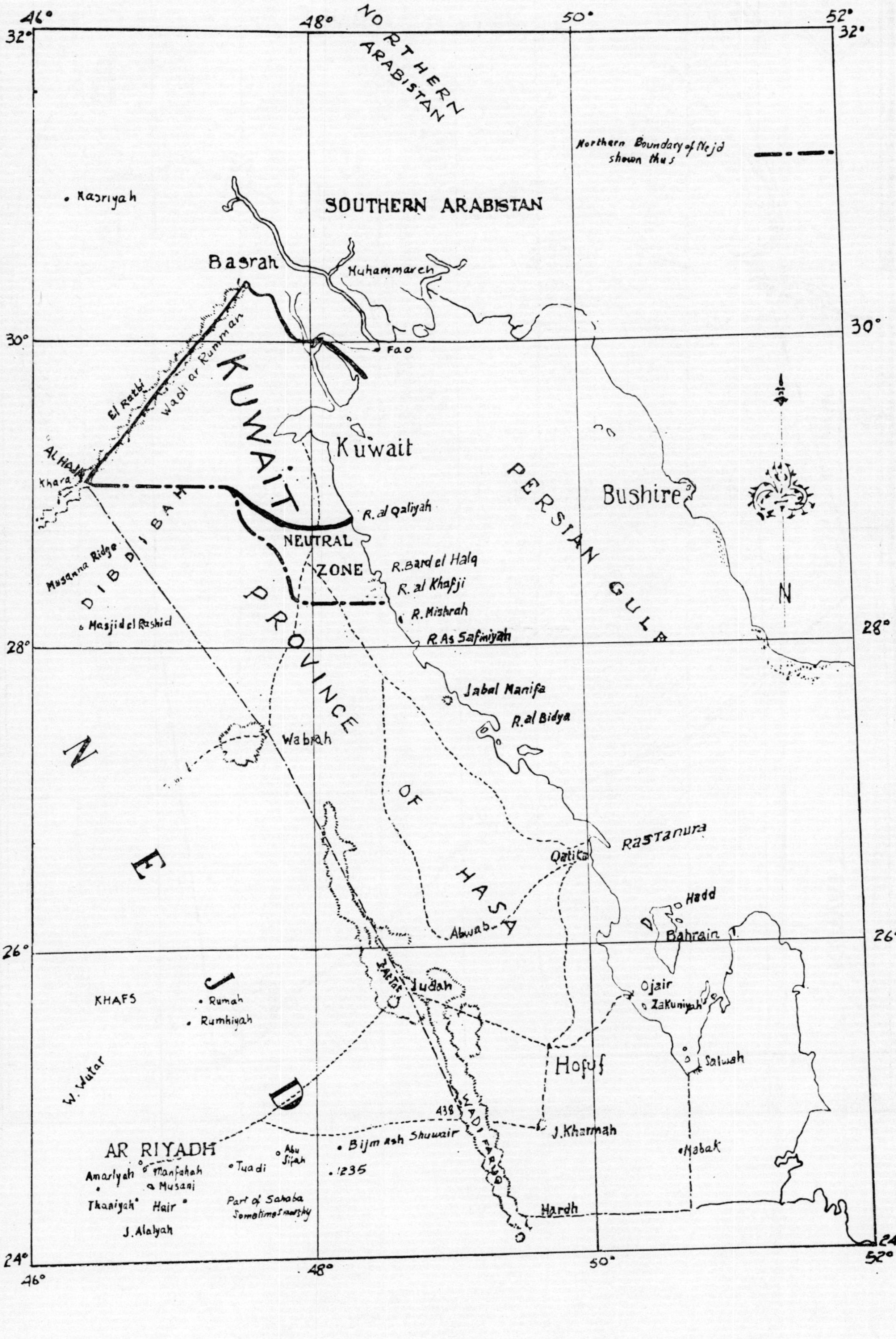
NORTHERN ARABISTAN
SOUTHERN ARABISTAN
Northern Boundary of Nejd shown thus
Kasriyah
Basrah
Muhammareh
Fao
Kuwait
KUWAIT
El Ratki
Wadi ar Rumman
AL HAJR
Khara
DIBDIBAH
Musanna Ridge
Masjid el Rashid
NEUTRAL
ZONE
R. al Qaliyah
R. Bard el Halq
R. al Khafji
R. Mishrah
R. As Safiniyah
Jabal Manifa
R. al Bidya
PERSIAN GULF
Bushire
N
PROVINCE OF HASA
Wabrah
NEJD
Qatif
Ras Tanura
Abwab
Hadd
Bahrain
Judah
Ojair
Zakuniyah
Salwah
Hofuf
KHAFS
Rumah
Rumhiyah
W. Wutar
J. Kharmah
Bijm ash Shuwair
WADI FARUQ
AR RIYADH
Amariyah
Manfahah
Musani
Tuadi
Abu Sifah
Thaniyah
Hair
J. Alaiyah
Part of Sahaba Sometimes marshy
Habak
Hardh
1235
46°
48°
50°
52°
32°
30°
28°
26°
24°

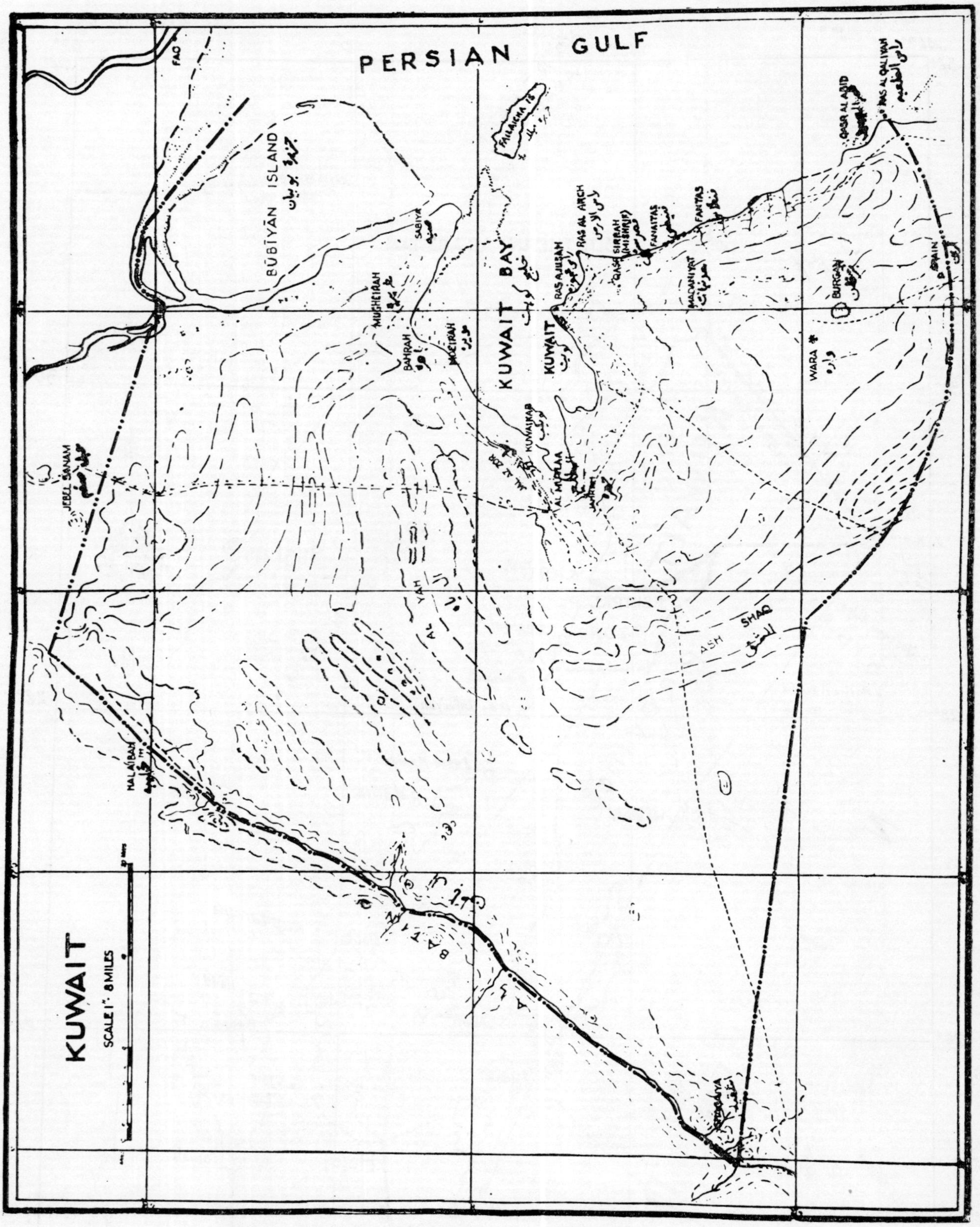
PERSIAN GULF
KUWAIT
SCALE 1" 8 MILES
FAO
BUBIYAN ISLAND
FAILAKA IS.
SABIYA
KUWAIT BAY
KUWAIT
JEBEL SANAM
BAHRAH
FANAITAS
FANTAS
BURGAN
WARA
QASR AL ABID
ASH SHAQ
AL BATIN

that party shall be in default only if that party has failed to conform to the decision prior to the expiry of that period and not otherwise.

(E) The place of arbitration shall be such as may be agreed by the parties and in default of agreement shall be London.

Article 19

The Company shall make all payments that become due to the Shaikh under this Agreement into the Shaikh's account at the Ottoman Bank in Basrah and the Bank's receipt shall be a full discharge for the Company in respect to the payment of the sum stated in the Bank's receipt. The Shaikh may from time to time designate in writing another Bank or Banks for the purpose of this Article.

Article 20

For the purpose of royalty payments the Company shall measure by a method customarily used in good technical practice all petroleum on which royalty is payable and the Shaikh by his representative duly authorised by him shall have the right to observe such measuring and to examine and test whatever appliances may be used for such measuring. Such representative shall comply with all necessary and usual safeguards for the prevention of fire or other accident; and shall make all examinations and tests at such times and in such manner as will cause the minimum of interference with the Company's operations. If upon such examination or testing any such appliance shall be found to be out of order the Company will cause the same to be put in order at its own expense within a reasonable time, and if upon any such examination as aforesaid any error shall be discovered in any such appliance, such error shall if the Shaikh so decide after hearing the Company's explanation be considered to have existed for three (3) calendar months previous to the discovery thereof or from the last occasion of examining the same in case such occasion shall be within such period of three (3) calendar months and the royalty shall be adjusted accordingly. If the Company should find it necessary to alter repair or replace any measuring appliance it shall give reasonable notice to the Shaikh or his representative to enable a representative of the Shaikh to be present during such alteration, repair or replacement.

The Company shall keep full and correct records of all measurements as aforesaid and the said representative of the Shaikh shall have access at all reasonable times to such records and shall be at liberty to make extracts from them. Such records shall be treated as confidential by the Shaikh and his representatives with the exception of such figures therein as the Shaikh may be required by law to publish.

Article 21

This Agreement is written in English and translated into Arabic. If there should at any time be disagreement as to the meaning or interpretation of any clause in this Agreement the English text shall prevail.

In witness whereof the parties to this Agreement have set their hands the day and year first above written.

ON BEHALF OF THE KUWAIT OIL COMPANY LIMITED		SHAIKH OF KUWAIT
Frank Holmes (signed)		Ahmad al Jabir as Subah
A. Chisholm (signed)		(signed)
IN THE PRESENCE OF		IN THE PRESENCE OF
H. R. P. Dickson, Lt.-Col.		H. R. P. Dickson, Lt.-Col.
(signed)		(signed)
23–12–34	(SEAL—Political Agency—Kuwait)	23–12–34
Lieut.-Colonel,		Lieut.-Colonel
H.B.M.'s Political Agent, Kuwait.		H.B.M.'s Political Agent, Kuwait.

---

156. *After his return from his London visit in 1935 Shaikh Ahmad wrote as follows to Mr Fraser of KOC and APOC:*

Ahmed Al Jabir Al Sabah
Oct. 22, 1935.

My dear Mr. Fraser,

Thank you very much for your letter of Oct 8th. Yes: I am very happy indeed to be home again, but I can assure you at the same time that I can never forget the happy days when my friends in London entertained me and organized such a glorious programme of receptions in my honour. Believe me those happy meetings are not oblivious memoirs and shall dwell in my heart for ever.

I am further glad to hear of your placing my photograph in your office.

I have great hopes of meeting some of my London friends again in Kuwait or London in course of time, when oil is found in Kuwait and Company's business prospers.

My kindest regards to you and all friends.

Yours sincerely,

157. *Shaikh Ahmad appointed Major Holmes his representative in London in the following letter:*

Kuwait 5th January, 1935.

PERSONAL

Major Frank Holmes,

As I have great confidence in your friendship and depend upon your honesty I have decided to appoint you my representative in London, and shall be pleased if you kindly inform me of your acceptance at your earliest convenience.

I also shall be much obliged if you ask your company to commence geological exploration at their earliest as possible, as the weather of these days is comfortable and cold.

Please inform your company my wishes and oblige.

Yours sincerely,
Sgd. Ahmed al Jabir al Subah.

---

158. *The arrival of KOC's Gulf and APOC geologists in Kuwait in March 1935 was recorded thus in an APOC memorandum from Abadan to London:*

From Abadan (General Management)
Subject Kuwait Survey

To London
Memo No. 34149
Date 31st March, 1935

CONFIDENTIAL

Reference Abadan/London Memo No. 34131

In continuation of the above memorandum we have to report that Messrs. Rhoades and Crowl duly arrived at Basrah on the 15th March and were met by Mr. G. M. Shaw, After a preliminary discussion they agreed to proceed to Kuwait by car the following day, and arrangements were made accordingly. On arrival they were met by Major Holmes and Abdul Mullah s/o Jabir, the Sheikh's London and Kuwait representatives respectively.

On the 17th instant they had an interview with the Sheikh of Kuwait, who invited them to dine with him on the 19th.

From the 18th to the 21st Messrs. Shaw, Rhoades and Crowl spent in going over the concessional territory, and Mr. P. T. Cox, who was on his return journey from the Trucial Coast, joined them from the 19th onwards.

The 22nd was taken up in discussing Messrs. Rhoades and Crowl's geological programme, and it was agreed unnecessary for Mr. Cox to remain in Kuwait whilst they were engaged in their preliminary independent observations, but that he should return immediately they were ready for him. Mr. Cox will therefore pay a further visit to Kuwait in approximately one month's time unless a message requesting his earlier return is received from Mr. Rhoades in the meantime.

Messrs. Shaw and Cox left Kuwait on the 23rd instant on their return journey and reached Abadan the same day.

We also take this opportunity to confirm the interchange of telegrams on the above subject, No. 36 being received from you on the 25th and replied to by our No. 28 on the 26th instant.

Sgd. E. H. O. Elkington

---

159. *Shortly after K.O.C.'s Burgan No. 1 well struck oil on the night of 23rd–24th February 1938 it was 'burnt off', a routine operation necessary to measure the well's productive capacity and involving its crude oil being allowed to flow briefly at full throttle into an adjoining reservoir where it was ignited to dispose of it. On this occasion the reservoir was a banked area of the surrounding desert, whose sand was converted by an overflow of the flaming oil into a sheet of glass, a fragment of which the author of this Record keeps as a souvenir of that historic moment for Kuwait.*

Addendum to Note 127 (pages 213–216)

(1) *On 9th January 1935 Lt. Col. Dickson (Political Agent, Kuwait) wrote as follows to the Political Resident (at Bushire):—*

Will you please refer to my Secret 'Oil' D.O. No. C-377 of 28th December 1934.

It may interest you to know that I have found out that Lord Glenconner was in Bagdad in the middle of June late (1934), and I am almost certain that he was present with a friend of his (who I suspect was R. C. Holmes) at the meeting which Chisholm and his friends had in the Maude Hotel, Bagdad, on the night of 17th June. See paragraph 10 of my D.O. referred to above.

This letter of Dickson's was just as erroneous, as regards any meeting of the author of this Record with Lord Glenconner, as were all his other suspicions regarding an 'APOC plot' described in his letter of 28th December 1934 (see Note 127). Neither in June 1934, nor before or since that date, has the author of this Record ever met Lord Glenconner.

(2) *The following is the text of a handwritten note initialled by Mr (later Sir J. G.) Laithwaite, addressed to his then India Office superior Mr Walton and dated 28th January 1935, which is attached to Lt. Col. Dickson's letter dated 28th December 1934 (see Note 127) now filed in the London archives of the British Foreign and Commonwealth Office:—*

Mr. W.

I am very doubtful as to the desirability of distributing this, though I might perhaps let Mr Randel [see Notes 134, 135] see it privately. Col. Dickson may be right (it is a most discreditable, if not impossible, story if he is) but the general conduct of Traders seems to me consistent with their being on their own and we have really very little to associate Messrs Strich with the business.

As regards para. 4, if this report had been repeated to us at once we should have been on our guard. As it went via Bushire we received it only about 18 December. We might deal with this question of repetitions in dealing with the general question (on which I am preparing an official draft to F.O.).

J.G.L.
28/1

plot described in his letter of 28th December 1934 (see Note 127). Neither in June 1934, nor before or since that date, has the author of this Record ever met Lord Glenconner.

(2) *The following is the text of a handwritten note initialled by the (later Sir J. G.) Laithwaite, addressed to his then India Office superior Mr. Walton and dated 29th January 1935, which is attached to Lt. Col. Dickson's letter dated 28th December 1934 (see Note 127) now filed in the London archives of the British External and Commonwealth Office:*

Mr. W.

I am very doubtful as to the desirability of distributing this, though it might perhaps let Mr. Rendel [see Notes 198, 135] see it privately. Col. Dickson may be right in the main, discreditable if not impossible story (if he is) but the general conduct of Traders seems to me consistent with their being on their own and we have really very little to associate Messrs Strick with the business.

As regards para. 4, if this report had been repeated to us at once we should have been on our guard. As it went via Bushire we received it only about 15 December. We might deal with this question of repetitions in dealing with the general question on which I am preparing an official draft to F.O.?.

J.G.L.
29.1

Index